Regional Climate Studies

Series Editors: H.-J. Bolle, M. Menenti, I. Rasool

Springer-Verlag Berlin Heidelberg GmbH

Hans-Jürgen Bolle (Ed.)

Mediterranean Climate

Variability and Trends

With 103 Figures and 24 Tables

 Springer

Series Editors:

PROFESSOR A.D. DR. HANS-JÜRGEN BOLLE
Stücklenstrasse 18c
81247 München
Germany

DR. MASSIMO MENENTI
Laboratoire des Sciences de
l'Image, de l'Informatique et
de la Télédétection (LSIIT)
Université Louis Pasteur
5 Blvd. Sebastian Brant
67400 Illkirch
France

Dr. Ichtiaque Rasool
60 Quai Louis Bleriot
75016 Paris
France

Editor: PROFESSOR A.D. DR. HANS-JÜRGEN BOLLE

ISBN 978-3-642-62862-7

Library of Congress Cataloging-in-Publication Data

Mediterranean climate - variability and trends / Hans-Jürgen Bolle (ed.).
 p.cm.--(Regional climate studies)
 Includes bibliographical references.
 ISBN 978-3-642-62862-7   ISBN 978-3-642-55657-9 (eBook)
 DOI 10.1007/978-3-642-55657-9
 1. Mediterranean Region--Climate. 2. Climatic changes--Mediterranean Region. I.
Bolle, H.-J. (Hans-Jürgen) II. Series.

http://www.springer.de

© Springer-Verlag Berlin Heidelberg 2003
Originally published by Springer-Verlag Berlin Heidelberg New York in 2003
Softcover reprint of the hardcover 1st edition 2003

Camera ready by authors
Cover design: E. Kirchner, Heidelberg
Printed on acid-free paper SPIN 10850295 32/3030/as 5 4 3 2 1 0

This publication results from two workshops organized back to back in Casablanca, Morocco, in February 2001. The workshop on the *"Assessment, assimilation, and validation of data for 'Global Change' related research in the Mediterranean area"* was organized in the context of the project on "Research In global ChAnge in the Mediterranean: A REgional network" (RICAMARE) coordinated by Gérard Begni (France) and Jose Moreno (Spain). RICAMARE is a project of the European Commission (FP5, ENRICH) and was supervised by Julia-Maria Kunderman, EC-DG Research, with finacial and ideal support of HassanVirji, IGBP/IHDP/WCRP START.

The workshop *"Development of Priority Climate Indices for Africa: A CCl/CLIVAR Workshop of the WorldMeteorological Organization and the World Climate Research Programme"* was organized by Valerie Detemmerman on behalf of WMO, and David Easterling on behalf of the chair of *the*

CCl/CLIVAR Working Group on Climate Change Detection, Tom Peterson, jointly with the Regional CCDAS Rapporteur Abdalah Mokssit, of the *Direction Nationale de la Météorologie of Morocco* which hosted both workshops.

The scientific organizers of the workshop, Hans-Jürgen Bolle and Emin Ozsoy as well as the participants would like to express their sincere thanks to these organizations and persons involved for hospitality and support. Special thanks are going to Ms. Chantal Le Scouarnec, *MEDIAS-FRANCE*, who was responsible for the technical and administrative arrangements. Thanks are furthermore extended for additional financial support to the Head of Earth Observation Delegation, Program Directorate, *CNES*, Jean-Louis Fellous, who was one of the founders of RICAMARE, and to *MEDIAS-FRANCE* for logistic support.

Preface

Mohamed Larbi Selassi
Deputy Director of the National Meteorology, Morocco
Welcome address (translated from French)

WMO, WCRP, Medias-France and scientific institutions representatives,
ladies and gentlemen,

I want first to thank WMO and MEDIAS-France, who have honoured us by organizing the two workshops, climate indices in Africa and data assessment for global change research in the Mediterranean region, in Casablanca and I welcome all of you here in Morocco.

It is with great pleasure that I open these two workshops on behalf of myself and on behalf of the Direction of the Meteorologie Nationale of Morocco.

Climate change is becoming the focus of the international community because of its global scale and unpredictable effects, the numerous impacts it causes, its global feature and the complexity of the solutions that can mitigate its impacts.

Global warming and the greenhouse effect became a subject of study at the international level since the United Nations Conference on the human environment that was held in Stockholm in 1972. The research and coordination efforts that have been made in this area have led to an "International Scientific consensus". High level meetings like those held in Toronto in 1988, in Lahaye in 1989 and in Geneva in 1990, did confirm the greenhouse threat and the emergency to treat it.

Climate variability in general and climate change in particular are becoming a permanent worry of all countries. They can cause serious impacts on human socio-economical activities. The international community has been mobilized by creating the scientific and technical instruments such as IPCC created by WMO and the UNDP in 1988 and the juridical instruments such as the Intergovernmental Negotiation Committee (INC) created by the United Nations as well as subsidiary organs of the United Nations Framework Convention on climate change (Rio de Janeiro 1992, Kyoto 97, Buenos Aires 98, Bonn 99 and La Haye 2000). Their common main goal is to develop scientific knowledge essential to help policymakers to do their duty.

Morocco ratified in 1995 the convention of Rio and affirms its commitment to build progressively an environmental policy that consists of developing suitable and integrated plans for water resources management, for the environment protection and for the conservation and protection of the ecological systems. Components of these efforts are to undertake scientific and socioeconomic research studies and to develop data sets of the climate system that will allow a better understanding of the causes and effects of climate changes.

Hence, the "earth summit" conclusions urged the international community to preserve the climate equilibrium for the welfare of the current and future generations.

Since the industrial revolution, changes in the atmospheric composition have been observed. The greenhouse gases concentration has increased by more than 25% and the air aerosol content has became more important. As you know, the main human activities that cause climate change are: industrialization and deforestation, through the use of fossil fuels respectively the destruction of 15 millions hectares of forests each year. The increasing concentration of greenhouse gases strengthens the greenhouse effect which is closely related to climate change.

Being worried about these problems is documented by the study of all the possible impacts of man's activities in our environment and through the determined intention to preserve the climate equilibrium that will ensure a welfare for future generations.

Climate change detection requires long time series of high quality data sets. This is why the collaboration between the meteorological services of different countries is encouraged and their active participation ensure the success of such kind of workshops.

Indeed, under the care of WMO, the meteorological services ensure collection, treatment and disemmination of data in order to set up a complete and reliable climatological data set at an international level.

With the support of all countries climate change research has been started at the global scale but most countries expect regional or even local studies. In this context the World Climate Research Programme has been encouraged to develop research and studies at the regional level and to focus on the extreme climate phenomena such as desertification, droughts, floods, etc.

These two workshops will be, I hope, a good opportunity to strengthen the collaboration between national meteorological services that are present here and to share their knowledge.

We are very glad that such kind of workshops are held in Morocco and I want to thank again WMO, WCRP, Medias-France and the staff of the DMN involved in the organization of these two workshops. I wish, finally, all the success for both workshops.

Thank you very much.

Authors

Marijan Ahel
Center for Marine and
Environmental Research
Ruder Baskovic Institute
POB 180
Bijenićka 54
HR-10 000 Zagreb
Croatia
ahel@rudjer.irbhr

Jean-Pierre Bethoux
Laboratoire de Physique et Chimie
Marines
CNRS/LPCM
Observatoire Océanologique
Quai de la Darse
BP 8
F-06238 Villefranche sur Mer
France
bethoux@obs-vlfr.fr

Marco Cacciani
University of Rome "La Sapienza"
Deptm. of Physics
Piazzale Aldo Moro 2
I-00185 Roma
Italy

Y. Casto-Diez
Departamento de Física Universidad
de Jaén
E-23071 Jaén
Spain

John DeLuisi
NOAA, SRRB
325 Broadway
80303 Boulder, Co
USA

Christina Anagnostopoulou
University of Thessaloniki
Deptm. of Meteorology and
Climatology
GR-54006 Thessaloniki
Greece

Hans-Jürgen Bolle
Stücklenstrasse 18c
D-81247 München
Germany
hansj.bolle@lrz.badw-muenchen.de

Massimo Candelori
Secretariat to the United Nations
Convention to Combat Desertification
Martin-Luther-King Str. 8
D-53175 Bonn
Germany
mcandelori@unccd.int

Michele Colacino
Istituto di Fisica dell' Atmosfera -
CNR
Via del Fosso del Cavaliere, 100
I-00133 Rome
Italy
m.colacino@rm.cnr.it

Lorenzo De Silvestri
ENEA, Global Environment and
Climate Laboratory
Via Anguillarese 301
I-00060 S. Maria di Galeria (Roma)
Italy

Tatiana Di Iorio
University of Rome "La Sapienza"
Deptm. of Physics
Piazzale Aldo Moro 2
I-00185 Roma
Italy

Alcide Giorgio di Sarra
Department of Physics
University "La Sapienza"
P.le A. Moro 2
I-00185 Roma
Italy
disarra@g24ux.phys.uniroma1.it

Jean-Marc d'Herbes
Coordinateur ROSELT/OSS
Laboratoire IRD (ex-ORSTOM)
Maison de la Teledetection
500, rue J-F Breton
34093 Montpellier Cedex 05
France
dherbes@teledetection.fr

Annick Douguédroit
Université de Provence
Institut de Géographie
29, Av. Robert Schuman
F-13621 Aix-en-Provence
France
annick.douguedroit@up.univ-aix.fr

Giuseppe Enne
Centro Interdipartimentale di Ateneo
Nucleo di Ricerca sulla
Desertificazione (NRD)
Università degli Studi di Sassari
Via Enrico de Nicola, 9
I-07100 Sassari
Italy
nrd@uniss.it

Maria Jesús Esteban Parra
Dpto Física Aplicada
Universidad de Granada
E-18071 Granada
Spain
esteban@ugr.es

Giorgio Fiocco
University of Rome "La Sapienza"
Deptm. of Physics
Piazzale Aldo Moro 2
I-00185 Roma
Italy
fiocco@g24ux.phys.uniroma1.it

Lorenzo Genesio
Applied Meteorology Foundation
F.M.A.
Via Caproni 8
I-50145 Florence
Italy
genesio@iata.fi.cnr.it

Anna Rita Gentile
European Environment Agency
(EEA)
Kogens Nytorv 6
DK-1050 Copenhagen
Denmark
anna.rita.gentile@eea.eu.int

Frank-M. Goettsche
Forschungszentrum Karlsruhe (FZK)
Institut fuer Meteorologie und
Klimaforschung (IMK)
Postfach 3640
D-76021 Karlsruhe
Germany
frank.goettsche@imk.fzk.de

Paolo Grigioni
ENEA, Global Environment and
Climate Laboratory
Via Anguillarese 301
I-00060 S. Maria di Galeria (Roma)
Italy

Ana Iglesias
Dept. of Agricultural Economics and
Social Sciences
ETSIA- Universidad Politecnica de
Madrid
Avenida de la Complutense sn
E-28040 Madrid
Spain
iglesias@ppr.etsia.upm.es

Birgit Klein
Institut für Umweltphysik
Universität Bremen
Postfach 330440
D-28344 Bremen
Germany
bklein@physik.uni-bremen.de

Dirk Koslowsky
Freie Universität Berlin
Institut für Meteorologie
Carl-Heinrich-Becker Weg 6-8
D-12165 Berlin
Germany
kosze@zedat.fu-berlin.de

Tarzan Legović
Rudjer Bošković Institute
P.O.B. 180
Bijenička 54
HR-10 002 Zagreb
Croatia

Jürg Luterbacher
University of Bern
Institute of Geography, Climatology
and Meteorology
Hallerstrasse 12
CH-3012 Bern
Switzerland
juerg@giub.unibe.ch

Panagiotis Maheras
University of Thessaloniki
Deptm. of Meteorology and
Climatology
GR-54006 Thessaloniki
Greece
maheras@geo.auth.gr

Abdellah Mokssit
Direction de la Météorologie
Nationale
Hay Hassani, en face Prefecture Hay
Hassani Ain Chock
Casablanca
Morocco
mokssit@mtpnet.gov.ma

Caroline Norrant
Université de Provence
Institut de Géographie
29, Av. Robert Schuman
F-13621 Aix-en-Provence
France
cnorrant@hotmail.com

Folke - S. Olesen
Forschungszentrum Karlsruhe (FZK)
Institut fuer Meteorologie und
Klimaforschung (IMK)
Postfach 3640
D-76021 Karlsruhe
Germany
folke.olesen@imk.fzk.de

XII

Emanuela Piervitali
Consorzio CRATI
Università della Calabria
Rende (CS)
Italy
emanuela@ifa.rm.cnr.it

D Pozo-Vázquez
Departamento de Física Universidad
de Jaén
E-23071 Jaén, Spain

Fabio Raicich
CNR
Istituto Sperimentale Talassografico
Viale Romolo Gessi 2
I-34123 Trieste
Italy
fabio.raicich@itt.ts.cnr.it

F S Rodrigo
Departamento de Física Aplicada
Universidad de Almería
E-04120 Almería, Spain

Paul D. Try
International GEWEX Project Office
1010 Wayne Ave, Suite 450
Silver Spring, MD 20910
USA
gewex@cais.com

Monique Viel
CTM-ERS/RAC
2, Via G. Giusti
I-90144 Palermo
Italy
mailto:monique.viel@ctmnet.it
monique.viel@ctmnet.it

Jean Palutikof
Climatic Research Unit
University of East Anglia
NR4 7TJ Norwich
United Kingdom
j.palutikof@uea.ac.uk

Ivan Raev
Bulgarian Academy of Sciences
Forest Research Institute
132, St. Kliment Ohridski Blvd.
1756 Sofia
Bulgaria
forestin@bulnet.bg

S. Ichtiaque Rasool
60, Quai Louis Blériot
F-75016 Paris
France
sirasool@compuserve.com

Wolfgang Roether
Institut für Umweltphysik
Universität Bremen
Postfach 330440
D-28344 Bremen
Germany
wroether@physik.uni-bremen.de

Murat Türkeş
Turkish State Meteorological Service
Department of Research
PO Box 401
Ankara
Turkey
E-mail: mturkes@meteor.gov.tr

Elena Xoplaki
University of Thessaloniki
Dep. of Meteorology and Climatology
GR-54006 Thessaloniki
Greece
xoplaki@giub.unibe.ch

Chiara Zanolla
Centro Interdipartimentale di Ateneo -
Nucleo di Ricerca sulla
Desertificazione (NRD)
Università degli Studi di Sassari Dip.
Di Scienze Zootecniche
Via Enrico de Nicola, 9
I-07100 Sassari
Italy
nrd@uniss.it

Claudio Zucca
Centro Interdipartimentale di Ateneo -
Nucleo di Ricerca sulla
Desertificazione (NRD)
Università degli Studi di Sassari Dip.
Di Scienze Zootecniche
Via Enrico de Nicola, 9
I-07100 Sassari
Italy
nrd@uniss.it

Contents

Chapter 1
Introduction

H.-J. Bolle

The project of the European Commission on "Research into 'Global Change' in the Mediterranean: A regional network" (RICAMARE) is part of the programme "European Network for Research into Global Change" (ENRICH). In February 2001 it organized with the support of IGPB-START and CNES a workshop on the *"Assessment, assimilation, and validation of data for 'Global Change' related research in the Mediterranean area"*. Part of this workshop was dedicated to the analysis of long time series of meteorological data to document the variability of climate in the Mediterranean region and to quantify trends.

In parallel the workshop on *"Development of Priority Climate Indices for Africa: A CCI/CLIVAR Workshop of the WorldMeteorological Organization and the World Climate Research Programme"* was held, organized by WMO and *the CCI/CLIVAR Working Group on Climate Change Detection* jointly with the *Direction Nationale de la Météorologie of Morocco* which hosted both workshops. Thanks the contribution from this workshop in addition the present trends at many African stations could be included. This provides a unique data base for the whole Mediterranean and African area to diagnose present climate variability and climatic trends and to measure the accuracy of its simulation by climate models.

·Because the results of these numerical simulations by climate models are called "data" as well, it is emphasized that the part of the workshop which dealt with climate variability primarily focussed on measured (and in one case proxy) data. In fact the Earth system can be described in different ways: by empirical data, to which since a few decades also observations from space must be counted, and by numerical simulations which produce an alternative data set.

The different ways to look at the Earth system generate different worlds: The model world, in which data are generated according to mathematical rules in strictly predefined geographical projections, the world based upon physical and chemical *in situ* measurements, where data are generated more or less at random at the ground and in the air, the world that retrospectively is reconstructed from proxy data, and the world inferred from remotly sensed signals received in space with more or less perfect coverage but afflicted with measuring and interpretation errors. In operational networks meteorological quantities are measured under specific controlled conditions which may not be fully representative of the surrounding landscape. Measured data sets provide us with certain information about the behaviour of nature in specific spots and models provide us with a laboratory to study how nature would react to specific forcing if it would have been constructed like the models. Climate research thus results in four different descriptions of nature which are neither identical nor can they be identified with nature: Proxy data, ground based measurements, observations from space, and model simulations.

One interesting argument articulated at the workshop in Casablanca was that if various models come to nearly the same conclusion about the reaction of the climate system to external forcing, then the probability is high that they simulate the trend of the natural system correctly. This behaviour of models can, however, also be interpreted in a different way. It is possible that the closer model results come to each other the more likely it is that the physics implemented in the models equal each other, whether this reflects the response of the real nature or not. Of course modellers try hard to make the models a mirror of nature by continuous improvement of the physics implemented into the algorithms. This can only be done, if there is a standard against which the models can be tested. The only "standard" we have are measured and estimated data sets of a few state variables which are another abstraction of the real behaviour of nature. With the aid of these selected measured data models are initialized and these data can also be used to validate the output of the models. Numerical weather forecast models are moored to measured data every six hours, for climate models this procedure is not so easy, especially if they are running into the future.

For the initialization and validation procedures the quality of the measured data is essential. Since they are not more than random samples which are taken, maybe, at nearly the same time but at widely spaced locations, it is essential to check, whether they provide a consistent picture, to prove, how good is their precision, and to estimate, how large their error with respect to the natural state variables may be. This last estimate is difficult to perform because no absolute standard exists. As an example, if one measures precipitation at a few positions and calculates the precipitation of an area which can be compared to the grid size of a model, then there is no way to prove how close this calculation comes to reality.

Measured data are inevitable for the improvement of process descriptions in models as well for their initialization and validation. They provide a kind of irregular honeycomb like structure to stabilize model performance. But beyond their use with models there is a strong motivation to use measured data as an diagnostic tool. Even if they do not precisely reflect the natural processes and area averages, a series of data measured at the same location under identical environmental conditions provides information about changes and variability. Long term data series give more precise and reliable information about local changes than models can provide. The analysis of measured data allows to study, as an example, changes in the occurrence of extreme events which cannot be resolved in climate models. There are periods in the data series that represent such extreme climatic periods or climate "excursions". Studies of the connections of such events with large scale climate phenomena such as the North Atlantic Oscillation and monsoons may improve the understanding how the Mediterraneran area responds to changes of the global general circulation system. Such studies, to which it is believed the data presented here will make an important contribution, may then lead to an answer of the key question how Mediterraneran climate and with it Mediterranean ecosystems may react to global climate change.

The following contributions deal with various aspects of mostly empirical data analysis with the goal to provide a consistent picture of the Mediterranean climate variability and to identify trends. In chapter two it is tried to make a synthesis of earlier work and to set the scene for the more detailed contributions that follow. Chapter three presents a wider perspective looking at the relevance of climate data for other environmental research such as for desertification studies and in the agricultural area. In view of what has been said before about the linkage between empirical data and nature the question is addressed which data are best suited as "indicators" of changes. Indicators are measured state variables by which changes can be quantified. This concerns the whole range of environmental quantities, not exclusively climate variables but in addition those quantities which depend on climate respectively reflect the impact of climate on the environment. Next, in chapter four, large scale climatic phenomena are discussed. This is followed by a presentation of regional climatic variability in chapter five. Papers on the changes observed in the Mediterranean Sea and of the interaction between the surface and the atmosphere follow in chapter six. Finally examples are presented in the seventh chapter. Experimerntal approaches are addressed to monitor atmospheric state variables which are not operationally measured and networks are described which address specific environmental questions and gather data.

Chapter 2
Climate, Climate Variability, and Impacts in the Mediterranean Area: An Overview

H.-J. Bolle

1 Introduction

For many inhabitants in the Mediterranean basin - as in other parts of the world - the old millenium ended and the new one started with extreme meteorological events of in some cases disastrous dimensions. For the year 1999 the data are summarized in the Annual Bulletin on the Climate in WMO Region VI (1999). It started dry in Portugal as well as in the Middle East where this situation continued to early summer. In Portugal March was extremely rainy as it was in northern Spain (Coruña: 246 mm). Sicily had the first snow in February since 1981. The April to July temperatures of the Spanish mainland reached their highest value since 1961 and in Portugal new maximum temperature records were measured. August was hot and dry in southern Europe. On the 10th of August the temperatures climbed up to 45 °C in Catania Signorella, which is a new record. In September the weather was very unsettled. Portugal had great damage due to flooding because the rain was much above normal. In Valladolid it rained 106.8 mm which is the highest monthly total since 130 years. Segovia had the highest number of rainy days during the last century. Three severe tornados damaged Alicante, Murcia, and Granada. Up to 230 mm rain per day fell in north-eastern Italy, an event which has a statistical probability to return every 50 years. Extreme weather conditions continued in October. The central Mediterranean experienced temperature anomalies up to 2°C above normal (37.4 °C in Palermo, 38.6 °C in Catania). In the Southeast of Spain the monthly precipitation totals reached 400 % of normal and torrential rainfalls occurred in the mountain watersheds of the Venetian plains and along the northern Apennines (23 and 26 October). In November temperatures in Israel were 5 °C above normal for eight consecutive days. Heavy storms led to floods and landslides in Greece with high losses in infrastructure and crops, four people died. An exceptionally rainy period occurred in south-western France with rainfall up to 620.2 mm in 48 hours causing estimated losses of more than 500 million € and 196 mm in 24 hours in Cagliary. On 18/19 November rainstorms with strong winds and 100 mm precipitation/10 hours went over Tuscany, inundated industrial areas and settlements, uprooted trees and blocked roads. Also Turkey was affected by storms.

In December 324 mm of rain were recorded within 2 days in Campania, east of Naples. In the South-East it was again dry. Cyprus experienced the 5th consecutive year below normal precipitation. Israel reported one of the hottest summer of the last century (1998 was the record with even one degree more in Bet Dagan) and, as in 1958/9, one of the driest September to December periods. In the Southeast of Spain and the Balearic Islands the year 1999 ended with 50% less rain than normal.

Long lasting dry periods during summer 2000 caused severe stress on plants, fostered wild fires in many places, and led to losses in harvests. Following the dry period in a number of places such as at the east coast of Spain, the south coast of France and northern and middle Italy long term rainy periods occurred during late autumn and winter 2000/1 with partially disastrous torrential rains causing flash floods and land-slides which killed people, destroyed houses, eroded fertile soil, and obstructed farmers to bring out seed at the right time. Venice had "aqua alta" the first time for several weeks continuously.

The sequence of extreme events continued in 2001. In Italy the dry period lasted from early summer to winter 2001/02. As a consequence of the dryness and the permanent inversion situation the pollution in all major Italian cities reached extremely high values so that traffic had drastically to be reduced. After a long dry summer torrential rainfall occurred in November in Algeria. Nearly 400 people were killed and terrible destructions were caused in Algiers. Strong storms crossed Spain and caused became especially strong over Balearic Islands. The East in January 2002 suffered from a very cold spell that brought snow to Crete with severe frost damages.

Extreme events always happen from time to time but their intensity and frequency seem to increase during the last decennium. Apparently these extreme meteorological events were regionally as well as temporarily irregularly distributed. It is difficult to detect the pattern and to identify the "driving force behind the scene".

Is this a scenario to which the region has to get accustomed to in the future? Are these events a regional manifestation of a global climate change? Do they indicate an instability of the climate, maybe a transient state, or are they merely statistical excursions? This to know is important for future planning because preventive measures would be different in the different cases. In view of the most recent new and alarming IPCC assessment the question can even be aggravated: *Will the amplitude and frequency of extreme climatic events be further enhanced in the future or will a general trend lead to a different "Mediterranean climate" within this century?*

To clarify the situation, it has to be asked for the causes of these events. The connections between the regional impacts and the large scale changes have to be analysed and the question has to be answered, how economical and ecological losses due to climate change respectively increased climate variability can be reduced. These "natural disasters" are intimately interwoven with man's actions on land. In fact some of the impact of torrential rains can only be understood because of the transformation of the land surfaces in the Mediterranean area has a long history and led to deforested, over-exploited and in some regions overpopulated landscapes which have not much to offer to mitigate the effect of strong rains. In addition agriculture, industrialization, urbanization and tourism draw on the water resources which causes sinking ground water levels in these areas to an extent that in some areas rivers draw water out of the surrounding landscapes rather then fed them with water (EFEDA 1992).

The Mediterranean basin is unique in the world because of its geographical position which brings it under the descending branch of the Hadley circulation in summer while the Westerlies prevail during the winter season. The alternation between these two regimes and the strong influence of the Mediterranean Sea, which constitute a more conservative but, as recent investigations have shown, nevertheless changing element, makes the region vulnerable to large scale climate changes. The problems arising in this area therefor differ considerably from those faced in other areas of the world at which the World Climate Research Programme (WCRP) and the International Geosphere-Biosphere Programme (IGBP) so far focussed their research activities. The Mediterranean constitutes a counterpart of the Baltic Sea which, as the Mediterranean, is surrounded by land masses but is dominated by the Westerlies and the surrounding landscapes are mostly wet. This makes a comparison of the hydrological cycles and the role of the different types of vegetation in both areas highly desirable.

The complex structure of the land masses surrounding the Mediterranean Sea makes it difficult to generalize the climatic variability, to assess the impact of vegetation on the hydrological cycle, and to model the Mediterranean climate. It requires high resolution models to resolve at least to a certain degree the coastal features and the variety of ecosystems which is necessary to include the boundary layer effects of both the influence of the sea and of the topography. At the same time the topping of the basin by the large scale circulation pattern has to be accounted for. This complexity makes it difficult to predict how the Mediterranean climate may respond to global climate changes and explains the weakness of global climate models in this region.

In this chapter some aspects of the climate variability and related issues in the Mediterranean basin are discussed from which research needs result to better understand (i) the role of the Mediterranean area in the global climate system, (ii) the dependency of the Mediterranean climate and its vegetation on the large scale circulation systems of both the ocean and the atmosphere, (iii) the internal processes which couple the Mediterranean Sea through the sea-land circulation systems with the inland hydrology and vegetation, and, finally, (iv) the possible responses of the Mediterranean biosphere and water resources to climate variability and the future development of global climate.

In the following sections firstly the present climatic situation in the Mediterranean area is sketched. This prompts the question how the present situation is related to past climatic developments and whether an inspection of past cycles may give us a hint about what succeeding generations may have to expect. The expectation is large that climate models may provide us with more reliable scenarios about future developments. This requires a better resolution of the Mediterranean topography and an improved representation of the land-surface processes in the models. These models have to simulate interacting processes from the scale of the general circulation down to regional and even local processes at or near the Earth surface. Without going into modelling details an overview of the interactions between the various scales involved is presented in the following section. Land surfaces are coupled to the atmospheric climate system by the exchange of energy, momentum, and fluxes of substances. These processes are dealt with in the next two sections under very specific points of view. Firstly, the exchange of carbon in its volatile compounds is considered, because the global carbon cycle is of fundamental importance for the estimation of the future evolution of the greenhouse effect. Some of these volatile compounds as well as particles raised from the surface furthermore interact with atmospheric chemistry and radiative energy transfer and therefor are of equal importance for the climate system at least in a regional sense. A central question for the Mediterranean area is, whether a trend is superimposed to the presently observed climate variability or whether "simply" an increase of the amplitude of the variability is experienced. The water cycle is discussed in the light of data analysis and model predictions. Conclusions are summarized in the final section.

2 Characteristics of the Mediterranean Environment

2.1 Climatological Mean Values

The Mediterranean type of climate is found mainly at the west coasts of the continents between $30°$ and $45°$ latitude in the European/African/Levantine Mediterranean basin, South Africa, Chile, Mexico, USA, and South Australia. It is characterized by winter rains and summer droughts with a strong soil water deficit in summer. It is a temperate rainy, humid meso-thermal climate with dry subtropical warm to hot summers. The average temperature of the hottest month can be $>22°C$. The strong difference between the wet winters and the dry summers is caused by the seasonal alternation of the dominance of cyclonic storms in winter and subtropical high pressure cells over the adjacent ocean in summer with subsiding maritime tropical air causing dry conditions in summer. The Mediterranean climate in most regions is confined to narrow coastal belts. Only in the European/African/Near East area, because of the Mediterranean Sea, it covers about 10 Mkm^2 reaching from Portugal to south of the Caspian Sea with a core area that stretches over 46 degrees longitude and covers about 6 Mkm^2. Of this area $2.496 * 10^6$ km^2 belong to the Mediterranean sea surface. If the Black and Marmara Seas are counted as well, the sea surface covers $3.1 * 10^6$ km^2.

Within the belt of Mediterranean climate exists a strong gradient of the meteorological parameters. As an example, in Monterey, California ($36.5°$ N), the annual monthly temperatures range between $10°C$ and $16.7°C$, the precipitation between 90 and zero mm/month with a total of 424 mm/year. The temperature range is only 6.7°C. The respective numbers are in the slightly further northern Naples (40.5° N) 9°C to 25°C (a range of 16°C), 122 - 15.2 mm/month, 860 mm/year, and in the slightly more southern Benghasi (32° N) 13 - 26°C (a range of 13°C), 94 - 0 mm/month, and 302 mm/year (Strahler, 1975).

The general temperature and precipitation distribution in the European/African Mediterranean basin is presented in Table 1 according to Legates and Willmot (1990a and b). The cloud frequency in winter is of the order of 40 - 60 % for European countries, 20 - 40% for north African coastal zones, and up to 70% over the sea. In summer the respective numbers are 20 - 40% (down to 10% in the Aegean), 10 - 40%, and 10 - 40%. This results in annual sunshine hours between 2500 and 3000.

Table 1. Mean surface air temperature and precipitation regime of the Mediterranean basin according to the European Climate Support Network (1995). Reference period 1960 - 1990

Quantity	Season	European lowlands	European highlands	North African coastal zone	Sea
Temperature (°C)	winter	7.5 ± 2.5	2.5 ± 2.5	12.5 ± 2.5	15 ± 5
	spring	12.5 ± 2.5	7.5 ± 2.5	17.5 ± 5	15 ± 5
	summer	22.5 ± 2.5	15 ± 5	32.5 ± 10	22.5 ± 5
	autumn	17.5 ± 2.5	10 ± 5	20 ± 5	20 ± 5
Precipitation (mm/day)	winter	1 - 3	3 - 10	0.25 - 2	2 - 10
	spring	1 - 3	3 - 10	0 - 2	0 - 2
	summer	0 - 1	1 - 3	0 - 0.5	0 - 1
	autumn	1 - 5	2 - 10	0 - 2	0.5 - 5

2.2 Spatial Variability of Mediterranean Climate

The climate gradient across the Mediterranean basin is extreme. It ranges from the cold mountainous areas of the Alps with annual mean temperatures below zero (-2 °C at 2500 m elevation) to the hot plains of Africa with annual mean temperatures up to about 22 °C even near the coast and monthly average summer temperatures of 26 °C near the coast and 32.5 °C inland (25.5° N). Inland the maximum annual temperatures are reached in July while at the coast it is delayed to August which indicates the effect of the sea on the climate (Fig. 1). The amplitude of the temperature wave is larger at the dry inland stations as compared to Mediterranean coastal sites and the differences between the stations are larger in summer than in winter. The Mediterranean basin also contains the sites with highest European annual precipitation of much above 1000 mm at the eastern Adriatic coast and its hinterland (4600 mm near 42.8 °N, 22.3° E) and less than 100 mm in North Africa at 32° N. A differentiated picture of the climate at a selection of operational stations around the Mediterranean Sea can e.g. be looked up in Wagner (2001).

In Fig. 2 the annual course of average rainfall across Tunisia is presented. It shows the gradient from the highest winter precipitation at the northwestern coast to the eastern coast and the inland. Remarkable is that at the more southern stations often the winter precipitation maximum is not in December or January but in October and March.

The spatial variability of the Mediterranean climate is high even at relatively small spatial scales. As an example we look at the records 1901 - 1970 of a triangle in Castilla - La Mancha, Spain, where in 1991 and 1994 the experiment EFEDA[1] took place (Roldan Fernández, 1988). At the station of Albacete (699 m) in the central plateau of Spain, 150 km off the Mediterranean coast, the annual rainfall varies between 185 mm (1950) and 657 mm (1959). The absolute maximum and minimum temperatures in August are 42 °C and 8 °C. The maximum average rainfall (50 mm/month) occurs in

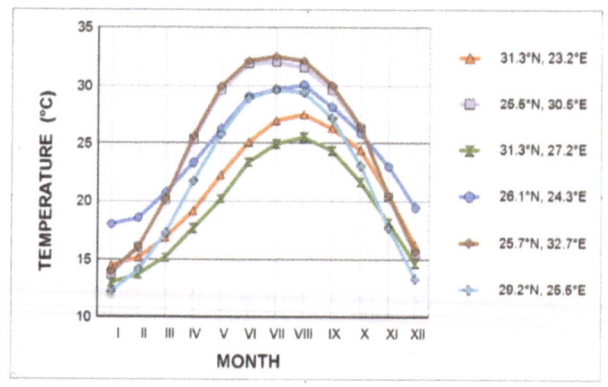

Fig. 1. Annual temperature wave at some sites in North Africa. The graph shows monthly mean values according to WMO (1999)

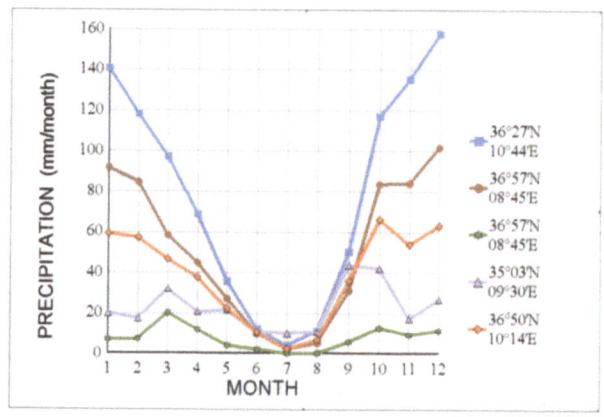

Fig. 2. Annual course of precipitation at some stations in Tunisia after WMO (1999)

[1] European International Project on Climatic and Hydrological Interactions between Vegetation, Atmosphere, and Land-surfaces (ECHIVAL) Field Experiment in Desertification-threatened Areas

May and October at average temperatures of 15 °C. Minimum precipitation occurs in July (9 mm) at 24 °C accompanied by south-easterly winds, especially during nighttime, and the winter rain is of the order of 26 mm/month at temperatures of 5 °C accompanied by westerly winds. In a distance of only 120 km to the north, in Cuenca (955 m) at the fringe of a mountainous range, the rainfall lies between 40 and 70 mm/month at temperatures between 3 and 13 °C during most of the year. Only in July and August the precipitation drops to 20 mm at temperatures of 22 °C. Total precipitation lies between 322 mm (1918) and 984 mm (1933). Mostly westerly winds blow all over the year except during nighttime in summer and autumn with low speed. 215 km to the north-west of Albacete, in Toledo (551 m), winds and the Mediterranean influence are much weaker than at the other two stations, in July and August 10 mm precipitation is measured per month when the temperatures are at 26 °C. In summer the air is much drier than at the other stations. Most of the rain (30 - 40 mm/month) falls during the rest of the year with totals between 191 (1950) and 575 mm/year (1955).

A number of studies have been made during recent years to assess the variability of the Mediterranean climate and synergies between the different parts of the basin. Because of its overriding importance for the vegetation, major attention is focussed here on the precipitation regime. Goossens (1985) applied a principal component analysis on a large set of data and found five different rainfall regimes in the European part of the Mediterranean:

- In the north-west of Spain, northern Portugal, northern Italy and the eastern part of the Mediterranean France rainfall exceeds 700 mm more or less equally distributed over the year.
- In the north-east and south of Spain, southern Portugal, and the western Mediterranean part of France the rainfall is mainly during the winter half-year and very little rain (50 mm) falls during summer.
- On the Balearic Islands, in the central part of Italy, northern Greece, and part of the former area of Yugoslavia precipitation falls mostly during winter but the summer is not without rain: 80 mm in the northern parts of this region but amount and duration progressively decreasing towards 20 mm in the south.
- In the south-east of Italy, Croatia, Serbia, and Albania occurs a single precipitation maximum in winter while the summer is almost but not completely rainless (e.g. Tirane 150 mm). This, in fact, is the area of highest annual precipitation (Crkvice: 4600 mm).
- Considerably less precipitation falls in other regions of dry and dusty lands throughout the long summer, e.g. Athens 14 mm, Hiraklion 3.2 mm, the Marathon area with 5-10 mm in both July and August according to Amanatidis et al. (1993).

Corte-Real et al. (1995) went one step further by correlating atmospheric circulation anomalies in the 500 hPa respectively sea-level pressure fields with monthly mean temperature and rainfall. The pattern of the canonical correlation analysis between the large scale pressure anomalies and the temperature anomalies are more coherent than those with precipitation. This indicates that the precipitation events are governed by smaller time-scale systems than those of the temperature. Between the eastern and western part of the Mediterranean exists an anti-phase relationship which can be attributed to blocking situations over the Atlantic Ocean. Strong Atlantic blocking accompanied by frequent depressions result in positive temperature anomalies in the east and negative in the west and *vice versa*.

The temporal variability of Mediterranean climate will be discussed in connection with trend analysis in section 6 with special emphasis on the hydrological regime.

2.3 Vegetation and Soils

Mediterranean woodlands are dominated by evergreen hardwood forest termed *sclerophyll forest*. Its major species are holm oak (*Quercus ilex*), cork oak (*Quercus suber*), Aleppo pine (*Pinus halipensis*), stone pine (*Pinus pinea*), black pine (*Pinus nigra*) and olive trees (*Olea europaea*). The olive appeared as a wild species after 10.000 BP and are widely grown since about 4.000 BP. In addition Eucalyptus grows in coastal areas.

Where the forests were destroyed either to obtain construction and burning materials or by wildfires, sparse to dense scrub (*low macchie*) regrows, termed *maquis* or *macchia* (Spanish *monte bajo*, in Greek *Xerovumi*). Most of the low maquis species root easily along cracks and layer surfaces and regenerate by root suckers which make its distribution less vulnerable by fires. Under favourable conditions holm oaks and a wealth of other plants (see e.g. Paccalet, 1981) regrow, forming the *high macchie*. Forest may regrow, but sometimes with different species, if the climatic and soil conditions are favourable and if there is no further intervention for a longer time period.

Under less favourable conditions on calcareous soils the vegetation may degrade to *garrigue*, Spanish *matorral* or *tomillares* which is characterized by less dense and lower vegetation. "Dwarf forest", chaparral, may develop. Heath species (*Ericaceae*: *Erica arborea* - tree heath, *erica multiflora*) are often found on degraded soils. In its extreme, the Spanish *matorral* vegetation sometimes covers only 2-4% of the surface (Albaladejo, 1995). Major species are rosemary (*Rosmarinus officialis*), specific gras species (*Stipa tenacissima*), sun-roses (*Helianthemum pilosum*), clover (*Anthyllis citisoides*), mastix (*Pistacia lentiscus*), broom (*Genista acanthoclada*), *Juniperus phoenicea*, *Thymelaeaceae* (in dry areas), *Cistus monspeliensis* and *salvifolius*, *Smilex aspera*, Lonicera etrusca, *Opuntia ficus-indica*, *Phillyrea angustifolia* (not in Greece) and many other species. *Garrigue* is resistant against arid conditions but during hot and dry periods the leaf area index and weight decreases.

The vegetation-climate relationship can most evidently be studied Turkey (Atalay, 2001). At the climate climax after the last glaciation the vegetation was governed by *Pinus brutia* (red pine) in the lower belt, and Pinus nigra (black pine), Cedrus libani (cedar), and Abies cilicica (Taurus fir) in the higher regions. The lower belt starts at the coast with around 800 mm precipitation and reaches up to

400 to 1500 m where the lowest winter temperatures are -15 °C. After destruction of the forests maquis substitutes the pines and if seeds of *Pinus brutia* are still present the red pine can regrow and form local woods. If further destruction goes on, the maquis degrades to garrique with mostly thorny vegetation of less than 1 m height. In the higher belt, starting at approximately 400 m in the Marmara region, at 750 m in the Aegean region, and at 1200 m in the Taurus area cedar black pine and Taurus fir are still present mainly at northern slopes. At southern slopes cedar and black pine are more destructed. In the areas, were the forests are destroyed, *Juniperus* communities develop.

It should be reminded here that the present Mediterranean vegetation is a mixture of "original" reminiscent and many "imported" species such as the olive tree and palm trees in the European Mediterranean.

Reddish brown and reddish chestnut soils are typical for the Mediterranean. Frequently Luvisols (15.6%), Cambisols (16.2%), Calcisols (20.3%), Vertisols (3.5%), and Regosols are found (Ibáñez et al., 1996). Red soil, *terra rossa*, forms on

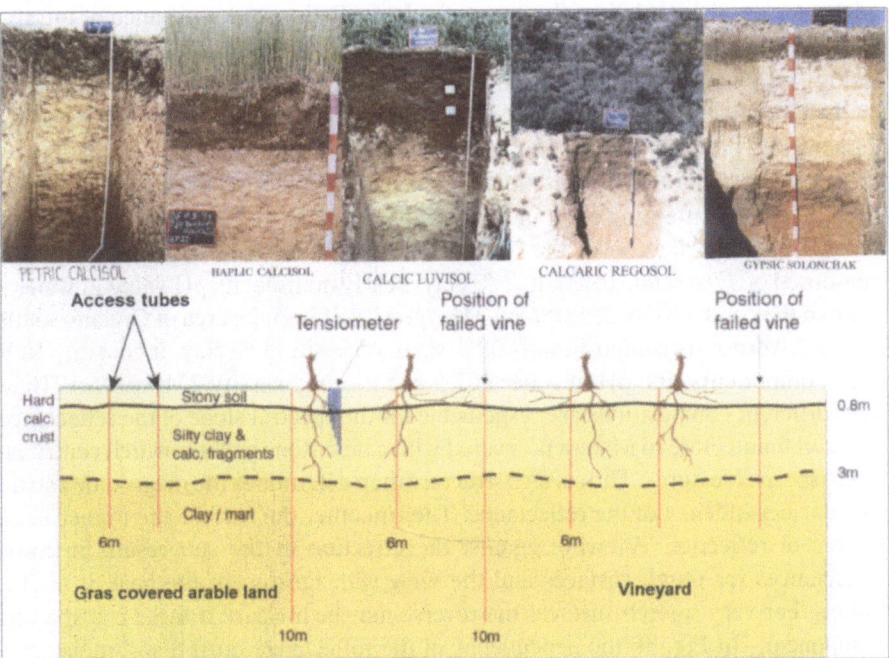

Fig. 3. Soils according to Diaz *et al.* (1992 and 1995) and a measuring profile across arable land and grapevine in the EFEDA experimental area of Castilla-La Mancha, Spain after Bromley (1995). Under dry conditions (29 September 1994) the moisture volume fraction (MVF) was close to zero in the surface layer, around 0.1 from 0.1 - 1.5 m depth with a first narrow maximum of 0.17 at 0.3 m. A second maximum up to 0.3 under grapevine was reached around 2.2 m and below the MVF was reduced to 0.13 and a minimum at 8m of 0.05. In the layer at 1 - 2 m depth the water storage was about half as large as under the arable land which indicates that the grapevine draws its water mainly from this layer which consists of a spongy soil layer that can keep the water. During the measuring period from August 1994 to July 1995 rain was only one third of the long term annual mean. Most of it evaporated from the layers above 0.4 m

limestone with iron oxides. Under dry conditions with capillary water rise from C-levels calcium carbonate is brought up and calcification takes place with the precipitated calcium carbonate in the B horizon below the root zone. A description of the soil classification and soil properties is given by Bouwman (1990).

Fig. 3 gives a view at soils found in the Castilla-La Mancha area in Spain, where the EFEDA (Bolle et al. 1993 and Bolle 1996) experiment took place. In near costal plains and hilly areas (colline metallifere) of Tuscany often iron rich reddish and sludge rich gray soils are found.

The constitution of the soils is an inseparable component of the water cycle over Mediterranean land surfaces. It determines whether or not roots can reach some water stored in deeper soil levels (Fig. 3) or have to rely upon rain water which they catch with their shallow roots in the upper soil layer. Consequently, in the major Mediterranean field experiments like EFEDA the structure of and water transfer through the unsaturated zone of soils is of fundamental importance for the investigation of water fluxes.

Soils as well as the vegetation have an important function for the energy budget of the surface due to their reflectance of solar radiation and emittance of infrared radiation. A few examples of the variety of observed spectral reflectances are given in Fig. 4a-d which demonstrate the radiative characteristics of the surface.

In Fig. 4a the reflectance of some characteristic soils is reproduced which demonstrate the dependence of the reflectance of both the chemical composition and the surface structure. Two soils which often appear in close neighbourhood are the red iron oxide containing soils and grey silty soils..The "red" soil the reflectance of which is shown here is of the Maremma near Venturina, Tuscany, Italy. It is composed of 77% sand, 16% silt, 7 % clay, nearly no lime, its pH value in water is 7.1 and the conductivity 235 µS/cm. The "grey" soil is of the area of Orciano south-east of Livorno. Its composition is 33% sand, 65% silt, 2 % clay, it contains 18 % lime components, its pH in water is 7.5 and its conductivity 2110 µS/cm. These very different compositions are responsible for the spectral slope of the reflectance. The soil found close to Matera is covered with calcic stone pebbles which contribute to the spectral features. The surface structure then determines the magnitude and the angular dependency of the reflectance. The smoother the surface the higher is the degree of reflection. Viewing against the direction of the sun results in lower reflectances for rough surfaces and the view with the sun in the back in higher values. For very smooth surfaces the reverse may be the case if there is a specular component. In Fig. 4b the dependency of the reflectance on the soil moisture is documented. Irrigation reduces the reflectance as long as the soil is not too wet and one does not look into the sunglint. Due to evaporation and suction the moisture dissipates from the surface and after a few hours the soil returns into the dry state and the surface may now have a different, smoother, structure. The figure shows the appearance of the water absorption bands at 1.4 and 1.8 µm and that changes in the reflectivity are measurable which happen within a few minutes.

If the surface is covered with gradually denser and biologically active vegetation the reflectance in the visible part of the spectrum is reduced due to chlorophyll absorption and the reflectance in the near infrared part of the spectrum is enhanced

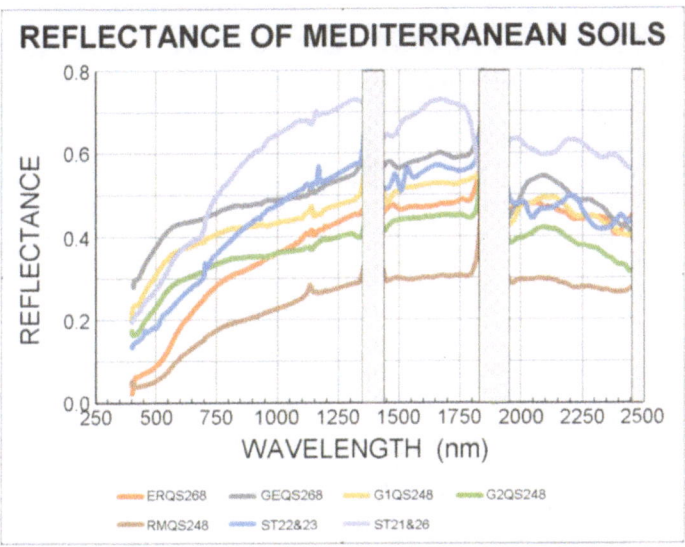

Fig. 4a. Reflectance of three soil types. The violet and blue spectra represent thinly vegetated soil with white calcic stone pebbles viewed with the sun in the back (violet) and with the sun in front (blue). The measurements were taken near Matera, Italy. The grey, yellow, and green lines are spectra of smooth respectively grainy "grey" soil surfaces. The red and brown lines stem from a smooth respectively grainy "red" soil surface. Further details see in the text

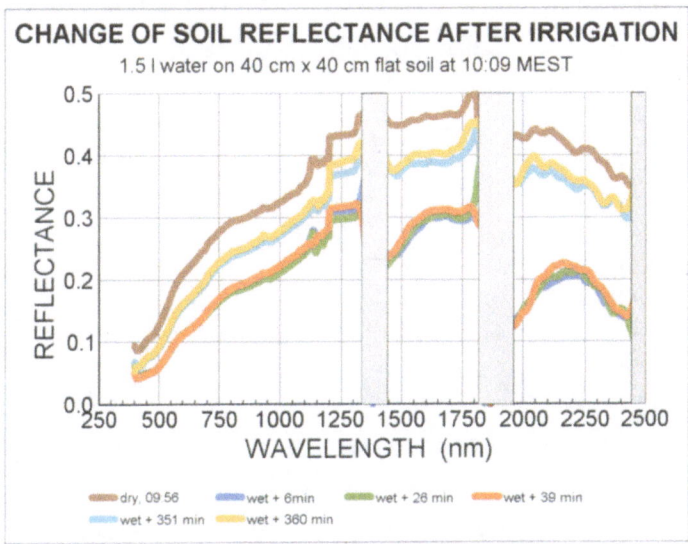

Fig. 4b. Dependence of spectral reflectance on the soil moisture. The in this case brounish soil (Tuscany, Italy) was irrigated at 10:09 hours with approximately 9 mm water. The liquid water absorption bands become a prominent feature after the irrigation and disappear again as the soil dries

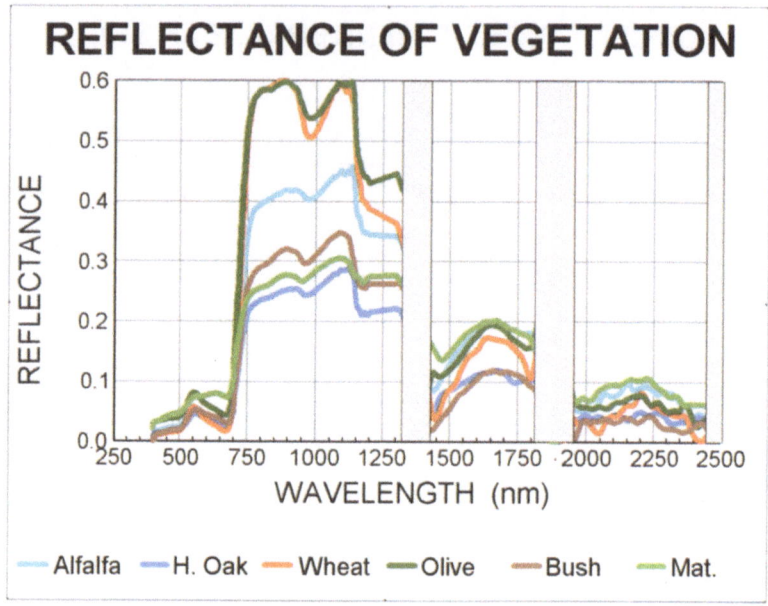

Fig. 4c. Reflectance of different plants. The olive tree (Tuscany) and the wheat (Matera) reach the highest values while the matorral (Castilla-La Mancha) and macchie vegetation remains at lower levels in the near infrared. The matorral vegetation is sparse. The underlying soil enhances the reflectance at 650 nm

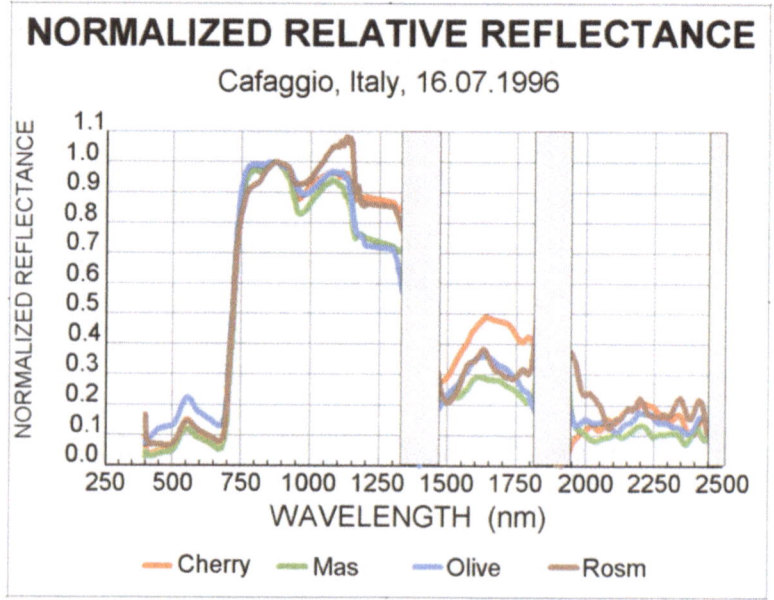

Fig. 4d. At 800 nm normalized spectral reflectance of some plants showing a very similar normalized spectral behaviour. The coloured lines show the spectra of cherry leaves, pistachio lentiscus (Mas), olive, and rosemary.

due to scattering of solar radiation by the leave and needle structures. The more active the vegetation is the stronger are also the water absorption bands of which even the weaker one at 0.94 µm appears. Fig. 4d finally shows that the general slopes of the spectra of different plants are not very different from each other though the amplitudes and the angular dependency of the reflectances are.

The emittance of vegetated surfaces follows, because of the water content of the plants, nearly the Planck formula for Black Body radiation. Integrated over the whole wavelength range it is proportional to about $(0.985 \pm 0.005)\,\sigma T^4$, where T is the absolute thermodynamic surface temperature in Kelvin and $\sigma = 5.6703 * 10^{-8}$ W m^{-2} K^{-4} the Stefan-Boltzmann constant. The emittance of bare soils is smaller and is proportional to $\varepsilon\sigma T^4$. ε, the emissivity, takes values around 0.9 and is very variable with soil type and moisture.

2.4 Contemporary Human Drivers

Land management and irrigation are major human impacts on land surfaces which may affect land surface processes and in turn the Mediterranean water cycle and regional climate. While in ancient times regional climates have been affected by human activities such as large scale deforestation (the wood was needed for constructions, heating and cooking) and the intensification of agriculture (see section 3) the modification of land-surface processes with the potential to affect regional climates has shifted to more complex causes and actions during the last century among which economical reasons and more recently the European subvention policy can be identified. Only few of them are discussed here.

During the first half of the century in southern Portugal forests have been cleared in favour of crops to make the country more autarkical. The result now is dramatically increased erosion due to flash floods. In most Mediterranean countries the introduction of heavy machines for land treatment has changed the agricultural practice from the former terracing to agriculture on flat or soft hilly land especially in estuaries which are preferred for growing agricultural products because here the land is flat and the soils are fertile. Land which does not meet the conditions for machine work is abandoned. In Italy the Pontine swamps have been drained since 1935 and converted to agricultural land. Here and in other coastal zones, the demand for irrigation is growing firstly for competitive reasons - higher harvests - and secondly because apparently the dry spells seem to extend throughout the summer months. Most of the irrigated water remains in the uppermost soil layers and is very quickly evaporated thus not being much helpful to reduce the soil water deficit. Because of the deep pumping in these regions salty sea water is entrained into near coastal aquifers which makes the water useless for agriculture. In addition industry is preferably settling in coastal areas and draws on the fresh water resources.

Especially in Portugal, Spain and Egypt rivers are dammed for large water reservoirs. In Morocco an irrigation system was constructed for about 1 million hectare using the water from nearly 25 smaller dams which have been constructed since 1950 (Wagner, 2000). This is also the practise in Tunisia and Algeria but in a smaller extent. The irrigation of 1.6 million hectare is the aim of the large Atatürk dam in Turkey. In Lybia a 600 - 800 km long water tube of 4 m diameter is constructed to irrigate 200 000 - 400 000 ha with prehistoric water from underneath the Sahara. It is expected that this not renewable water reservoir will spend the necessary water for 400 years. The use of smaller water reservoirs down to private pond sizes is becoming very popular to bridge the dry summer months with winter rain water. In Spain a system of dams and channels was build to provide for balanced supply. Over long distances water is transported from the central parts of the Iberian Peninsula through the Canal de Trasvase Tajo Segura to the South. The new Spanish water plan even attempts to branch off Ebro water to Murcia and Almeria in the South. The area around Almeria, were water is very scarce, is a major provider of vegetables already in very early spring. For this purpose the whole peninsula of Llanos de Almeria as well as other places are covered by greenhouses made of plastic covers. Not very far away and adjacent to the desert of Almeria near Jaén a monoculture of olive trees under which all undergrowth is removed stretches over a terrain of about 10^3 - 10^4 km^2. Olive trees already in ancient times served as a stabilizer of threatened landscapes.

The ground water level in several parts of central and south-eastern Spain declines up to one metre per year. This being also the case in many other regions of the Mediterranean area. In the Messara valley of Crete, as another example, the groundwater level fell about 25 m from 1985 to 1996 (Grapes, 2000). In total 22 000 ha are estimated to be under irrigation by 1997 in the Mediterranean basin (Wagner 2001) which is an increase of 39 % since 1975. One of the consequence of these irrigation systems is that albedo, evaporation and surface temperature are regionally changed. An additional factor to agriculture is the construction of tourist facilities with pools near the coast but also in the hinterland which draws on the water resources.

In Greece, e.g. east of Athens, new villages which will draw on water resources are build into a treeless landscape completely degraded by wild fires and overgrazing (Schultze-Westrum, 1994, Papanastasis, 2000). Whether man-made or caused by natural events like lightening, forests and macchia in the Mediterranean are repetitively damaged by wild fires which either lead to degraded land or to the regrowth of unpretentious vegetation, mostly macchia, or of species the seeds of which better survive the fire. Thus it has been observed in Tuscany that a forest of *Pinus pinea* once planted by man regrew after burned down as *Pinus halipensis* forest. Reforestation efforts are large, especially in Spain, but this is a slow process, sometimes interrupted by new fires.

The pattern of human actions on land throughout the Mediterranean basin is very differentiated and regionally controversial. Some areas became dryer during the last century, some areas received more water due to human actions, sometimes on the expense of other areas or of non-renewable reservoirs . It is difficult to judge to which climatic effects the sum of all these activities finally will result. The extent of the various regional impacts have not yet been established. It may well be that the overall effect is small because the various actions compensate each other, but it is also very likely that regionally the water cycle may be enhanced where water is spend and aridity will increase where water is withdrawn. There is no doubt that in many areas the soil moisture deficit increases during the dry summers due to overexploitation of the water resources. Because of the long distance water transfers made possible by modern techniques (or the revival of the old practice of aqueducts) a change of regional precipitation often is not the alone a measure for a sustainable development. Therefore the prediction of precipitation changes by models without an efficient soil parametrization and without taking into account water transfers in natural as well as artificial aquifers do not represent adequately regional idiosyncrasies.

The growing European Union with its diversity of agricultural conditions enforces a redistribution of food production and a new structuring of the land-use. An overriding issue is the conservation and useful distribution of water. The storage of water of the wet seasons for use during dry intervals got large priority and the constructions for this purpose range from big dams, drilling projects and long aqueducts to small private ponds. The impact these new water surfaces have on regional climate is difficult to estimate. At the one hand these water resources allow the irrigation of large areas in which the land surface properties are changed and on the other hand the ground water level in adjacent areas is lowered by the concentration of the water. A study was carried out by the Turkish Department of Meteorology of the impact of the southeastern Anatolian irrigation project on regional climate. The project is composed of 22 dams (19 hydroelectric power plants) with a total irrigation area of $1.7 \ 10^6$ ha which would affect an area of 74.000 km^2 when completed in 2010. So far no significant signals could be detected in both long term temperature and precipitation data. The statistics of extreme events on the other hand shows an increase of high temperatures and an decrease of low temperature but no effect was detected in the precipitation. In 1992 the low temperature periods were at an extreme of 33 cases (Halpert et al., 1993). Model simulations indicate an increase of the precipitable water in the lower levels of the atmosphere (Ünal, Karaca, and Dalfes, 2000).

Another factor is the competition in the production of field fruits. In many places cultures are planted to obtain an economic advantage though looking at the totality of the European production and the import from elsewhere in the world these products are partly superfluous. They are also quickly changed if the production becomes unattractive respectively if subventions are cancelled. In this connection it is interesting to note that in parts of the Ottoman empire, e.g. Palestine, because of the taxation of trees these by the time vanished completely (Wagner, 2001, p. 230): "Political ecology" obviously played an important role since ever. In Italy some areas such as le Crete (SE of Siena) which were abandoned are taken under the plough again to grow grain. In Spain as well as in Italy rocky land is made available for agriculture again by smashing even large stones with big machines. Urbanization is growing in coastal areas and land is transformed to generate an environment for tourism. Fortunately a growing number of "natural parks" are conserved.

3 Past Climates and Vegetation in the Mediterranean Basin

3.1 Late Cretaceous Period to Pliocene Epoch (135 Ma - 1Ma B.P.)

Traces of vegetation, primarily pollen, and of animals, sediments, dunes or geo-
hydrological structures allow to draw conclusions with respect of the water, wind
and temperature regime in ancient times and the development of vegetation and
climate. 100 Ma B.P. the Sudan and Arabia were positioned near the equator and
southern Europe at 30 degree North. The African plate separated from the South
American one and moved towards the Eurasian plate from which it was separated
by the Tethys Sea (which was a broad seaway linking what we now call the North
Atlantic and Indian oceans until 20 million years ago). The Sahara at that time was
still in the tropics and southern Europe had a climate like the northern African coast
has nowadays. The first Saharian sand deposits, indicating its desertification, have
been found at the ground of the Atlantic Ocean from the Oligocen (33 - 28 Ma B.P.)
on (Sarnthein, 1978). 11 - 5 Ma old wood was found from trees which were adapted
to dry climate. Due to the northward movement of the African plate since 65 Ma
B.P. land masses slowly build up around the western part of the Tethys Sea and
formed the Mediterranean basin. In the East the brackish Paratethys Sea was left
over from which later on the Carpathian lakes, the Black Sea, the Caspian Sea, and
the Aral Sea formed. When the African continent collided with the Asian one the
Middle East mountains formed and interrupted the conection to the Indian Ocean.
The Balkan mountains blocked the drainage of the Paratethys Sea and northern
rivers into the Mediterranean. Around 15 Ma B.P. the Mediterranean basin had
approximately its present shape but its geographical position was still a few degrees
further South and the climate became dry. The continued northward movement of
the African plate build up the Iberian-Moroccan mountains which closed the
connection to the Atlantic Ocean between 10 and 6 Ma B.P. (Hsü, 1983). Due to this
complete isolation and the geographical position in the dry belt some four million
km^3 water evaporated and the basin fell dry and left a salt-covered surface with the
exception, maybe, of a few brackish lakes. Prior to the geological evidence that the
Mediterranean basin had been a desert, French and Italian paleontologists had
postulated a "salinity crisis" during the "Messinian" period between 6 and 5 Ma B.
P. to explain the extinguishment of species that cannot survive in sea water with
high salinity. It seems to be that the Balkan barrier got permeable for some rivers
around 5.5 Ma B.P. when the Paratethys broke up into the smaller lakes and seas
and some brackish lakes developed in the Mediterranean basin. Also major rivers
like the Rhone and Nile poured water (and with it gravel) into the dry basin. Around
5.2 Ma B.P. the North Atlantic Ocean broke through the Iberian-Moroccan barrier
and opened what we now call the Strait of Gibraltar. The much lower lying
Mediterranean basin regained its water volume by what must have been a gigantic
waterfall. Later on the Bosporus barrier eroded and the Mediterranean Sea was
connected with the Black Sea and northern rivers as well.

3.2 Pleistocene (1 Ma - 10 ka B.P.)

An overview of the climatic and vegetational development within the more recent period of the Earth's history was presented by Yassoglou (1998, 2000) based upon the work of Gilman and Thornes (1985), Verheye (1991), Tzedakis (1993), Runnels (1995), Grove (1996) and the ARCHAEOMEDES Project (van der Leeuw 1998, McGlade and van der Leeuw, 1998). A description of the succession of vegetation phases is given by Wagner (2001) as well.

About 1 Ma ago the Aegean and the Adriatic Seas opened. The era of the periodic glacials started during which fluvial terraces were formed from 300 ka B.P. on. Southern Europe was covered during the cold stages of the Quaternary by open *Artemisia* (mugwort) steppes with scattered forest stands, at least during 70% of the last 2 Ma. From temperate forests which developed during warmer and wetter periods the vegetation degraded to steppe forest, grassland steppe and finally desert steppe during the cold periods. During the Würmian, as an example, the Epirus area was arid and treeless except in favoured areas with a mosaic of grassland and open vegetation, seasonally with a fair amount of bare ground, unstable slopes and poor soil development. With the retreat of the glaciers and the advent of a wetter period around 15 ka B.P. Pine-Juniperus forests appeared followed by oak forests and as climate became wetter, 12.5 - 11 ka B.P., these forests expanded. Around 11000 B.P. the first settlement in Jericho is dated. At that time there was enough water in the Negev to support a larger population concentrated in one place without modern technology (Bloch, 1970).

During the suddenly cooler upper Dryas period (10.5 ka B.P.) with its reduced precipitation wind blown silt from Africa was deposited over wide areas of the Mediterranean.

3.3 Early Holocene

The Juniperus-Pinus-Amygdalus (almond)-Pistacea forests and steppe mostly survived the cold spell of the Dryas. After that phase wild olive trees appeared and a rich vegetation developed at the time of the climate optimum around 8 ka B.P. with oak, birch, hazel, elm, ash-tree, lime, pine *(Quercus, Betula, Corylus, Ulmus, Fraximus excelsior, Tilia, Abis)*.

The development at the end of the last glaciation is closely related to the latitudinal position of the Westerlies. The subtropical high pressure belt is assumed to be positioned until 10 000 years B.P. at approximately 21°N in winter and 33°N in summer. The subpolar low pressure system was at that time around 50° N with small seasonal variations. About 9000 B.P. its winter position suddenly moved northward. By the time of the optimum of the interglacial, about 9000 B.P., the polar front had reached its extreme northern position at 80° N. The subtropical high pressure system followed this development and reached its extreme northern position at 41° N in winter respectively 47°N in summer around 5000 years B.P.. This is consistent with summer monsoon rains penetrating much deeper northward into the Sahara than nowadays. It is estimated (Lamb, 1982), that at this time there was 200-400 mm annual precipitation in the now hyperdry regions of the Sahara. The current "Azorean" high was at that time probably close to the Biscay, north-west of Spain (Lamb, 1977). The Saharan drying period started about that time, though until about 5000 B.P. there are relics in the Sahara indicating that still enough water was present to sustain life. This era coincided with a remarkable change in the composition of forests in other parts of the world and advances of the alpine glaciers. Between 5.5 and 5 or 4.8 ka B.P. there has been a short (few centuries) lasting cold period that ended the warm period with its storm belt at high latitudes and had remarkable impact on the vegetation, which did never completely recover from this "Piora Oscillation". From that time on the high pressure belt moved back into its current position of about 30/32°N, and the positions of the Azorean high and the Icelandic low determine to a large extent the weather situation in Europe.

From 7 ka to 5 ka B.P. oak, hazel and birch slowly gave way to hornbeam, fir and beech. A decline of vegetation development was noted around 6 ka B.P., except for maple, accompanied by an increase of herbaceous taxa. This could have been caused by increased rain but also during this time the anthropogenic intervention, clearances and grazing, became evident, probably first in Greece. Small villages were established in lowlands near the coasts.

4.2 ka B.P. climate changed abruptly, it occurred the Mesopotamean drought (Otterman and Starr, 1995). Dryness increased in general but with imbedded wetter periods. Intensive agro-pastoral activities developed. Olive trees were widely grown which contributed to the protection of vulnerable land but coastal pine forest were cut for ship construction. First traces of desertification appeared and propagated towards extensive erosion on hilly land and salinisation of irrigated land. Deciduous woodland was successively replaced by garrigue. In Turkey, which is one of the palaeolithic and neolithic settlement areas, first clearance started around lake Beysehir 3 - 4 ka B.P. and continued up to the present time (Atalay, 2001). Around 3.2 ka B.P. the population decreased severely until about 2.5 ka B.P. when Greek colonies were established at the northern and southern coasts of the Mediterranean. The population moved from low lying settlements at rivers to higher fortified locations and concentrated more in centres.

3.4 From the Roman Period to the Little Ice Age

Reorganization of land use took place by the Phoenicians and Romans. The Roman domination started about 2.1 ka B.P. when also marginal land was cultivated and the environment modified by drainage systems, river damming and aqueducts. Associated with agricultural expansion and further deforestation around 2 ka B.P. the vegetation changed as documented in palynological records and reconstructed from historical sources (Reale and Dirmeyer, 2000).

These large scale changes of land use may have contributed to long term climate change in this area. To test this hypothesis a coupled land-atmosphere climate model recently was re-run with the present distribution of vegetation exchanged against the land cover as suggested by Reale and Dirmeyer with the vegetation for the end of the Roman classical period (Reale and Shukla, 2000). Many of the properties of summertime precipitation suggested by both the historical and archeological evidence are recreated in the simulations. For example, North Africa from the Nile valley to the Atlas mountains are simulated as significantly wetter with the vegetation at the time of the Roman classical period. Much of the southern Europe (e.g. Greece and the Iberian Peninsula) are also wetter, with reduced rainfall over the Mediterranean Sea itself. The model results also underline the interaction between the Mediterranean area and the Hadley circulation. There is a stronger northward propagation of the Inter-Tropical Convergence Zone and its associated rains into the Sahel during boreal summer in the Roman classical period case by which also rainfall over sub-Saharan Africa is affected. In these computations all other parameters except the vegetation cover remained unchanged. In view of the complex interactions between the surface and the atmosphere this may not be a fully adequate procedure to really *proof* that climate is changed by changes of land-use, but it gives a hint for future research needs.

Font Tullot (1988) reports about strong cold spells over the Iberian Peninsula around the years 570 and in the winter of 763/4 when large snow masses were recorded and a "Siberian" cold 859/60. During this period with variable climate large floods of the rivers Guadiana (620 and 686), Tajo (849) and Guadalquivir (974 and 1011) are reported and also very dry years such as 675 and between 650 and 1000 the climate in Galicia in principle was dry and relatively warm. Again around 1010 cold spells occurred.

The years 500 - 700 A.D. were marked as "dark age" because of the bubonic plague which again reduced population but brought resilience to the "natural" ecosystems. Rural estates were abandoned and hydraulic constructions destroyed. After 700 A.D. until about 1500 A.D. agriculture expanded with improved irrigation systems and terracing. New crops were introduced which were brought primarily into the Iberian Peninsula by the Arabic. Wheat, barley, olives, legumes, almonds, fruit trees were grown in semi-arid conditions, which must have been similar as today, and the ecosystem is believed to have been in equilibrium at that time.

Again, around 1350 A.D., a severe depopulation of at least 30% took place due to the Black Death. 1500 A.D. the Arabs were expelled from Spain, irrigation systems were abandoned and badlands formed. Mulberry were replaced by olive trees. From 1700 A.D. on there has been a steady increase of population and an intensification of agriculture. Forest clearing accompanied with soil erosion went on. The "little ice age" did not leave traces in the Mediterranean area. The more recent variability and trends are discussed in sections 2 and 6 and are the topic of the chapters 4 and 5.

4 The Complexity of the Mediterranean System: Interactions at Various Scales

4.1 Scales of Interaction

Different disciplines use different names and definitions for the scales which they are addressing. Therefore here some neutral terms are used. It is clear what the global scale means and "large or macro scale" addresses something of the order of a continent. The "synoptic scale" (also addressed as 'meso α', $4 * 10^4$ - $4 * 10^6$ km²) is that of travelling eddies, of the order of 10^6 km². "Mesoscale" covers an area of the order of 10^4 - 10^5 km² (sometimes "meso β" is used for $4 * 10^2$ - $4 * 10^4$ km²). The "regional scale" is that of a province or a small to medium sized watershed (10^3 - 10^4 km²) and "local" covers about 1- 10^3 km² (almost equivalent to "meso γ", 4 - 400 km²). The scale below 1 km² also sometimes is addressed as 'micro'.

The vertical extent of the layer in which interactions between the atmosphere and the ground is perceptible has different and variable dimensions. The surface extends its influence on the structure of and processes in the atmosphere up to the top of the atmospheric boundary layer which ranges between a few hundred metres to about 3 km, depending on the roughness and temperature of the surface. The lower boundary is determined by hydrological and thermal processes in the soils and aquifers as well as the structure and vegetation of the surface. The diurnal heat wave penetrates the soil to 30 - 40 cm depth. Evaporation occurs in the upper centimetres of the soil but plants can draw water from much deeper layers. To understand the interaction between the locally highly variable sub-surface hydrological regime and the atmospheric system the processes of heat and water transfers in soils are as important as the evaporation processes at the surface and the transport processes just above the surface.

4.2 Large Scale Processes

The specific situation of the European-African-Near East Mediterranean is that land masses with highly structured coastlines surround nearly completely the Mediterranean Sea at a latitude where the general atmospheric circulation regime has a large seasonal variability. This variability is caused by the shift of the wets wind belt (the Westerlies) between a more southward position during winter into a more northward position during summer. This brings the Mediterranean basin under the influence of the Westerlies during winter and the descending branch of the Hadley[2] circulation during summer. The influence of the Westerlies respectively the

---

[2] The Hadley circulation consists of rising wet air near the equator that descents as very dry air around

(continued...)

Hadley circulation on the Mediterranean Basin not only varies with the seasons but also from year to year depending on the large scale general atmospheric circulation and the North Atlantic temperature regime.

The global scale connections probably go even farther because the Hadley cell is coupled to the zonal tropical circulation (Walker Circulation, Southern Oscillation) which is interconnected with the Indian Monsoon the influence of which reaches out to the Eastern Mediterranean. Some meteorologist even do not exclude teleconnections between the Mediterranean climate variability and the El Niño phenomenon possibly because at least the equatorial Atlantic oscillation is influenced by the El Niño (Latif and Grötzner, 2000). In fact small correlations have been found for the western as well eastern parts of the Mediterranean (Pongráz et al. 2000) but Colacino et al. (2000) could clearly demonstrate by applying coherence analysis that for the central Mediterranean area (Italy) no correlation exist between El Niño and meteorological parameters respectively events.

The wave-like pattern ("Rossby waves") of the general circulation is in the first instance responsible for the influence which the Westerlies expose on the Mediterranean area. This zonal flow can be modified by the geographical position and magnitude of North Atlantic sea surface temperature (SST) anomalies which are coupled with the semi-permanent pressure systems that develop over the ocean (Malberg and Frattesi, 1995). The Azorean high and Icelandic low pressure cells according to their positions direct the atmospheric flow either eastward into the Mediterranean area or north-eastward to northern Europe (Fig. 5). The pattern of the steering mechanism over the Atlantic Ocean results from both the general global atmospheric circulation and the intensity of the oceanic circulation system, the conveyor belt, which is driven by the formation of cold bottom waters in the arctic ocean and the influx of heavy salty water from the Mediterranean Sea through the Strait of Gibraltar. The production of the cold bottom water switches in the course of the years between a westerly position (Drake Strait) to an easterly position (Frahm Strait) and causes a variability of the strength of the conveyor belt including its upper level back-flow, the Gulf Stream. The warm surface water arriving from the Caribbean affects the SSTs of the North Atlantic Ocean which couples with the atmospheric pressure systems on top. The variability of the North Atlantic pressure field is known as the North Atlantic Oscillation (NAO). The variability of the general circulation system, which drives the sea surface layer, is called, according to Wallace (2000), the "Northern Hemisphere Annual Mode (NAM)". Its strength is defined by the zonal index cycle, which alternatively can be represented by the Eurasian circulation index (EU), see Luterbacher et al. (1999). Marsh (2000) identified the following phases of the winter NAO: A positive phase (Azores high strong and Icelandic low deep) was observed 1865, 1905-11, 1920-25, 1945, 1980-1995, and a negative phase 1905 (nearly neutral), 1918, 1940, 1964-75, and 1996.

---

[2](...continued)
30° latitude after leaving behind its water vapour in condensation processes at the Inter-Tropical-Convergence-Zone (ITCZ).

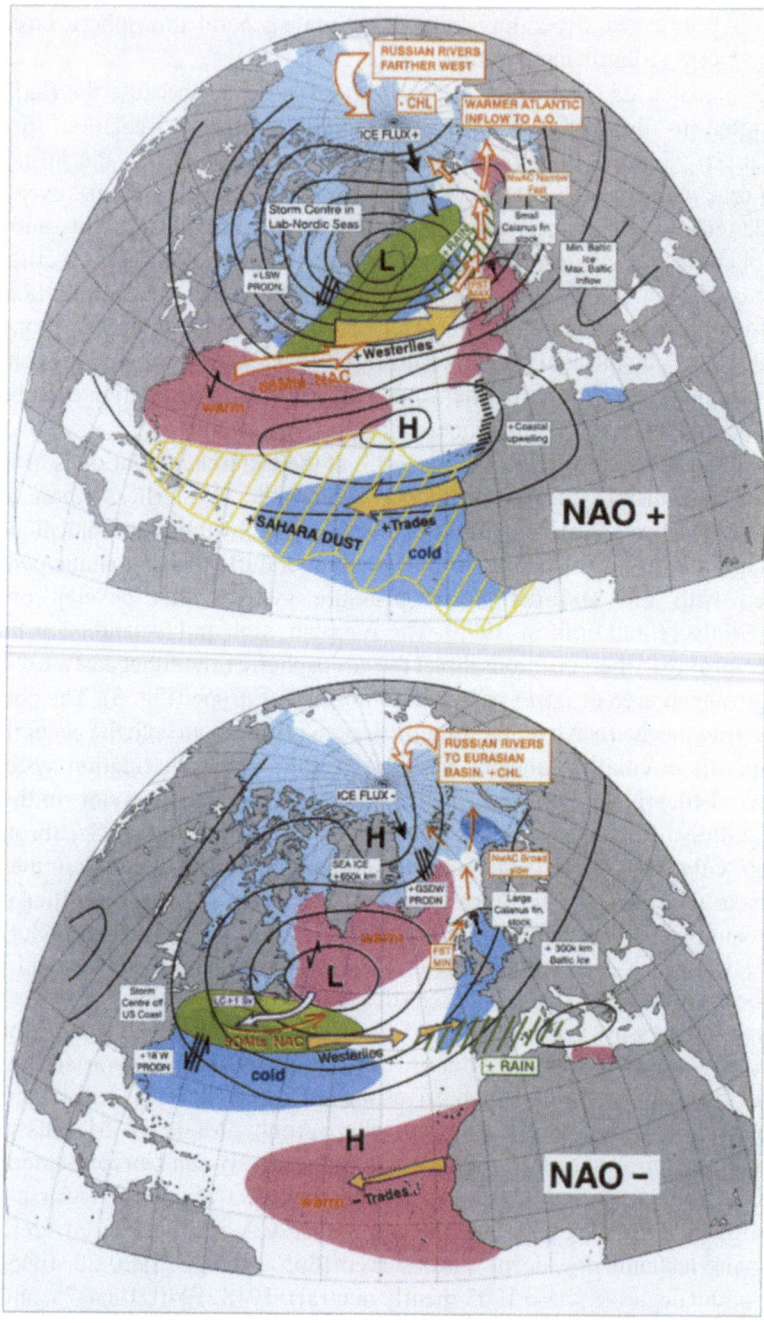

Fig. 5. Schematic representations of the (a) positive and (b) negative phases of the North Atlantic Oscillation. The figure originates from H. Wanner and J. Luterbacher (Wanner et al. 2001) and was reproduced from Sarachik and Alverson (2000) after Dickson (1996), CEFAS. By courtesy of H. Wanner and PAGES International Project Office

The general trend currently is increasing towards the positive phase.

Flohn (1993a, 1993b) reported, that between the periods of 1971-1991 and 1951-1971 the mean pressure in winter of the Icelandic low deepened by 3.5 hPa and the mean pressure of the Azorean high and the pressure over the Mediterranean increased in the average by 2-2.5 hPa. Thus the pressure gradient between about 30°N and Iceland steepened by about 4 hPa. Malberg and Bökens (1993) as well come to the conclusion that the pressure gradient over the north Atlantic and Europe increased over the time period from 1960 to 1990. They *inter alia* investigated the seasonal changes of the pressure gradient over the traverse from the Azorean high and the Icelandic low (2810 km) for the sixties, seventies and eighties, which is covering a full cycle of the periodic variability. During winter the gradient changed in 20 years from 12.9 hPa to 22.5 hPa, or +9.6 hPa, while during summer the change was rather in the reverse direction, from 14.9 hPa to 13.9 hPa or -1 hPa[3].

4.3 Synoptic Scales

The interaction between the general circulation and the semi-permanent North Atlantic pressure systems determine the path of the disturbances imbedded in the Westerlies primarily during winter and cause an interannual variability of the Mediterranean climate, specifically its precipitation regime.

During winter vigorous synoptic scale low pressure systems detaching from the Rossby waves determine weather and climate. High mountain barriers such as the Atlas, the Pyrenees and the Alps are necessary to modify or generate these systems which are developing in the middle troposphere and gain their momentum by internal energy transfer processes. Well known is the Genova cyclone generated by the interaction of westerly airstream and the bow of the Alps. Behind the mountains often chinook-like pattern develop. Smaller topographic obstacles are less important for these processes. The temperature during this season over the whole Mediterranean area including the north-western African coasts to a large degree is determined by the sea surface temperature. Topographic temperature features almost disappear in satellite images with the exception of high mountains and the Sahara (Bolle 1999).

4.4 Processes and Interactions At Meso- to Local Scales

During summer and early autumn the Mediterranean Basin is more decoupled from the zonal circulation. The westerlies pass further north and only seldom (in "blocking" situations) affect the Mediterranean basin. Its atmosphere now develops

---

[3] For the traverse Malta to north of Norway (3350 km) the respective numbers are: winter change from 9.4 hPa to 16.2 hPa, or +6.8 hPa, during summer from 4.1 hPa to 3.9 hPa with a minimum during the seventies of 3.6 hPa, or in the average, maybe, -0.2 hPa

internal dynamics which to a large degree is determined by the temperature gradients between the sea and the surrounding more or less vegetated land surfaces as well as their large topographic and ecological manifold. The topographic effects gain importance and interact with the much smoother large scale pressure systems in the Mediterranean area. Mesoscale phenomena now prevail, which result from topography, specifically the land-sea distribution. Because of its mountains and land cover differences local to regional phenomena develop like the "pumping" on top of elevated hot surfaces which create thermal lows in the high reaching Atmospheric Boundary Layer (ABL), mountain induced wind systems or topographically induced precipitation pattern. During summer semi-permanent high pressure systems with sinking motion of the dry upper tropospheric air impede moist warm boundary layer air to reach the condensation level.

A marked difference exists between topographically highly structured areas like the central Mediterranean and the Aegis and the large nearly quadratic landmass of the Iberian Peninsula or the long stretched North African coast. The circulation systems over Italy and Greece are strongly modified by coastlines, mountains and hills. The zonally stretched North African coast is positioned parallel to the boundary between the subtropics and the mid latitudes. During winter north of the coastline dominate the Westerlies and south of it in the lower troposphere - the atmospheric boundary layer - north-easterly winds which are directed towards the Inter Tropical Convergence Zone (ITCZ) at 10 °N. During summer high pressure dominates in the North of the coastline and the ITCZ is positioned near 20° North. Easterly winds prevail throughout the troposphere over the Sahara. The subtropical jetstream develops at the position of the maximum temperature gradient near 12 km height at 30 °N in winter and 35-40 °N in summer. Further South, near the ITCZ, two easterly jetstreams develop, the African Easterly Jet at 3 km height and the Tropical Easterly Jet at 12 km height. The eastward directed winds north of the Sahara and the westward directed winds south of the Sahel lead to a depression over North-western Africa.

Water shortage, wild fires and tilling generate treeless or sparsely vegetated areas which in summertime heat up under the intensive insolation. This causes dry convection and "heat lows" in the lower atmosphere which entrain the surrounding air like a chimney (see e.g. Smith, 1986). Nevertheless in most cases no clouds develop for two reasons. Firstly the air is quickly warming so that the relative humidity even of entrained primarily wet maritime air becomes low and secondly because the sinking motion of the overlying high pressure system prevents the air to reach its condensation level. The situation is different if there are forested mountains in the hinterland which push the sea breeze upward and even feed it with additional water evaporated from the woods. In these cases at the front of the sea breeze clouds develop over the mountains which can lead to rain and even to strong summertime thunderstorms. In contrast the coastal zones, where agriculture is concentrated, remain dry because here the land is heated up and dry convection dominates.

Over the Iberian Peninsula often a low pressure system develops during summer in the boundary layer (Gaertner et al., 1993). The centre of this thermal low lies over the southern central part of the Estremadura. It can appear from April to September with a maximum of occurrence in July and August (above 50% of the days). Portella and Castro (1996) summarise its characteristics in the following way: (i) During the early morning hours winds run parallel to the coast and turn to the interior as the low intensifies moving up pollutants through river valleys and mountain passes. (ii) In the Mediterranean zone stronger surface pressure gradients almost coincide with the orographic line which divide the Atlantic and Mediterranean basins. (iii) A zone with strong surface convergence in the NE of the peninsula coincides with the maximum of the occurrence of summer storms. (iv) The vertical structure of the low shows strong thermal influence below the 850 hPa level, convergent flow up to the 700 hPa level, and strong divergence just above. In the 500 hPa level geostrophic winds prevail. Fiedler and Adrian (see Fiedler et al. 1996 and Bolle 1998) computed the progression of the sea-land circulation front during one day on the basis of EFEDA data and demonstrated that the Mediterranean sea-land circulation front just reached in the evening the area of Albacete but did not go beyond.

The vertical convection over the Iberian Peninsula is supported by the higher temperatures of non-forested land surfaces and the contrast between the surface temperature of the surrounding sea and the land. Maximum surface temperatures range between 52 and 60 °C over bare soils and 49 - 56 °C over a fallow field, depending on the wind speed. Over a fallow field with short herbs three to four degree lower temperatures were measured as compared to the bare soil, and 10 °C lower temperatures over alfalfa. This demonstrates the effect of even sparse vegetation in this hot environment (RESRAPS Final Report 1994). Measurements at leaves of grapevine resulted in maximum temperatures of 35 - 38 °C while temperatures of nearby bare soil reached 60 °C. The nighttime temperatures were very close (Giordani *et al.* 1996). The maximum temperature differences between the surface and the air at 2m height are in the average 26 °C over bare soil and 22 °C over fallow land. The high temperatures develop because of the deforestation in many places, the low matorral vegetation, harvested crops, fallow land and non-irrigated agriculture which create a desert-like situation during the summer months as reflected in the vegetation index derived from satellite data (compare section 6.3).

In the EFEDA experimental area near the centre of the heat low during the intensive measuring periods of June 1991 and July 1994 the noon values of the albedo over fallow land were 27±1% and despite this high albedo the maximal values of the net radiation fluxes were 500 to 525 Wm^{-2} of which about 300 Wm^{-2} were returned to the atmosphere as sensible heat flux and approximately 75 Wm^{-2} as latent heat flux. The residuum of the fluxes measured at the top of the surface suggest a soil heat flux of 125 - 150 Wm^{-2} at noon.

The COMPARE/PYREX exercise has emphasized the fact that the orography representation still not is sufficiently close to reality for accurately forecasting an orographic flow at the mesoscale (Georgelin et al., 2000). It is mainly the land-sea circulation that becomes responsible for the exchange of dry and humid air between land and sea. Already during spring large thermal contrasts build up during daytime between sea and land causing warm air to rise in coastal zones. This generates low surface pressures which entrain the cooler, wetter, and heavier air from the sea which warms up rapidly. The air on top of the during summer 22 - 27°C warm sea can take up large amounts of water vapor but, because of the much higher temperatures over land, it often needs additional lifts by near coastal mountains or by mixing with dry convection to move this air up to the condensation level. Especially warm air moving from south to north over the sea can be loaded up with large amounts of water vapour. Often this *scirocco* carries as well desert dust with it which is released together with the condensing water.

The evaporation processes are determined by the availability of water in the soil, the pumping through the vegetation and the structure of the relatively thin Atmospheric Boundary Layer (ABL). The top of the ABL is the altitude at which the direct influence of the surface on the dynamics of the atmosphere ceases. The ABL is intimately connected to the surface by the vegetation which protrudes into this layer thus contributing to the roughness of the surface. Its thickness and its structure depends on many parameters such as turbulence, dry convection, moisture, and irradiance. In the case of the Iberian Peninsula ABL height up to 3 km were experienced, often filled with dust raised from the surface which causes large optical depths (see section 5.4). Locally it may as well be influenced by lakes, dammed rivers and small water ponds in which the water of winter rains is stored.

Though there is little rain or dew during summer, the evaporation of about 1 mm/day, as measured during the EFEDA field experiments in June, obviously is daily handed over through the boundary layer into the free troposphere and carried away. It results from the transpiration of the remaining matorral vegetation and evaporation from deeper levels of the soil due to the diurnal heat wave by which a fraction of the soil heat flux is transformed into latent heat. The exchange processes between the atmospheric boundary layer and the free troposphere on top of it are of completely different nature than those at the surface. Since there always is even during dry summers a small amount of evaporation at the surface this amount of water must somehow disappear if it is not raining. This is very probably due to the divergence at the top of the heat low and the variable height of the ABL: During daytime it is expanding filling its upper parts with water vapour and during nighttime because of the decreasing temperatures it is shrinking leaving some of the water vapour behind in the free troposphere, where it is transported away with the large scale wind systems (Fiedler 1994).

In other regions the land-sea circulation leads to cumulus convection over mountain chains such as the Italian Apennines (Bolle, 1999) but local circulation systems may also develop into vigorous thunderstorms if the moisture is available which feeds these systems by its latent heat. As can be seen in satellite imagery, comparable high temperatures develop in summer in coastal areas where in flat lands most of the agricultural areas are situated. These areas are in summer only sparsely vegetated, many fields are bare after the harvest of crops. These hot areas support the raise of wet air entrained from the sea, they are even spending some water vapour from irrigated fields though water for permanent and area covering irrigation becomes rare. Over the mountains of the hinterland then convective clouds and thunderstorms may develop which support the growths of the forests that remained there.

The situation is different in areas where enough water is available like in the large irrigation systems which partly use fossil water resources from underneath of the Sahara. Here the surface temperature is kept low because a substantial amount of the net radiation can be converted into latent heat which is exported from the region.

In early autumn the situation may change insofar as now the sea remains relatively warm in comparison to the atmosphere and the land. At this time of the year convection may start already over sea and strong latent heat fluxes lead to thunderstorms that enter the land. Though the surface temperature of large land areas at this time of the year tend towards sea surface temperatures the contrast between mountainous areas and plains is still considerable. This locally leads to complex valley-mountain circulation systems and even catabatic winds known as "Bora" at the eastern Adriatic coast.

Topography also gives rise to small scale climate anomalies such as dry lowland "islands" surrounded by elevated areas with higher rainfall (Guzzi, 1981). The opposite is experienced if in a dry environment ground water reaches the surface e.g. in depressions and creates the well known "oasis effect" with reduced temperatures and higher air humidity.

4.5 Experimental Strategies to Deal with the Hierarchy of Scales

The discrepancy between the wide range of scales involved in the land surface - atmosphere interactions and the limited possibilities to measure the quantities which govern the land-surface processes and the surface - atmosphere interaction enforces to rationalize the number of stations which have to be distributed over an area compatible with the resolution of global climate models. Ideally the measurements should be carried out for time periods the individual processes need to go through at least one complete cycle. This would be of the order of days if the diurnal wave of the energy fluxes is considered, of the order of a year if the annual hydrological or vegetation cycles are investigated, of the order of a decennium if the climate variability is taken into account, and of the order of hundred years, if the life cycle of a forest is under observation. It is clear that the long time scales can only be covered by operational observation systems with in most cases spatially scattered measuring sites.

EFEDA (see Appendix 1) was an experiment designed to study the shorter end of the time scales and specifically the transition from the wet to the dry period during the month of June in the desertification threatened area of Castilla - La Mancha in Spain. To bridge nevertheless over to the longer time scales, some measurements were prolonged to three months, some groups revisited the area after three years and satellite data have been evaluated for ten years. Furthermore it could be relied upon data of the meteorological and hydrological networks.

The goal was to investigate simultaneously all interacting processes from local scale to an area of 10^4 km^2 and from sub-surface water transports to the free troposphere. In order to cover such an large area three sites typical for the mixture of land cover in that area were selected which form a triangle of 70 km side length. At the centre of each of these sites heavily instrumented central stations were established and surrounded by simpler equipment at distances up to five kilometres to assess local inhomogeneities.

The participating teams grouped into seven task forces (see Fig. 6). The Hydrology Group investigated the geo-hydrological sub-surface structure, the variability of the ground water level and sub surface water flows. The Soil Physics and Vegetation Group made an inventory and analysis of the soil and vegetation types during the drying period. The Group on Impacts supplied the socio-economic boundary conditions and studied the impact of dryness on vegetation. The Boundary Layer group measured the soil - vegetation - atmosphere energy and water transfer and determined the energy budgets. The Aeronomy Group conducted radiosonde ascents and flights with instrumented aircraft and motorgliders. The Modelling Group used mesoscale models to integrate measured quantities and conducted a thorough comparison of SVAT models with the measured data. From these activities result data sets for the parametrization of processes. Finally the Remote Sensing Group analysed various types of satellite data and thus extended the observed area

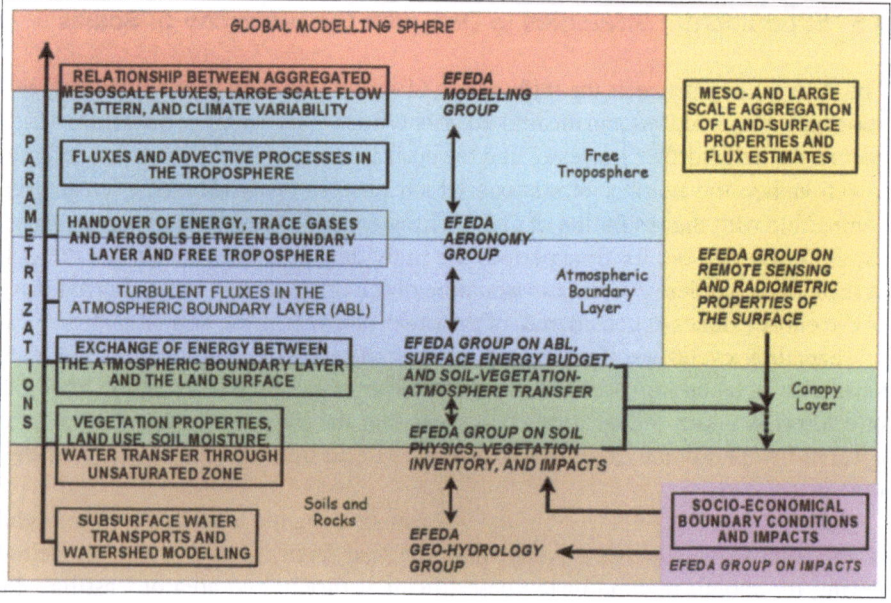

Fig. 6. Experimental activities required to investigate at various scales the interacting processes which are directly or indirectly coupled with the land surface and reach from sub-surface water and heat flows to the free atmosphere into which evaporated water escapes as demonstrated by the EFEDA project. Remote sensing is used to explore the water vapour and aerosol distribution in the atmosphere from aircraft and the surface conditions from satellites. Socio-economical processes were only considered as far as water resources and land use are concerned such as irrigation needs.

both in space to the whole Mediterranean basin and in time over ten years.

A specific program was set up to validate the information derived from the measurements in space with quantities directly observed at the ground. These necessarily spotty data served the validation of the few available high resolution satellite data. These in turn were aggregated to the areas sizes resolved by meteorological satellites to transfer the validation on to these daily (NOAA satellites) and hourly data sets of medium spatial resolution from METEOSAT.

4.6 The Role of the Mediterranean Sea

It would be incomplete not to refer to the overriding role of the Mediterranean Sea for the European-African Mediterranean climate and vegetation. Because of this enclosed sea the area of Mediterranean climate is so extended in this region and not, as on other continents, restricted to a narrow coastal strip. The importance of the sea is present in almost all areas of the Mediterranean basin. At the larger scale the sea gains more solar energy than it loses as infrared radiation to space. It is surrounded in almost all directions by areas where the net radiative flux is negative (Fig. 7). The only exception is a narrow strip to the West that connects the Mediterranean Sea with the sub-tropical North Atlantic Ocean. Because of this positive radiation budget

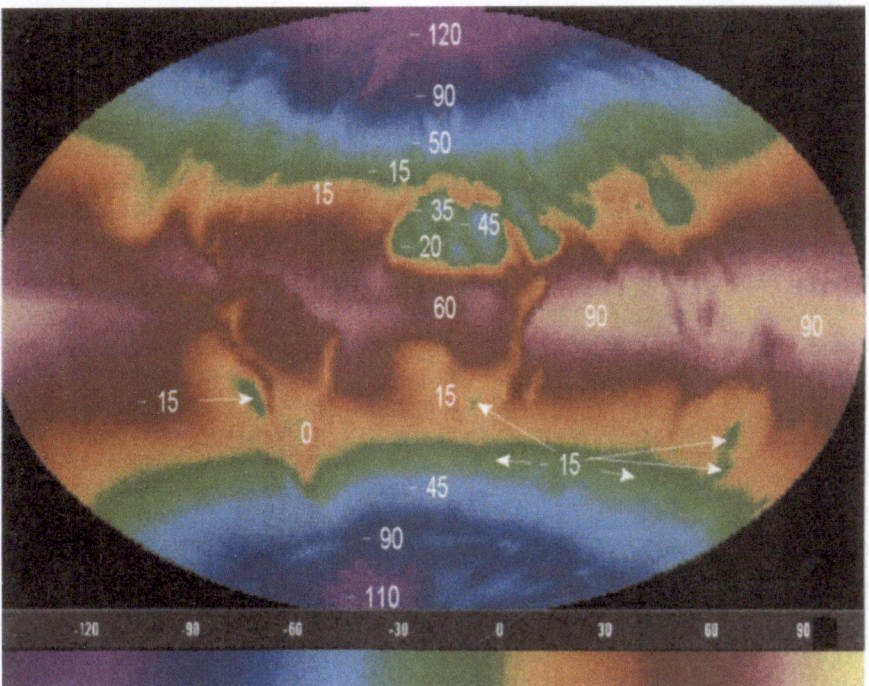

Fig. 7. Net radiation flux in W m^{-2} at the top of the atmosphere averaged over five years after Graßl (1995)

the Mediterranean Sea in the annual average is a heat source surrounded by areas that require an influx of heat to compensate for their negative radiative energy budget. Furthermore strong albedo differences exist in south-north direction. Over land the albedo decreases from South to North but this gradient is interrupted by the low albedo of the Sea.

In the south, over North Africa, the energy budget is maintained by the adiabatic warming of dry air which descends in the northern branch of the Hadley cell until it meets at daytime the uprising convective flow of sensible heat from the hot desert. In the North the loss of radiant energy is compensated by wet maritime air from the North Atlantic Ocean that is transported north-eastward within the warm sectors of travelling eddies. The coastal zones in addition obtain energy from the Mediterranean Sea which is transported inland with the regional sea-land circulation systems and released by condensation processes.

The Mediterranean Sea provides, especially in its eastern part, most of the water necessary for the vegetation by evaporation. It either is transported inland by the sea-land circulation systems which exist during summer on all coasts or it is handed over to the free troposphere and is entrained into disturbances in which rain is generated (Kossman et al. 1999). The sea breeze in addition cools in two ways the coastal zones over which it wafts. Firstly the air arriving from the sea at daytime in summer is cooler than the air over the at that time hot continent and secondly the wind increases evaporation since the air which warms up over land is not saturated with water vapour. This, of course, may as well have an unwanted by-effect because it enhances the drying effect which is energized by the strong irradiance.

In the northern part of the Mediterranean Sea the average sea surface temperatures are about five degrees lower than in its southern part. But the pattern is not uniform. The central and eastern parts of the Mediterranean Sea are mostly warmer than the western and Aegean sectors. There are different reasons which leads to this pattern: The reduced solar irradiance in the North, upwelling of cooler waters due to internal circulations, run-off of cooler river water into the sea, and the intrusion of cold water from the Black Sea through the Sea of Marmora. This situation is indicated already in spring, when the Black Sea often has the lowest temperatures in the region and the Peloponnese can be much cooler (nearly at sea surface temperature: one cannot see the coastlines in satellite images, see Bolle 1999) than e.g. the south of Spain - a situation which varies in strength from year to year.

The question is, whether the marine system remains unaffected if global climate changes or whether it can be expected that changes occur in the sea which affect the exchange with the atmosphere thus feeding back to the climate system. The circulation in the Mediterranean Sea is a response to the acting driving forces. The basic view at the Mediterranean circulation system is according to Wüst (1961) that through the Strait of Gibraltar warm and fresh Atlantic water enters the Mediterranean and is transformed into cold and salty deep water in the Eastern Mediterranean which is exchanged against the entering Atlantic water. The circulation is driven during winter by the Westerlies, which move the surface water masses eastward, the outflow of heavy salty water through the Strait of Gibraltar, and the influx of fresh water from rivers. The resulting circulation, also called the "Mediterranean conveyor belt", is an inflow of Atlantic surface waters through the Strait of Gibraltar which remains near the African coast and heads eastwards. At its southward fringe it generates by interaction with the coast small anti-cyclic gyres which often can be visualized in satellite imagery. In the eastern Mediterranean the "Mid Mediterranean Jet" dissolves under the influence of the high salinity of this region into a number of gyres, mixes, and the heavy water is drawn down and joins into the Levantine Intermediate Water (LIW) which returns to the Strait of Gibraltar at depth down to 500 m. It generates two sidearms. One is directed into the Adriatic Sea from which the Western Adriatic Coastal Current returns near the surface. The other one flows into the Gulf of Lion where it surfaces in the gyre of the Lion and the water joins the Ligurian-Provencial Current which flows towards the Spanish east coast. Several other gyres develop as well due to the complex topography of the Mediterranean sea and its shores.

New empirical evidence (Roether et al. 1996) has shown that obviously during the last two decennia increasingly more deep water was formed through dense water outflow from the Aegean Sea and simultaneously in context with the LIW in the Rhodes Gyre core (Özsoy, 2000). Around 1987 the production of deep water switched from the Adriatic to the Aegean which increased the temperature of the Eastern Mediterranean Deep Water by $0.5°C$ from $13.3°C$ to $13.8°C$ and its salinity from 38.66 psu to 38.8 psu[4]. The potential density increased from below 29.18 to above 29.2. Coupled to this new deep water production its nutrient content seems to increase with the temperature.

Pursuing this discovery of Roether a number of empirical and model studies have been carried out to explain the observed phenomenon and investigate more closely the energy and water budgets of the Mediterranean Sea. An review of the progress made in understanding the variability of the Mediterranean circulation systems starting from the classical view and interpreting the new discoveries was given by Pinardi and Masetti (2000). The starting point of these investigations are estimates of the annual heat gain of the Mediterranean through the Strait of Gibraltar lying between 5.2 ± 1 W m^{-2} (Bethoux 1979) and 7 ± 3 W m^{-2} (Macdonald et al. 1994) distributed over the whole area of the Mediterranean Sea. The evaporation was estimated to be 1320 - 1570 mm year^{-1} (Castellaro et al. 1997) corresponding to a latent heat flux of 105 - 124 W m^{-2}.

The increased salinity in the deep waters of the eastern Mediterranean in 1986/7 would need an evaporation increase of 20 cm/year over the entire Eastern Mediterranean for seven consecutive years (Wu et al.2000) which is a rather unlikely mechanism. The alternative approach to explain this phenomenon is that the new deep water was formed by anomalous cooling and a moderate increase of the net water exchange with the atmosphere, evaporation (E) minus precipitation

---

[4] psu = Practical Salinity Unit defined by the International Practical Salinity Scale 1978. It is measured as ratio of the electrical conductivity of the sea water to that one of a standard KCl solution. A mass ratio of salt to sea water of 0.035 corresponds approximately to $S = 35$ psu.

(P).

While the wind provides the stress which moves the water masses, the surface energy budget is responsible for the thermohaline circulation. Pinardi and Masetti (2000) emphasized the importance of the wind stress as a major driving force. Korres et al. (2000a and b) showed that it delivers more than 50 % of the kinetic energy of the currents. The variability of its curl and amplitude has been investigated by Myers et al. (1998). To assess the role of the energy budget, oceanographers need to know the net energy flux at the sea surface which determines cooling respectively warming of the surface water with the related changes of density that initiate thermohaline circulation. Furthermore the net influx of fresh water, run-off from rivers (R) plus precipitation (P) minus evaporation (E), determines the salinity of the water body. The net flux of water (E - P) at the surface is important for the buoyancy in the upper level of the sea and adds to the thermohaline circulation. Meteorologists on the other hand are interested in the individual heat fluxes between the sea surface and the atmosphere because the sensible heat flux (H) is important for the temperature structure, convection, and height of the boundary layer of the atmosphere, while the latent heat flux transports the water into the atmosphere which may be precipitated locally or at distant locations.

One crucial point in explaining the increase of salinity is, whether such extreme events occur often and strong enough. Another one is, how the changes set off in the sea may feed back to the atmospheric and hence affect regional climate. As we have seen, precipitation over land depends, especially during summertime, to a great deal on the evaporation processes and labilization of the atmosphere and land-sea circulation systems. The problem to be solved therefor is, to what degree the processes which alter the internal structure of the sea are also affecting the surface processes.

The system is highly variable and it needs long term data series to isolate these effects. Fortunately there have been two strong events in the nearer past which gave a large signal and are satisfactorily documented to study this problem. These were the extreme "Mistral" event of 1981 and the cold spill over Greece, 1986/7. The 1981 period is characterized by a strong winter cooling related to northern to north-western winds starting from the French south coast, blowing over the Gulf of Lions mainly during wintertime and extending from there to the eastern Mediterranean as westerly winds. The cooling over the Aegean, 1986/7 was caused by an extremely cold and dry surge of continental winds which occurred over the Greek peninsula with temperatures as low as -4 °C over the northern Aegean and below 10 °C at Crete. It caused a drop of the SST by 2 °C below the climatological average. It was followed by a hot summer in 1987. The hypothesis behind the investigations, which was supported by the results of model studies, is the following: The cold but initially not extreme salty water of the Aegean is pushed southward by strong Etesian winds which accompany the cold surge and sinks down because it is losing energy to the atmosphere. The outflow from the Aegean entrains the more saline Levantine Intermediate Water (LIW) into the Aegean. This would be a positive feedback increasing the salinity in the Aegean and favouring the production of deep and now more saline water. From the bottom of the Cretan Sea the new deep water spills over to the Levantine Basin towards the end of the year. Because the LIW is drawn into the Aegean and afterwards dumped into the deep sea, the outflow of intermediate water from the Adriatic Sea and through the Sicilian channel is blocked and reduced (Wu et al. 2000). A simulation of the surface heat flux with ECMWF reanalysis data indicates a number of cold events over the Aegean Sea also between 1990 and 1994 which would keep this process going.

To explore the involved processes the energy budget at the surface has to be computed. For the large Mediterranean area this is only possible by using bulk formulas which are either fed with assimilated operational data respectively are used in models which are initiated by these data. Assimilated data are data which, with the aid of models, are gridded into a predefined system so that a dense network of equally distributed data points is obtained. For model simulations the Mediterranean version of the Geophysical Fluid Dynamics Laboratory modular ocean model (MOM) has been used. Neither the state parameters measured by the operational network nor those produced by models are precisely those one would like to have for accurate flux determinations of which meteorologists anyhow think that only with the application of eddy correlation methods one can come close to the real fluxes. It is therefore not astonishing that between the different approaches discrepancies exist and one is left with estimates the absolute accuracy of which is not exactly known.

Castellari et al. (1998) investigated the air-sea interactions with a nine years data set (1980 - 1988) of the 12 hourly NMC atmospheric analysis combined with SST data of Reynolds (1988) and cloud coverage of COADS (Comprehensive Ocean-Atmosphere Data Set, da Silva et al. 1994). The 1981 extreme event caused extraordinary strong sensible (maximum 80 W m^{-2}), as well as latent (maximum 280 W m^{-2}) heat fluxes during the event in winter followed by a year of extremely small wintery fluxes (maxima 34 respectively 140 W m^{-2}). The Aegean event was followed by two summers with very low evaporation (minimum 20 W m^{-2} instead of about 70 W m^{-2}). The surface temperature and total heat flux fields are correlated with a time lag of 2 - 3 months with the total heat flux leading the SST field.

Korres et al. (2000a and b) investigated in great spatial detail the ocean response to low frequency atmospheric forcing for the years 1980 - 1989. Their results with respect to the surface heat budget and transports are reproduced in Table 2. The strong interannual variability and the response to the extreme events in 1981 and 1986 can clearly be seen. Their general conclusion, underpinned by detailed results about wind stress, energy budgets, and kinetic energy, is, that the circulation of the Mediterranean indeed reacts to anomalous atmospheric forcing which occurred in the years 1981 and 1986. Largest amplitudes occur along the Spanish continental shelf, the western coasts of Corsica and Sardinia as well as in the central Levantine and Ionian.

The model simulated a surface heat loss of the eastern Mediterranean of 5.3 W m^{-2} in "normal" years which increased to 7.7 W m^{-2} in cold years. At the same time E - P changed from 55 to 58 cm year^{-1} for the whole Mediterranean while over the Aegean the increase was 6% and over the eastern Mediterranean 11% corresponding to 6.6 cm year^{-1} which is much less than needed for the formation of the additional saline deep water (which was estimated to be 20 cm year^{-1}). In winter high losses of heat occur primarily in regions where water mass formation occurs like in the North Adriatic, the Liguro - Procencal and Levantine - Aegean region (Table 3).

Table 2. Variability of the annual mean surface heat budget (AMSHB) and the net heat transports at three transects after Korres et al. (2000). The heat transports (HT) are presented as follows: HT(Gibraltar)/Area Mediterranean, [HT(Gibraltar) - HT(Sicily)]/area Western Mediterranean (WM), and HT(Sicily)/area Eastern Mediterranean (EM) respectively

Year	AMSHB (W m^{-2})			Heat Transport (W m^{-2})		
	Whole Mediterranean (2.48E12 m^2)	Western Mediterranean (0.84E12 m^2)	Eastern Mediterranean (1.65E12 m^2)	Across Gibraltar Strait	Net Western Mediterranean	Across the Sicily Strait
1980	0.076	12.894	-6.804	1.38	-8.608	6.716
1981	-1.591	14.012	-9.96	1.425	-9.429	7.224
1982	11.234	24.054	4.459	1.227	-13.229	8.948
1983	7.175	16.356	2.257	1.107	-9.177	6.6
1984	7.681	18.408	1.926	1.02	-11.244	7.571
1985	5.94	23.147	-3.306	0.901	-11.318	7.428
1986	-3.946	6.205	-9.396	0.873	-7.744	5.476
1987	14.553	23.037	9.999	1.144	-10.416	7.379
Mean	5.14	17.139	-1.353	1.135	-10.146	7.175

Table 3. Monthly 1987 - 1993 climatological surface heat fluxes in W m^{-2} computed from model experiment by Angelucci et al. (1998) extracted for some geographical positions: NA: North Adriatic, LP: Liguro-Provencial, LA: Levantine-Aegean, SES: South-east Spain, SSicili: South of Sicily

Season	NA	LP	LA	SES	SSicili
Winter	-200	-160	-150	-80	-120
Spring	-40	0	-40	40	20
Summer	160	180	200	120	200

The response time of the sea to these extreme events is of the order of years and probably the event experienced in 1987 was not the beginning of the formation of additional deep salty water. The 1981 event had a similar effect. Its impact can be traced at 420 m depth until 1985 in the amplitude of time series of the first EOF mode of the 1981 cooling (Korres et al 2000b).

Long term deep water warming was also measured in the Alghero-Provencial basin between 1000 and 2700 m (Bethoux and Gentili, 1996). Between 1959 and 1995 a warming of 0.13°C and an increase of salinity of 0.04 psu were measured. Twice as high values are reported (see Zodiatis and Gasparini 1966) for the intermediate waters in the Ligurian Sea and the Tyrrhenian Sea which may have been imported from the eastern Mediterranean basin. There are different hypotheses about the causes of the warming in the western deep waters. Temporarily the beginning of these effects seem to coincide with the damming of the Nile and of Black sea rivers as well as a decrease of rainfall around the northwestern Mediterranean Sea.

Evaporation computed from latent heat fluxes obtained from ECMWF budgets result in 1100 mm year^{-1} which is out of the range suggested by Castellari. The precipitation would be 450 mm year^{-1}, which is smaller than the 590 mm year^{-1} of the climatology. The long term means of E - P is 650 mm year^{-1} (climatology: 950 mm year^{-1}). However this "climatology" was performed for two years only in which a heat accumulation occurred in the Mediterranean basin (1987 and 88).

The Adriatic was in more detail studied by Artegiani et al. (1997) and Maggiore et al. (1998). The obtained mean surface heat budget indicates a loss of 19 - 22 W m^{-2}. This should be compensated by heat advection through the channel of Otranto. The surface water flux E - P varies between 60 and 520 mm year^{-1}: E ranges between 1080 and 1340 mm year^{-1}, P between 820 and 1020 mm year^{-1}, and the river run-off is 1170 mm year^{-1}. Thus there is a gain of water of the order of 650 to 1100 mm year^{-1} which means that the Adriatic is a dilution basin with respect to salinity. Its winter surface heat loss leads to deep water formation.

There is a lack of results to confirm that a relationship exists between the variability of the sea surface heat fluxes over the sea and climate variability - such as precipitation changes - over land. In rainfall statistics no marked effect of the 1981 and 1986 events have yet been found. Maracchi et al. (2000) investigated the relationship between the SST of the northern Thyrrenian Sea according to COADS data and the precipitation in Tuscany. They found a correlation e. g. between the September SST and extreme rainfall events in October as well as a correlation between extreme rainfall events in December and SSTs in June and conclude that an increase of the SST by 1.8 °C may cause an increase between 40% and 65% of extreme monthly rainfalls. They investigated the time series of SSTs from 1950 to 1997 and came to interesting results. Firstly, monthly SSTs deviate from the 27 years mean by about ± 2 °C. Secondly, the trend varies considerably with the length of the interval. For the whole period it is only 0.02°C/100 years which is not significant. But for shorter intervals, such as 1970 - 1997, it can amount to 3.3 °C/100 years with largest values in May and August. This is due to the fact that the

SST was in the average low between 1975 and 1985 and high between 1960 and 1970 as well as between 1987 and 1997. No trend is recognized in these data since 1988.

Recently Andersen and Knudsen (2001) published a long term analysis of SST and sea level data obtained from satellites. SST trends have been deduced from ERS-Along Track Scanning Radiometer (ATSR) as well as NOAA-AVHRR measurements. The altimeter data of the ERS satellites and the TOPEX/POSEIDON mission were evaluated from September 1992 to March 2000. The SST data set ranges from September 1992 to July 2000. In Fig. 8 the trends obtained for the Mediterranean section of the world wide data set are reproduced. It can be seen that the two pairs of data sets do not result in identical pattern but the general trends are clearly visible in both of them: In the eastern and southern Mediterranean Sea the

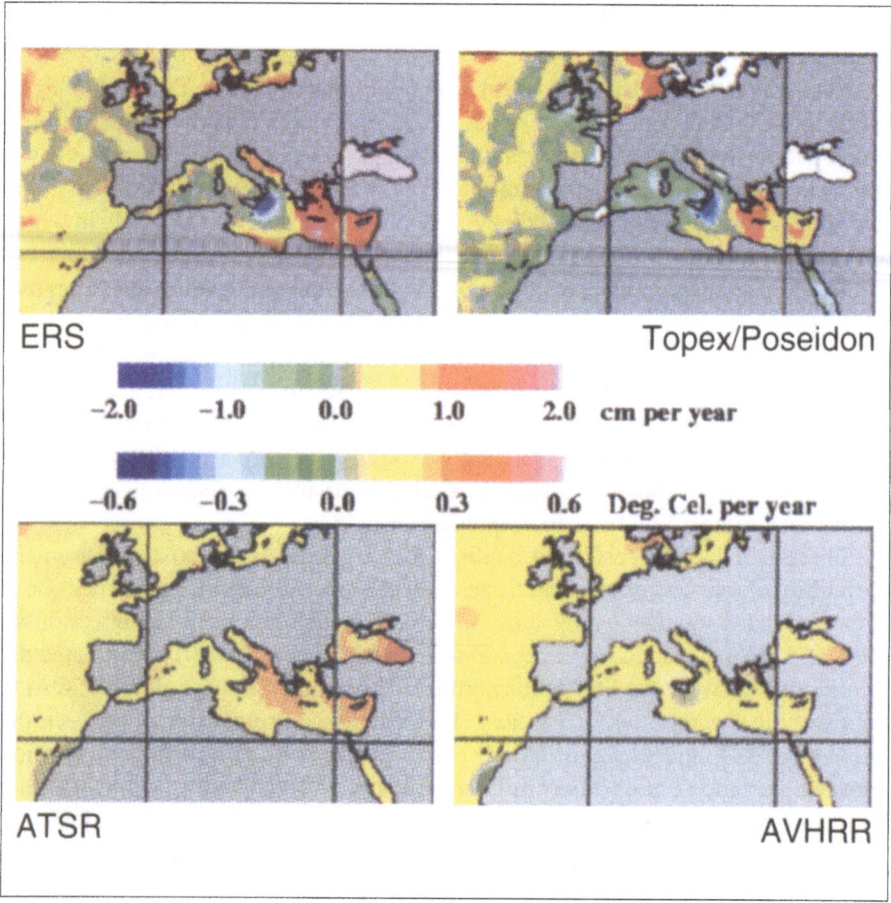

Fig. 8. Sea level and sea surface temperature changes between September 1992 and March respectively July 2000 after Anderson and Knudsen (2001). The sea level change bases on measurements of the ERS respectively Topex/Poseidon altimeters and the SST change was calculated from ERS-ATSR and AVHRR measurements. The scales are in cm respectively °C per year. After ESA-ESTEC, EOQ No. 69

sea level increased while around Sicily a strong and in most parts of the western Mediterranean Sea a much smaller decrease is observed. The SST is rising everywhere at a pace of at least 0.1 °C/year but in its north-eastern parts according to the ATSR measurements regionally up to 0.3 °C per year. In the AVHRR data, and only here, a small increase of the SST is observed south of Sicily which would not explain the depression in the sea level. This indicates that the sea level changes in the Mediterranean must have dynamical reasons.

In the average these measurements suggest that the Mediterranean SST increases in the order of 0.15 °C/year, which is much more than any land surface temperature. To assess the impact of SST changes on the water cycle one has to take into account that evaporation not only depends on the temperature of the evaporating surface but on the gradient of the vapour pressure above the surface and the wind speed as well. The latent heat flux is computed according to the formula (Korres et al. 2000a):

$$E = \rho_a C_e |W| [e_{sat}(T_s) - r_h e_{sat}(T_a)] \frac{0.622}{p_a} \tag{1}$$

where
ρ_a air density (average value 1.27 kg m^{-3} at sea level)
C_e turbulent exchange coefficient (estimated to be 1100)
W monthly mean 1000 hPa wind vector
e_{sat} saturation water pressure (17 hPa at 15°C, 25 hPa at 21°C, 41 hPa at 30°C, 74 hPa at 40°C)
T_s SST
T_a monthly mean atmospheric temperature
r_h monthly mean relative humidity at 1000 hPa
p_a surface air pressure (1013 hPa).

This give an impression of the complexity of the task to assess magnitude and causes of climate variability throughout the Mediterranean basin. It needs accurate information about the humidity and wind velocity fields on top of the sea surface and it must be questioned whether the computations results in the same monthly latent heat flux average if monthly mean values are used in the equation or if, say, half hourly instantaneous values are averaged.

5 What Does the Mediterranean Basin Contribute to the Global Carbon Budget and Atmospheric Pollution?

5.1 Fluxes of Biogene Volatile Organic Compounds

Land-surface processes are linked to the global climate system through the exchange with the atmosphere of energy, momentum and matter. Of overriding importance for vegetation growth and its sustainable development is the water cycle which will be dealt with in the next section. Of equal importance for the global system are the production respectively exchanges of carbon dioxide, nitrous oxides, dust, and Biogene Volatile Organic Compounds (BVOCs). Worldwide it is estimated that the biogene component of the VOCs exceeds with 127 - 480 Tg C yr^{-1} (Kesselmeier and Staudt, 1999) the anthropogenic emission of non-methan hydrocarbones of 100 Tg C yr^{-1} (Singh and Zimmerman, 1992). There have been so far two major activities in the Mediterranean area to determine fluxes between the vegetation and the atmosphere, the MEDEFLU network (Miglietta and Peressotti, 1999) which concentrated on CO_2 and the Biogenic Emissions in the Mediterranean Area (BEMA) project which concentrated on BVOCs (Seufert et al. 1997).

The Mediterranean area is rich of aromatic plants with high emission of BVOCs and the photochemical activity that transforms these compounds after they have been released is high. Monoterpenes, which constitute one component of the BVOCs, react with OH, NO_3, and O_3. Depending on the NO/NO_x ratio they either destroy or produce ozone. (Lerdau 1991; Fehsenfeld et al. 1992; Meixner 1994). Due to these reactions BVOCs, according to Seufert et al. (1997), may increase - by depletion of OH and producing CO - the lifetime of radiatively active gases such as methan, foster the formation of aerosols and of cloud condensation nuclei, enhance the acid deposition in remote areas by forming organic acids, and control the tropospheric ozone formation. The assessment of the BVOC source strength up to now is limited to few investigations of a limited number of species. More often investigated trees are *Quercus ilex* and *Pinus pinea*, which have been identified as a major monoterpene sources.

The carbon skeletons of terpenes are composed of characteristic C_5 units. Isoprene belongs to the group with five carbon atoms. Monoterpenes consist of C_{10} structures and are described in Kesselmeier and Staudt (1999). Kesselmeier et al. (1996) list 18 Monoterpenes which are emitted by *Quercus ilex* of which the five most frequent are α-Pinene, β-Pinene, Sabinene, 1,8 Cineole, and p-Cymene. Less volatile terpenes consist of C_{15}, C_{30}, C_{45} or higher carbon structures. Other volatile compounds are aldehydes and organic acids. Kesselmeier and Staudt list in total 37 volatile non-terpenoid compounds of which emission rates of different plants have been estimated. The emission rates generally depend on the photosynthetic active radiation (PAR), temperature, leaf position, plant age, season, and stress. Consequently the individual measurements vary in wide limits. In the case of *Quercus ilex*, as an example, at a PAR rate of 2000 µmol photons m^{-2} s^{-1} an emission rate of acetic acid of only about 10 nmol m^{-2} (projected leaf area)$^{-1}$ was measured in May while during a hot and dry while during a hot and dry day in August maximum values were 30-40 nmol m^{-2} (projected leaf area)$^{-1}$ and negative values occurred at eight o'clock in the morning. The respective numbers for formic acid are 20 nmol m^{-2} (projected leaf area)$^{-1}$ in May and 50 nmol m^{-2} (projected leaf area)$^{-1}$ in August (Gabriel et al. 1999). Some data relevant for Mediterranean species are compiled in Table 4a and 4b.

From data gained during a large field experiment in the natural reservation of Castelporziano near Rome probably the first time it was attempted to estimate BVOC emission fluxes for some major Mediterranean land cover types. The results

Table 4a. Isoprene and Monoterpene emission rates of some selected Mediterranean species after Seufert et al. (1997), Kesselmeier et al. (1998), Kesselmeier and Staudt (1999), and Gabriel et al. (1999).

Species	Emission rates normalized at 1000 μmol m^{-2} s^{-1} PAR, 30°C			
	Isoprene μg (g d.w.)$^{-1}$ h^{-1}		Monoterpenes μg (g d.w.)$^{-1}$ h^{-1}	
	high	low	high	low
Arbutus unedo (strawberry tree)				0.12
Ceanothus leucodermis (chaparral)		0	5	
Cistus				✓
Citrus sinensis (orange)		✓	1 - 3	
Cypress		0		0.1
Cytisus sp. (broom)	6 - 27			0.5
Erica multiflora (heath)	2			0.03
Erica arborea (tree heath)	6.2 - 20			
Eucalyptus	8 - 57		3 - 48	
Fig	27		0.2 - 1.6	
Helianthus annuus (Sunflowerus)		<0.1		0.7
Lycopersicon esculentum (tomato)		✓	13 - 60	
Medicago sativa (alfalfa)		✓		✓
Myrtus communis	25.2 - 137			✓
Oleander, Magnolia		✓		✓
Olive, Cedar		0	0 - 9	
Palmae (various palm trees)	5 - 170			✓
Pinus halepensis		✓		0.2 - 1
Pinus pinea (and pinaster)			2 - 15	
Pistacia lentiscus (mastix)				0.4
Pistacia vera		0		9
Populus (different species)	37 - 100			✓
Quercus frainetto	134			
Quercus fagina	111			✓
Quercus ilex		✓	2 - 58	
Quercus cerris		✓		
Quercus rotundifolia			15	
Quercus suber		✓		
Quercus calliprinos			3.1	
Quercus petraea	45			
Querscus pubescens	91			
rice, wheat, grape, cypress		0		0
Rosmarinus officialis		✓	2.2	
Salvia mellifera (sage)			5 - 27	

Table 4b. Acetic and formic acid and aldehyde emission rates of Mediterranean species after Seufert et al. (1997), Kesselmeier et al. (1997, 1998), Kesselmeier and Staudt (1999), Staudt et al. (2000)

Species	Emission rates in ng (g d.w.)$^{-1}$ min^{-1}			
	Acetic acid	Formic acid	Acetaldehyde	Formaldehyde
Citrus sinensis (orange)	1 - 1.3	2 - 3		
Pinus pinea (and pinaster)	0 - 5	3.3 - 8.3	6.7 - 25	1.6 - 15
Quercus ilex	0 - 6.6	1.6 - 8.3	6.6 - 15	5 - 13

Table 5. BVOC emission fluxes after Seufert et al. (1997) calculated from enclosure measurements for the main land cover types of Castelporziono (Manes et al., 1997)

Land cover type	LAI (projected)	Specific Leaf Weight g m^{-2}	Emission flux ng m^{-1} s^{-1}	
			Isoprene	Monoterpene
Pine plantation	3 - 4.5	200	70	1000
Mixed oak forest	4.5 - 8	75	3113	1006
Mediterranean maquis	2 - 6	175	30	750
Cork oak forest	4 - 6	100	50	0.1
Pseudosteppe with trees	2.5 - 4.5	100	450	4
Pseudosteppe	2 - 3	100	150	2
Suppressed macchia	3.5 - 4	175	10	750
Dunes	0 - 3.5	175	50	300

reported by Seufert et. al. (1997) are presented in Table 5. One of the major results is the important role oaks play for Isoprene (deciduous oaks) as well as Monoterpene (Quercus ilex) emissions.

5.2 Carbon Fluxes

During daytime plants are taking up carbon dioxide from the air through their leaves or needles due to the photosynthesis process which exceeds the losses by respiration that takes place at the same time. Carbon dioxide emerges also from soils resulting from respiration by soil organisms, roots, humus decay and myccorhizal activity. It is replaced by organic matter input to the surface (surface litter) and root detritus. Part of the CO_2 released from the soil during daytime will again be taken up by the leaves and only the rest returns to the air outside the canopy. Therefor the net amount of carbon dioxide integrated into the biomass through the photosynthesis process, the net assimilation, is higher than the net CO_2 flux measured in the atmosphere (Calvet et al. 1999) provided there is enough photosynthetic radiation (PAR), water, phosphor and nitrogen.

The total input to the ecosystem due to photosynthesis is called gross ecosystem exchange (GEE), gross photosynthesis or gross primary production (GPP). This leads to the production of biomass and drives the output of the system, the CO_2 respiration (RE). Respiration occurs directly from the living plants or at the end of an internal cycle by which litter and humus is produced from the primary production of biomass and decays after some time (Aber, 1998).

The atmospheric CO_2 net flux measured at the top of the canopy represents the difference between GEE and RE and are called Net Ecosystem Exchange (NEE = GEE - RE). It represents the uptake respectively loss of carbon by the ecosystem during a specified time period. The *instantaneous* CO_2 net flux measured at the top of the canopy provides no information about the carbon budget because of the time lag between the inhalation of carbon dioxide which is restricted to the vegetation period and daytime and its release by decomposition of litter which occurs any time. The *annual net flux* of carbon dioxide at the top of an ecosystem is calculated from the summation of carbon dioxide fluxes at the top of the canopy and leads to budgets. It includes the CO_2 consumed for production of sustaining biomass, the direct re-emission into the atmosphere (due to respiration from leaves, fires, decomposition of dead timber above the surface), and the respiration from the soil of which a fraction is taken up again by the vegetation.

Raich and Schlesinger (1992) estimated the soil respiration rates for different ecosystems on the basis of CO_2 fluxes measured at the surface and not at the top of the canopy. From the CO_2 fluxes they calculated the following carbon fluxes relevant for the Mediterranean ecosystems in units of g C m^{-2} year^{-1}: Temperate coniferous forests 681±95, temperate deciduous forests 647±51, Mediterranean woodlands and heath 713±88, croplands, fields etc. 544±80, and desert scrub 224±38. If one assumes that in the long run the soil carbon flux cannot exceed the biomass production on top of the soil, these values provide an upper limit for the net primary productivity (NPP). A least square regression analysis resulted in an average relationship between the NPP and the soil respiration (SR) of SR = 1.24 NPP + 24.5. For the Mediterranean woodland and heath the relationship would even be SR = 1.75 NPP (all units in g C m^{-2} year^{-1}).

IGBP in 1996 launched the international initiative FLUXNET (Baldocchi et al.

1996; Valentini et al. 1997; Valentini et al. 1998) to measure carbon, water, and energy fluxes in important terrestrial ecosystems. The European components of this international activity, EUROFLUX (Valentini et al. 1999) and MEDEFLU (Miglietta and Peressotti 1999), are concentrating on the carbon exchange in European and specifically Mediterranean ecosystems, primarily forests.

Measurements conducted by the MEDFLUX project (Miglietta and Peressotti 1999) funded by the EC provide for specific ecosystems long term carbon dioxide flux data. The CO_2 emission above *Quercus ilex* according to flux measurements for a first annual cycle made on top of a holm oak stand in Southern France seems to have a double annual period being a sink in the average of 5 g m^{-2} d^{-1} during late autumn and spring/early summer and a small source (during major rain events) respectively close to zero in August/September and January.

The result of one year measurements in a Mediterranean *macchia* forest in Sardinia was that it is a small CO_2 sink of 70 g m^{-2} year^{-1}. Only from mid August to mid November and especially during a rainy period from mid September to October the *macchia* was a small CO_2 source of the order of 2 - 3 g m^{-2} d^{-1}. The net carbon sequestration was less than 80 kgC ha^{-1} year^{-1} which probably is due to the low vegetation cover and arid environment.

The carbon sequestration of an average EUROFLUX forests (Valentini, Baldocchi, and Olson, 1999) is 2 - 4 t C ha^{-1} year^{-1} but increases from low to high latitudes. According to Valentini et al. (2000) the NEE of four tested Italian forests range between 4.5 and 6.7 t C ha^{-1} yr^{-1} and at the same time the respiration is between 4.45 and 6.4 t C ha^{-1} yr^{-1} resulting in a NEE/RE rate of about one. The respiration (RE), to which also the comparably small flux of BVOC's belongs, is calculated from the nighttime fluxes, whole day fluxes during leafless periods, and daytime respiration which is estimated by extrapolation of nighttime fluxes to the rest of the day taking into account the functional relationships with soil and air temperature. While the NEE depends on the latitude, the GEE is nearly constant at 12 t C ha^{-1} yr^{-1} for1 the investigated European forests. Younger stands accumulate carbon to build up their biomass. Older stands are close to equilibrium and can switch from being a carbon source during one year and a sink in another. Thus the budget seems to depend on climate variability.

Another approach was made by Chirici et al. (1999) who combined the ecophysical forest model (FOREST BGC) of Running and Coughlan (1988, see Running et al. 1989) with Landsat-TM data for deciduous forest stands of *Quercus cerris* near Radicondoli, Tuscany, Italy. The model was validated with measured evapotranspiration fluxes (sap flow). According to this model the photosynthesis was of the order of 80-100 kg CO_2 ha^{-1} d^{-1} during the vegetation period between May and October 1997. This results in a total photosynthesis (GEE) of 8 - 10 t C ha^{-1} year^{-1} in the forest. It is estimated that a NEE of about half this figure results.

CO_2 fluxes also have been measured during the long term MUREX experiment (Calvet et al. 1999) near Toulouse of a dense herbaceus agricultural fallow site covered mainly with *Brachypodium sp, Potentilla reptans, Geranium rotundifolium, Erigeron canadesis*, and *Rumex acetosa*. On 1 September 1997 under a clear sky during daylight time (10 hours) average CO_2 flux densities of 0.24 mg CO_2 m^{-2} s^{-1}

soil respiration and 0.1 mg CO_2 m^{-2} s^{-1} flux at 2.7 m height in the atmosphere have been measured. This results in a daytime net assimilation of about 0.34 mg CO_2 m^{-2} s^{-1} or 12.2 g CO_2 m^{-2} d^{-1}. The modelling of the net assimilation for the year 1995 resulted an annual average of approximately 20 g CO_2 m^{-2} day^{-1} with values up to 50 g CO_2 m^{-2} day^{-1} near day 140 and 30 g CO_2 m^{-2} day^{-1} around day 220 but with a dip down to 12 g CO_2 m^{-2} day^{-1} around day 165 when the herbs were cut and the LAI went down to zero for a few days. At the day 247 (1 September) the assimilation was according to two different models 12.4 respectively about 23 g CO_2 m^{-2} day^{-1} which is for the first model very close to the value of 1997 (Calvet, 2000). The annual assimilation equals in the average about $7.3 * 10^4$ kg CO_2 $ha^{-1}year^{-1}$ (or 20 t C ha^{-1} $year^{-1}$ according to Calvet 2000) which points at a very efficient ecosystem. Without the cutting even more could have been be expected.

5.3 Forest Fires

According to Moreno (1999) in Spain the number of fires increased rapidly after 1973 from 4 000 to a maximum of over 25 000 in 1995 and the burned area jumped up from about 5 000 to sometimes over 400 000 ha/year with large interannual variability. In 1993 the number of fires was 15 000 and 100 000 ha burned. In 1986 there have been 12 000 fires and 500 000 ha burned. The average burned surface per year presently is about 250 000 ha/year in Spain, 140 000 ha in Italy (prior to 1977 less that 70 000 in the average), 25 000 in France, 50 000 in Greece, and 90 000 in Portugal. In North African countries but also in Turkey the burned area per year is much smaller, e.g. around 20 000 - 30 000 ha/year in both Algeria and Turkey. Between 1989 and 1992 the total number of fires in the northern Mediterranean area seems to have been fairly constant at 60 000 per annum but the burned area decreased from 700 000 to 400 000 ha (Parry 2000).

Vélez (1990) counted 173 715 fires in Spain between 1974 and 1994 and 4.6 Mha were burned. 41 % were forests (33 % *Pinus*, a minor fraction *Quercus*), the rest treeless scrubland or afforestation areas in the East. 3.3 % may be caused by lightening. The financial losses are of the order of 300 M€ per year and the expense for fire fighting 310 M€ per year.

Wildfires are favoured by long dry and hot periods. They are as far as it is known only to 3.3% caused naturally such as by lightning. Another 2.5% results from management burning which are fires that escape if pasture or agricultural material is burned. Conese et al. (1999) investigated 140 fire events which took place at the western third of the island of Elba and determined the "risk level" for different ecosystems ranging from 8 for chestnut forests to 1 for pasture. Meteorological conditions have to be considered in addition. López-Bermúdez et al. (1999) reported a case in the Murcia region where after three years of extreme drought in 1994 a fire developed at air temperatures higher than 48 °C and strong winds with a fire front of 50 km and burned 29 000 ha.

Pinus species are easier inflammable then *Quercus* species and once burned do not recover from the roots (such as olive trees do) but only from seeds. Schultze-

Westrum (1999) point out that the natural regrowth of burned forests in Greece partially depends on the grazing goats which eat leaves and suppress the recovering of deciduous trees like *Quercus* species.

Often in burned areas monocultures of *Pinus* are re-forested rather then mixed cultures and the carpet of fallen needles underneath suppresses undergrowths and degrades soils by acidification.

The burning of agricultural debris as well as the wildfires are an important source for aerosols and cause a hazy atmospheric boundary layer over large areas. The burning thus have an impact on the radiative transfer but also an atmospheric chemistry since with the water vapour, carbon dioxide and aerosols also large amounts of the BVOCs are set free by the fires. These fires are concentrated at the end of the vegetation period and at the beginning of the rainy period. Some of the material therefore is washed out and deposited at the surface. Since these are partly acidic components it contributes to acidify the soils.

5.4 The Role of Aerosols

The aerosol content of the Mediterranean atmosphere varies largely. Few measurements indicate that the amount of aerosol increased since the fifties at least in some heavily industrialized areas. Aerosols interact in various ways with the atmospheric radiative transfer and constitute an important climate factor. Dry aerosols reduce the solar radiation at the surface and hygroscopic aerosols are responsible for cloud formation and rain, if the moisture is present. Aerosols furthermore interact with remote sensing from space. The limited knowledge of their radiative properties and concentrations hampers the correction of satellite data. One of the great deficits for coping with these different aspects is the lack of an aerosol climatology for the Mediterranean basin.

Four types of aerosols are of specific importance in the Mediterranean area: Industrial aerosols, sea-spray aerosols, sand particles, and aerosols of biogene origin. They act in different ways. Hygroscopic salt particles produced by sea spray foster the condensation as do many of the industrial aerosols. The presence of Saharan sand particles is an outstanding feature of the Mediterranean atmosphere. Desert sand particles are not hygroscopic and due to absorption of solar radiation they may even warm the dusty atmosphere, dissolve clouds, and thus prevent condensation, a process which sometimes can be observed in satellite images. Their climatic effect primarily is the backscattering of solar radiation to space and thus the reduction of the short wave radiant energy which reaches the surface. This counteracts the greenhouse effect since it causes cooling at the surface. The same effect have sulphate particles of industrial origin. On the other hand, in the thermal infrared, they make the atmosphere more opaque which increases the long wave atmospheric radiation. The ratio of their longwave to shortwave properties determine which of the effects dominates - cooling or warming. From the many fires in the Mediterranean area of partly wet material not only aerosols are emitted, at which immediately condensation takes place, but in addition biogene substances emerge from which aerosols build up by chemical reactions.

D'Almeida (1986) studied the Saharian dust load near its source in Agadez (Niger) and found mass concentrations of mineral dust varying between 40 μg m^{-3} under clear sky conditions up to 9300 μg m^{-3} during severe sand storms. This is a very high load as compared to global average background concentrations of 7 μg m^{-3} (Heintzenberg, 1980) and sea salt mass concentrations of 100 μg m^{-3} for a wind speed at 20 m/s (Lovett, 1978). Towards Europe the annual mass transport from the source region primarily between Air and Tamanrasset was estimated to be 8 * 10^{7} t in 1981 and 12 * 10^{7} t in 1982. The optical aerosol depth at 500 nm over Assekrem (2730 m, southern Algeria), as an example, varied in 1981/2 between 0.04 in winter and 0.8 during several months in summer (D'Almeida, 1987).

The optical properties of Saharan aerosol was investigated in 1980 (Fouquart et al. 1987a and b). During this 17 days lasting experiment (ECLATS) the optical depth at 520 nm even rose up to 1.4 while minimum values were 0.2. The single scattering albedo of the particles was estimated to be 0.95 and the asymmetry factor about 0.65 in the shortwave domain. The ratio of thermal infrared (8-14 μm) to visible (0.55 μm) optical depths were found to be close to 0.1. Calculations of the heating rate resulted in an additional shortwave heating between 0.7 K day^{-1} averaged over the most clear days and 1.8 K day^{-1} for the dustiest day. The infrared cooling is smaller, 0.2 K day^{-1} and 1 K day^{-1} respectively. The net effect of the desert

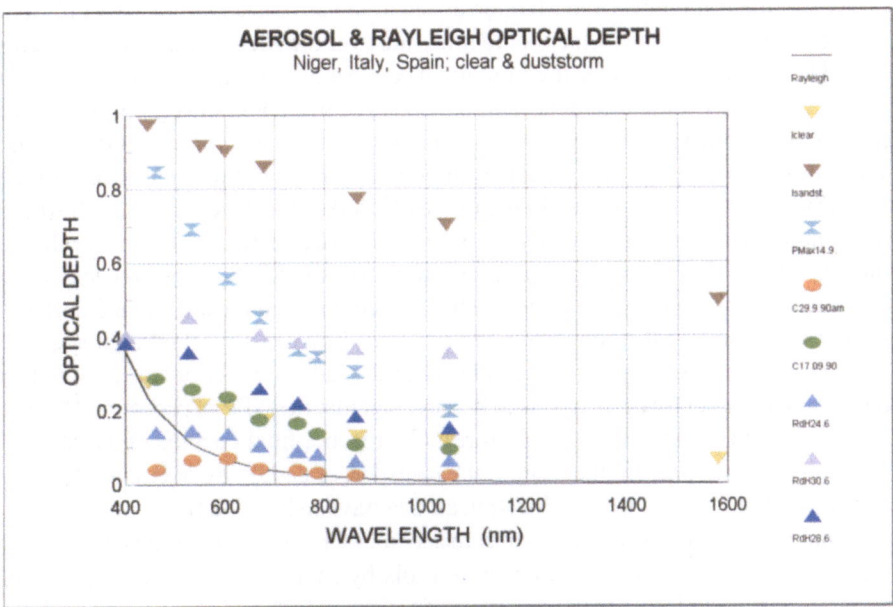

Fig. 9. Spectral dependence of the aerosol optical depths measured at different locations. I: Ibecetene, Niger, under "clear" and "sandstorm" conditions. P: Po valley, Italy, 14.09.90. C: Cafaggio, Italy, coastal area, 17.09.90 (V < 10 km) and 29.09.90 a.m. (V > 50 km), 29.09.90. RdH: Rada de Haro, Spain, (EFEDA experimental area) at three days with different atmospheric turbidity. Clearly to be seen is the effect of different size distributions. For comparison the Rayleigh optical depth is plotted as line

aerosol therefor is an atmospheric warming of a little bit more than half a degree per day. Model computations with aerosol parameters observed at a hazy day (optical depth at 550 nm around 0.7) as input data showed a marked effect on the downwelling shortwave flux. Without aerosols the calculated flux at the surface was 530 Wm^{-2} and with this aerosol load of the order of 400 Wm^{-2}. The infrared flux with aerosols was 390 Wm^{-2}, which agreed with measurements, while the calculated flux without aerosol was 370 Wm^{-2}. These results indicate that the Sahelian aerosol has a net cooling effect at the surface.

The spectral distribution of the optical depth during to a sand storm is presented in Fig. 9. The effect of large particles can clearly be seen in the nearly linear dependence of the optical depth on wavelength. In the Mediterranean basin the Saharan aerosol effects may be slightly smaller because of the thinning effect during the long transport. But even here during *scirocco* events the optical depth at 525 nm increases to values around 0.8 as compared to a very clear day when it is around 0.2. A selection of some characteristic cases of aerosol optical depths measured in the Mediterranean area is presented in Fig. 9.

5.5 Unknown Budgets

As can be seen from these mostly preliminary data, the span of carbon fluxes is large. An integration over the Mediterranean area has to account for very different source and sink situations. One tool to integrate over the Mediterranean ecosystems would be to apply models for the individual ecosystems and integrate over the area they occupy. According to a model comparison organized by GAIM (Hibbard, 2000) most of the applied models overestimate the NPP in the range of 0 to 1000 g C m^{-2} year^{-1} in single cases up to a factor of three while they tend to underestimate the higher values of NPP measured in the range of 1000 to 1700 g C m^{-2} year^{-1}. It would need more measurements in different parts of the Mediterranean basin to validate the available models and a careful integration over the Mediterranean ecosystems e.g. by means of satellite data to decide how much the Mediterranean land surfaces contribute to the global carbon budget respectively how sensitive the Mediterranean flora is with respect to climate variability and change. To establish a complete budget the effect of forest fires has to be rated against the NPP of regrowing forests which is higher than that of an old forest. To reach a final conclusion the carbon that remains after fires in the ground, the transport of carbon by rivers into the sea, and the carbon cycle of the Mediterranean Sea have to be added.

Analog problems exist with respect to aerosols. Emission from industrial plants, sea spray, and fires, the formation of aerosols by chemical processes, the flux of Saharan dust particles and the local raise of soil particles by wind from dry surfaces as well as the raining out of theses constituents are not well known. The high variability of the concentrations due to wind direction, wind velocity, sedimentation and precipitation makes it difficult to build reliable aerosol models. A key problem is how the cloud climatology is affected by the changing aerosol load. For this purpose also a cloud climatology separately for land and sea would be needed at higher spatial resolution than presently provided by the International Satellite Cloud Climatology Project (ISCCP). In different regions of the Mediterranean the spectral atmospheric transmission has been measured and also regional networks exist. It would be necessary to intensify these measurements with modern equipment at "Anchor Stations" distributed around the Mediterranean, to evaluate these data with a standardized procedure and to compile these data into a central archive to generate a more complete picture for the whole basin.

There may also exist another climate relevant link between carbon dioxide and desert aerosols which is not yet explored. It is known that the uptake of CO_2 by the oceans increases with its iron content. Into the Mediterranean Sea continuously Saharan minerals are precipitated among which are iron containing minerals. Does this lead to an additional dumping of CO_2 into the Mediterranean Sea?

6 Will Mediterranean Water Resources and Vegetation Sustain Dramatic Climate Changes?

6.1 The Search for Long Term Variability and Trends in Operational Precipitation Data

Some features of the Mediterranean climate and its spatial variability have been discussed in the second section of this chapter and in the third section the climate development prior to the establishment of the operational measuring network, the "instrumental era", was addressed. Here now more recent studies are reported of temporal variability primarily of precipitation. The latest results of Mediterranean climatologists are then presented in Chapters 4 and 5.

Maheras (1988) investigated the changes of precipitation conditions in the western Mediterranean basin from 1891 to 1985 by means of a principal component analysis. Two principal moist periods occurred from 1901 to 1921 and 1930 to 1941 with dry periods in between and 1942 to 1954. 1980 again a dry period began. This indicates a periodicity of about 20 years. The wet periods coincide with weaker zonal circulation and the dry periods with stronger zonal circulation. The analysis was expanded retrospectively by Rodrigo et al. (2000) for Andalusia to 1500 AD where three major long lasting dry periods, 1501-1589, 1650-1775, and 1938-1997 could be identified and periodicities of 2.1, 3.5, 7-9, and 16.7 years have been found. Droughts are related to high positive NAO indices and floods to negative NAO values.

With respect to the whole basin Kutiel and Maheras (1998) and Maheras, Xoplaki, and Kutiel (1999) concluded that there exists a complicated pattern of hydric anomalies throughout the Mediterranean basin. There can be positive as well as negative excursions from average (i) basin-wide, (ii) primarily in the west, (iii) strong in the west and east but weak in the central parts, or (iv) positive in the east and weaker in the western part. Temperatures of Malta, Athens and Jerusalem are positively correlated. Except for winter the correlation between the Iberian Peninsula and Greece/Israel is negative or zero.

The antiphase relationship between the western and the eastern part of the Mediterranean known as the *Mediterranean Oscillation* and mentioned already at the end of section 2.2 was investigated in more detail by Colacino and Conte (1993) in relation to the effect of global warming in the Mediterranean area. These authors found an upward trend in the 1865 to 1990 surface pressure field which started in the forties with an average slope of 0.04 hPa/year. This general increase of surface pressure is related to a trend of the geopotential at 500 hPa which is positive with a maximum of 0.7 m/year centred just west of the Strait of Bonifazio (between Corsica and Sardinia) and takes slightly negative values (0.1 - 0.2 m/year) in the eastern part of the Mediterranean. This trend is superimposed by a variability with a prevailing period of about 22 years with an opposing phase. If in the western part of the Mediterranean is a negative departure of the 500 hPa height from the mean as in 1953 and 1975 there is a positive excursion in Cairo. An interesting observation is that the mean values of both cycles, which were clearly separated by about 50 metres (lower in Algiers than in Cairo) in the forties, tend to approach each other in the eighties. Furthermore since 1970 both the number of blocking events and the number of blocking days increased while the number of strong cyclonic events with > 20 hPa/day deepening fell from in the average 5 to 1 - 2 events during the same period. The authors point out that it cannot yet be postulated with certainty that these changes are induced by the increasing greenhouse effect because in relation to the 22 years periodicity the analysed time period is still short (enough radiosonde data only became available after the second world war) and because of the complexity of the Mediterranean climatology.

Recently Piervitali, Colacino and Conte (1998 and 1999) published two papers on precipitation trends during the interval from 1951 to 1995 over the Central-Western basin. They found a decreasing trend in this region of the order of 3 mm/year with largest reductions during winter (the winter 2000/2001 may change this picture). The reduction increases from North to South. There seems to be a negative correlation of this trend with the increasing pressure during this period. The interannual variability is large, up to 200 mm from one year to the next around an average of 650 mm/year. The spectral analysis shows peaks at 3, 15 and 22 years. Only in the western part the correlation with the NAO is relatively strong with positive values in Oran, Tunis and Malta and negative values in Lisbon and Mertola (Portugal). The negative correlation with the Mediterranean Oscillation (MO) generally is higher. The index used for the MO is the height of the 500 mb surface over Algiers which is anticorrelated to that of Cairo.

Trigo and da Camera (2000) investigated the relationship between weather types and precipitation in Portugal. They confirmed the observation of other researchers that there is a significant decrease in March precipitation in southern Portugal from about 175 mm in 1946 to 60 mm in 1991 for the station Coimbra and from 110 mm (1946) to 40 mm (1991) for the station Mertola, both with large interannual variability ranging from zero (1966) to 250 in Coimbra. The extreme precipitation minima in the South correspond to anomalous low occurrence of wet weather circulation types with an Atlantic origin and negative sea level pressure anomalies at the western coastlines from England to Spain. Cyclonic weather types provide more equally distributed precipitation over Portugal.

By investigating tree rings of *Cedrus atlantica*, Chbouki et al. (1995) came to similar results for the Rif and Atlas region in central Morocco with respect to droughts that occurred between 1845 and 1974. They identified three longer drought periods, 1860-1900, 1925-1950, and around the 1970's. During these periods the Azores high was positioned slightly more to the North positions and shifted towards North-East thus blocking the disturbances to reach the region.

Mild summer rains in the Levant regions have been attributed to a weakening of the rain suppressing thermodynamic conditions (persistent subsidence aloft and lower level cool marine flow) rather then synoptic scale forcing (Saaroni and Ziv, 2000). This would imply that longer drought situations can be expected if the Hadley circulation is strong.

Kömüşçü (1998) found periodicities of 2.4-3.1, 3.3-4.3, and about 11 years in the temperature time series of 71 stations in Turkey. The inland regions exhibit a grey noise spectrum, significantly different at the 95% level from a white noise which is predominant in coastal regions. This indicates the strong influence of the sea-land circulation in coastal areas while the inland variability seems to be caused by large scale processes. Periods with similar frequencies have been reported by other authors as well. Piervitali, Colacino and Conte (1998 and 1999) made a spectral analysis for the years 1951 to 1995 over the Central-Western basin and found peaks at 3, 15 and 22 years.

Otterman et al. (1990) pointed out that vegetation cover influences convective processes. If an accelerated hydrological cycle produces more rain over land it may keep the surface cool due to increased evaporation.

6.2 The Water Cycle

As has been learned from the data analysis both the seasonal variability as well as the multi-annual variability of the water cycle are high. Very recently extreme summers with extended dry periods have been experienced in various parts of the Mediterranean area. These long dry spells have consequences for the dry agriculture - such as wheat, tomatoes or olive harvests - and for the water supply systems. Water reservoirs and water re-use systems to bridge several months without rain become essential under such climatic conditions. It must be emphasized that even an enhanced precipitation must not necessarily increase soil moisture and reduce the soil water deficit in summer. Firstly the enhancement of precipitation may be limited to the winter months during intensive rainfalls and secondly the soil moisture depends on three parameters: precipitation, evaporation, and run off. Under the present conditions summer rains are only wetting the uppermost layers of the soils. The water is evaporated before it can penetrate to deeper layers.

An illustration of the various components of the water cycle is given in Fig. 10. Depending on the distribution of land cover part of the rain feeds forests and fallow

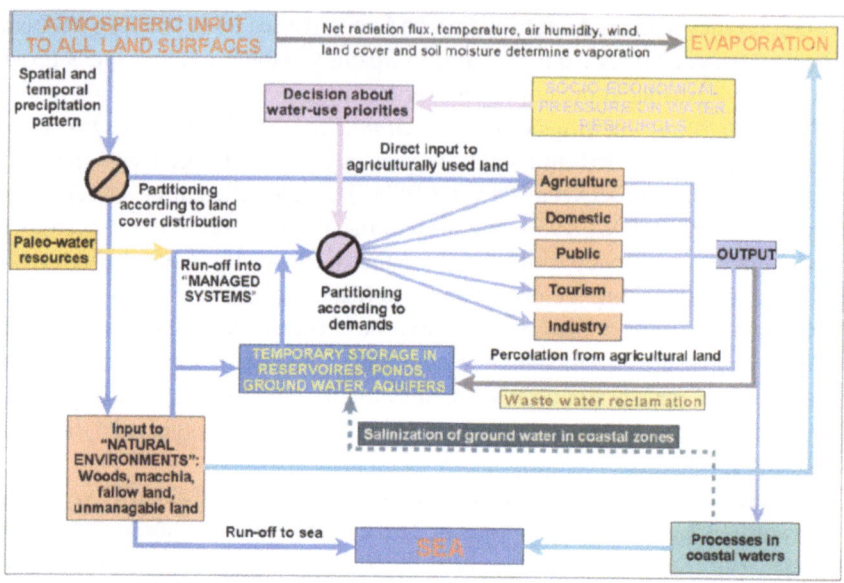

Fig. 10. Schematic of the water cycle. The water cycle over land is characterized by two major partition "switches": One is the distribution of rain into the agricultural areas where it is used to grow food and into the other land cover types where it serves to grow biomass or runs off either into the sea or into reservoirs. The second partition is under human control. It directs the water collected from rivers and aquifers to the five major user communities. Part of the water is evaporated back into the atmosphere, the rest finally and partly after having been recycled runs off into the sea.

land and the other fraction is partitioned into agricultural areas and serves the production of food. A large fraction is evaporated by the strong solar input which leads to the water deficit in summer. The run-off through rivers and sub-surface aquifers from all areas is partly directed into water reservoirs (to which also ground water belongs) and water management systems, to which in some countries in addition fossil reserves may be added. This water underlies the direct influence of water authorities and can be distributed according to the needs respectively the national water management policy to the main competing water use communities such as agriculture, industry, domestic, public, and tourism. Part of the used water can be recycled, the rest runs off to the sea.

The use which the different countries made in the eighties of their water resources is plotted in Fig. 11. In total about $18 * 10^6$ ha have been irrigated in 1989 which was about 14 % of the total cropland and the tendency is further increasing (Gleick 1993). This in some areas leads to a *circulus vitiosus*. To increase agriculture and by this the income of the rural population it is necessary to irrigate the fields. This can lead to a dramatic decrease of the groundwater level. In the 235 km^2 large Guadalentin watershed (Murcia, Spain) the pumping lied between 40 and 70 (100 m)^3/year. This corresponds to 200 - 300 mm and approaches the average annual precipitation sum of which only 14% is believed to infiltrate the aquifer. As a consequence the groundwater level between 1972 and 1996 fell from −50 m to

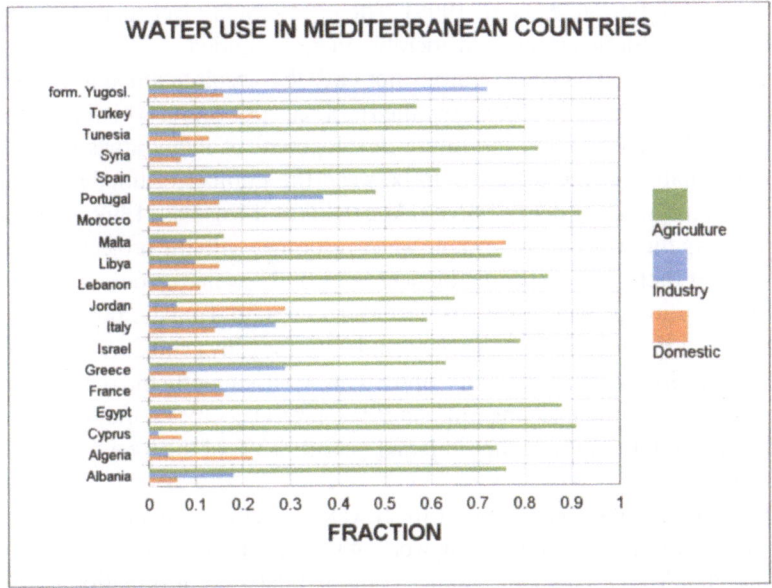

Fig. 11. Use in percent of the total water resources for agriculture, industry and domestic purposes respectively in some Mediterranean countries according to Nunes Correia (1999)

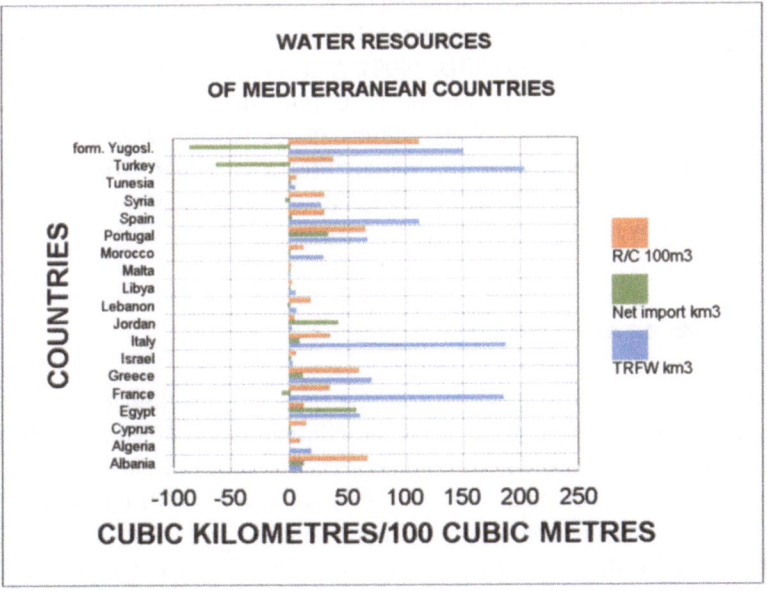

Fig. 12. Water resources of Mediterranean countries per capita after Nunes Correia (1999). Total Resources of Fresh Water (TRFW), net import across the boarder in km^3, and resources per capita (R/C) in 100 m^3

- 250 m (Lopez-Bermudez et al. 1999). The unsaturated layer increased by 200 m and the drilling for water had considerably to be expanded.

For the Mediterranean countries not only the absolute amount of renewable annual water resources is critical but the resources per capita which are presented in Fig. 12. As can be seen, the future development of many areas of the region becomes a matter of the relationship between the number of inhabitants and the amount of renewable water resources. Angelakis and Kosmas (1999) point out that by combining the projected renewable water resources with the projected population development there may be dramatic decreases of the water resources per capita in 2050 by 50% in Algeria, Egypt, Israel, Morocco, and Tunisia. 60% reduction is expected for Lebanon, 40% for Turkey, and 25% for Libya and Syria. Increases are expected for Greece, Portugal, Italy, and Spain. Consequently the recycling of used water and, maybe, the exchange of water resources becomes a major issue for the future. The authors estimate that in Israel already 13% of the total water resources is recycled.

Due to the rapid development of fresh water consume it is necessary to update these data continuously. In the framework of the UNEP Mediterranean Action Plan the Blue Plan Water Thematic Studies compiled detailed information about water availability and consume in the Mediterranean countries (Margat and Vallée, 1999). The WMO with the support of the World Bank established the *Mediterranean Hydrological Cycle Observing System (MED-HYCOS)* as part of a global effort with the goal to "contribute to the assessment of water resources and to their management by helping national Hydrological Services to strengthen their capacities and by promoting the exchange of information and skills among the countries participating in the project" (MED-HYCOS/IRD, 1999). At a wider scope under the guidance of the World Commission for Water in the 21st Century the "World Water Vision" was presented at the Second World Water Forum convened by the World Water Council in The Hague, The Netherlands, in March 2000. The *Global Water Partnership* prepared for this forum the basis for taking action which is documented under the title: "Towards Water Security: A Framework for Action" (GWP, 2000).

6.3 Vegetation and Climate

Vegetation is an indicator of land-surface processes and thus climate variability. The vegetation cover depends strongly on the climatic environment, specifically its hydrological properties, and it has a function in regulating the energy budget due to albedo and stomata resistance effects. Otterman et al. (1990) furthermore claim land-use change as a reason for precipitation change.

The from North to South decreasing vegetation cover is responsible for the albedo gradient across the Mediterranean basin which keeps the surface net radiation on both sides of the sea in narrow boundaries. In the South where the albedo is high because the soil is bleached out over long times (the organic component gets lost) a larger fraction of the solar energy is reflected and the radiation budget is comparable with the values measured at the northern more vegetated areas where the solar irradiance is smaller but its absorbed fraction higher.

Furthermore the vegetation, as long as water is available, keeps down the temperature and infrared emission because of the evaporation. Bare soils with their low emissivity heat up at daytime to high thermodynamic temperatures at which infrared emission compensates the absorbed solar and atmospheric radiation minus the mostly sensible heat fluxes and cools strongly at night into the mostly cloudless sky. These extreme temperature contrasts reduced the chance for survival of vegetation to species that withstand the high noon and low night temperatures with little water. In these sparsely vegetated areas the net radiation energy is nearly completely converted into sensibly heat. In the more vegetated areas of the North the temperature wave is dampened though the amount of absorbed radiation energy is not much less than in the South because the here smaller solar irradiance is compensated by the lower albedo. The surface furthermore does not need to compensate for the absorbed solar energy by infrared emission and sensible heat fluxes alone, the evaporation of water in addition cools the surface. Because of the high emissivity the surface emits more infrared energy than the desert at equal thermodynamic temperature. If the vegetation for a short time vanishes (e.g. after harvest or in dry summers) the soil is not immediately changing its composition and colour and the albedo might only be little affected by the change at the surface. This is because vegetation has a very low albedo in the visible part of the spectrum but a high albedo at near infrared wavelengths while the soil may have a medium high albedo both in the visible and the near infrared and thus nearly exhibits the same broadband albedo as the vegetation. This property, of course, varies regionally.

By making use of its spectral reflection properties the state of the vegetation can be traced by remote sensing techniques. There are many possible combinations of these spectral reflectances which lead to so-called "vegetation indices". The most simple combinations are ratio of visible to near infrared reflectances or the difference of these two reflectances normalized to their sum. The analysis of time series of these data provides information about the length and phase of the vegetation period, the intensity of the "greenness" as well as the interannual variability. As an example the 10 years time series for the Iberian Peninsula of Koslowsky (1999) is reproduced in Fig. 13. It clearly indicates the variable pattern of the dry areas. An interesting result is, that whatever happens during the early vegetation period (which is governed by the winter rain) it does not much affect the summer picture of which only one representative scene is reproduced in Fig. 15a. This reflects the large water deficit in summer which cannot be overcome by strong winter precipitation: The winter- and springtime precipitation is used up in summer mainly because of the high net radiation. If this picture is combined with the effective surface temperature (Fig. 14 and 15b) the impact of vegetation on surface temperature can be visualized: Vegetation covered areas remain cooler due to evaporation.

After a forest fire the albedo for a short time is reduced which means more solar energy can be absorbed and a black "hot spot" is visualized. In most cases vegetation reappears soon but it may be of a different type than prior to the fire because the survival of the seeds depends on the species and the intensity of the fire. An

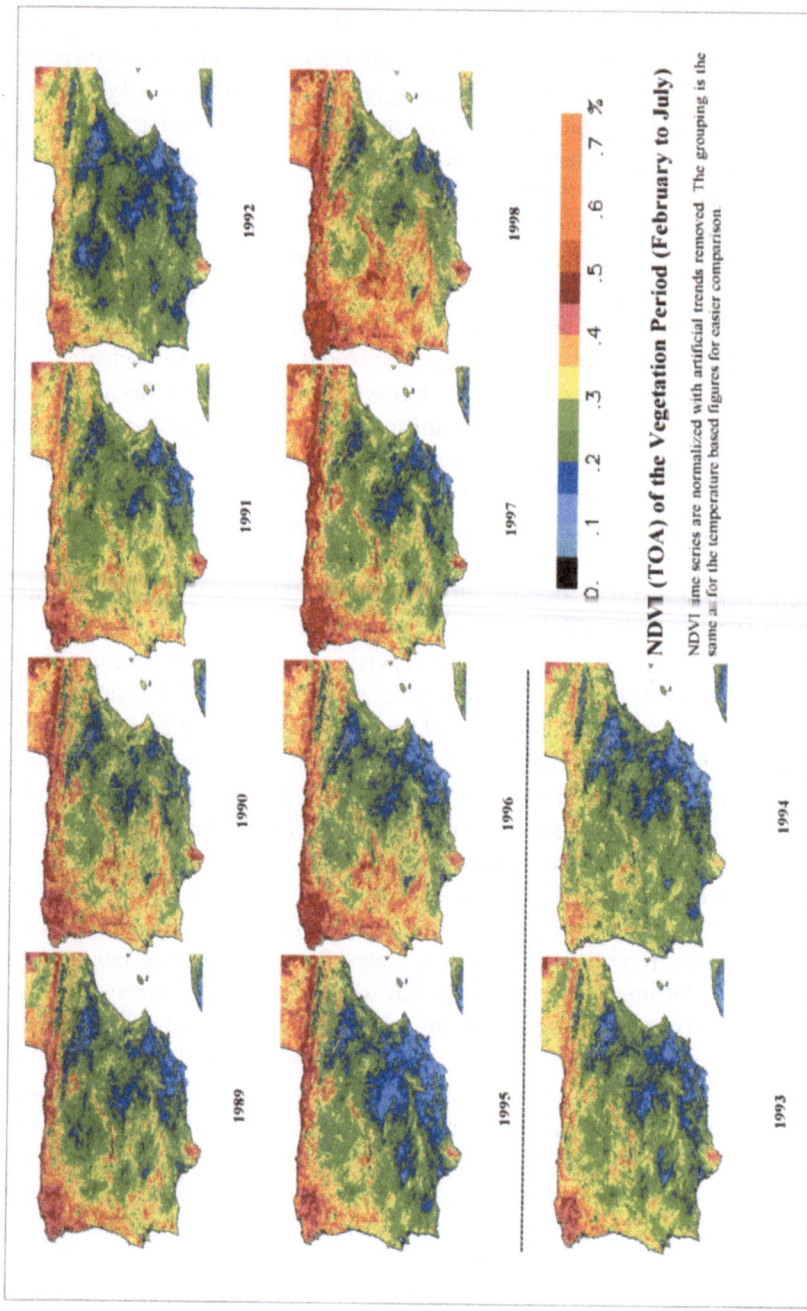

Fig. 13. Normalized time series 1989 - 1998 of the Normalized Difference Vegetation Index (NDVI) averaged over months February to July for the Iberian peninsula after Koslowsky (1999)

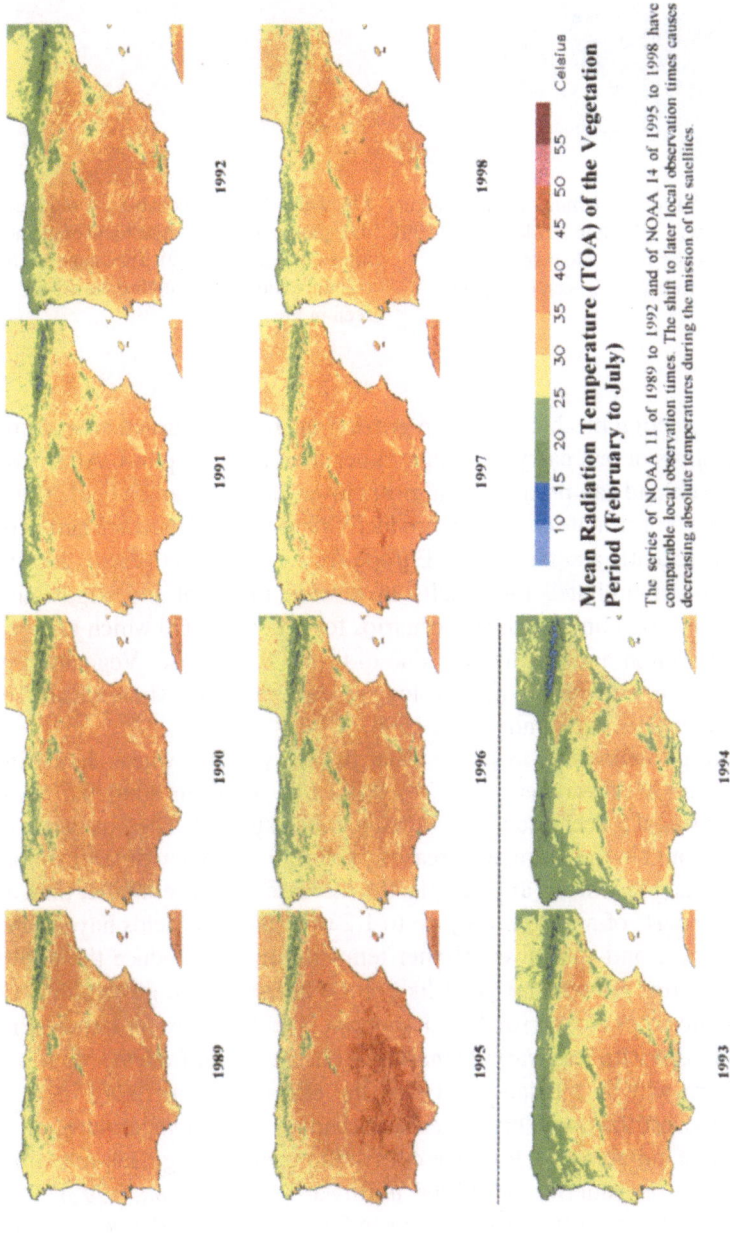

Fig. 14. Time series 1989 - 1998 of the mean effective radiation temperature at the top of the atmosphere for the Iberian peninsula averaged over months February to July after Koslowsky (1999)

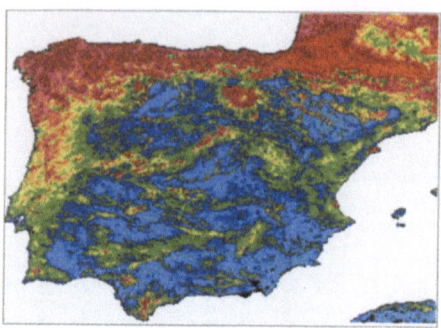 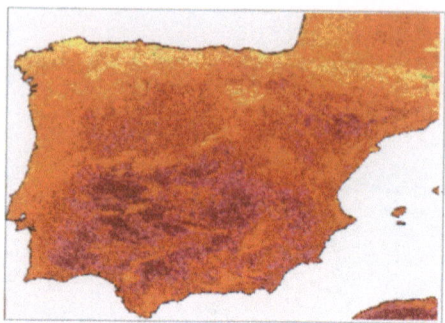

Fig. 15a. NDVI for the summer months August and September 1998 after Koslowsky (1999). The image has the same colour scale and is representative for all years seen in Fig. 11

Fig. 15b. Mean effective TOA temperature for the summer months August and September 1998 after Koslowsky (1999). The image has thecsame colour scale and is representative for all years seen in Fig. 12

additional selection mechanism is the grazing of sheep in burned areas as they eat the young shots of leaf carrying plants rather than those of pines. After five years regrown scrubland can produce an almost closed canopy (Faraco et al. 1993).

The impact of climate change on productivity, yields and the distribution of plant species was recently assessed by the Europe ACACIA Project (Parry, 2000) which gathered and assimilated the results of a wide range of investigations. This assessment is based upon climate scenarios for the year 2050 which are discussed in the next section and fertilization due to the CO_2 increase. Vegetation growth models are then fed with the meteorological state variables which result from the climate models (see e.g. Bindi et al., 1996).

According to this assessment the water limited yields of wheat should increase in most of the Mediterranean areas by 0 - 2 t/ha. Only in southern Spain and southern Portugal a decrease of 0 - 3 t/ha is to be expected. The potential yield for grapevine should in the average increase by 0.6 - 0.9 t/ha but with large differences. In Castilla - La Mancha, as an example, the increase would be smaller than 0.3 t/ha and in some parts of northern Italy up to 1.2 t/ha. Smaller yields have been found for grain maize and sunflower. Higher temperatures will reduce the duration of determinate crop growth and yield like onion but stimulate growth and yield in indeterminate crops like carrot. Less important is temperature for lettuce. Fruit dry matter increases with less warm scenarios but decrease with warm scenarios but in general there will be a balance between the effect of temperature increase and elevated CO_2 fertilization. Olive trees are sensitive to low temperatures and water shortage. The suitable area for olive cultivation could enlarge due to temperature and precipitation patterns, or shifted northward (Bindi et al., 1992). Extended summer droughts may affect Mediterranean forests growth negatively and increase the risk of fire.

The reported results are qualitative and of the kind "if temperature increases, our vegetation models tell us that ...". Presumed that the vegetation models are correct, the results derived from their application are only trustworthy if the underlying climate scenarios can be validated. This now has to be investigated.

6.4 Will the Mediterranean Water Cycle Change?

There are three ways to develop scenarios for the future. One is a simple extrapolation of observed regional trends which can completely be in error if the large scale circulation changes. The second way is a mixture of model simulation and empirical knowledge: If it can be demonstrated that precipitation correlates well to some large scale phenomenon which can be simulated with some reliability the future precipitation can be deduced from the development of these large scale phenomenon. The third way is to rely exclusively on climate model scenarios and to scale down the results of these computations at certain time slices into the regional scale. This would require a validated scaling method. The application of all three methods bears risks in the case of the Mediterranean area. Small shifts in the large scale flows respectively circulations not only brings an area under different climatic regimes but in addition generates changes in the interaction between motions at different scales down to land-sea circulation and mountain induced circulation systems which can destroy the present correlations. We shall inspect now to which conclusions some of the applied methods lead. Because water is the "life blood" of the Mediterranean we shall specifically address this parameter.

A simple extrapolation of the meteorological records to obtain an indication how the future may look like does not seem to have great merits in view of the complexity of the interactions with the NAO and the Hadley circulation. As an example, López-Bermúdez et al. (1999) deduced from the 130 years precipitation record of Murcia (Spain) a downward trend from 368 mm in 1860 to 241 in 1996. If one looks closely at the record this trend is strongly biassed by a very wet period from 1882 to 1895, when up to 800 mm were recorded within one year. Taking the time period 1910 - 1996 rather an upward trend would result due to maxima around 1950, 1973, and 1985. Generally one can conclude that since 1940 the situation in the average is pretty stable with a large interannual variability.

North African data series (WMO, 1999) show positive as well as negative temperature trends but also the time series are of different length (Table 6). Precipitation statistics has to fight with the large variability of the annual rainfall. In dry areas with a few mm average precipitation per month the standard deviation is of the order of the average precipitation. At 29° N, 16° E the temperature trend was 0.015 °C/a (1955 - 1989) , the average rainfall 2.56 mm/month, the standard deviation ±2.35 mm/month and the trend - 0.013 mm/a. At 27.0° N, 14.4° E the mean precipitation lies at 0.79 mm/month, the standard deviation is ± 0.93 (1930 - 1988). But also the spatial variability in areas of moderate precipitation can be large. Algiers had a positive precipitation trend of 0.1mm/a (1841 - 1972) biassed by the years 1950 - 1970 while Oran (as well as Constantine) had a negative one of the same magnitude (1841 - 1990) which is strongly biassed by the wet years around 1870. The temperature is very stable at 17.5 °C. Alexandria and Port Said have a very similar rainfall trend as Malta but quite different mean values (Table 6).

Malta between 1841 and 1986 had a positive precipitation trend of 0.046 mm/a (average 44.4 mm/month, standard deviation ±12.6 mm/month). No significant trends can be detected in the data of Palma de Mallorca, Nicosia and Limassol, which belong to other islands in the Mediterranean. Aleppo measured a negative temperature trend (1951-1986) of 0.084 °C/a.

A study by Türkeş et al. (1995) of the annual mean temperatures at several stations of Turkey came to the conclusion that in the different regions of Turkey no trends could be detected (up to 1994) which can be related to global warming. In some areas even a small cooling was observed. In the nine years running mean the maximum was reached in 1961 (14°C) and minima in 1932 (12°C) and 1992 (12.5°C).

Table 6. Mean values, standard deviation and trends of temperature (T) and precipitation (P) for some stations at the southern boarder of the Mediterranean

Site	Quantity and Period	Mean	Standard Deviation	Trend mm/year
Alexandria	P 1868-1973	16.2	± 5.4	+0.031
Alexandria/Nouzha	T 1945-1990	20.15		-0.0024
Alexandria/Port	T 1870-1952	20.5		+0.0040
Port Said	P1888-1987	6.7	± 3.3	+0.039
Port Said	T 1883-1990	21.3		-0.01
SIWA (29.2° N, 25.5° E)	P 1920-1989	0.74	± 0.74	-0.0015
SIWA	T 1950-1980	21.1	± 0.35	- 0.018
Malta	P 1841-1987	44.4	±12.6	+0.046

Schönwiese et al. (1994) analysed the precipitation data from European stations over the last 100 years and, more specifically, for the period 1960-1990. The conclusion of these authors is a decrease in the central part of the Mediterranean basin over the last 100 years of 50-100 mm annual precipitation (1891-1990), and a very variable situation during the last 30 years with a strong reduction of precipitation in autumn in the northern part of the Mediterranean (European coastline except South Italy) and an increase in spring. Most of the investigations indicate different trends in Portugal and northern Spain versus the Mediterranean coast of Spain.

A more sophisticated canonical correlation approach was applied by von Storch et al. (1993). They observed that the winter precipitation over the Iberian Peninsula correlate quite well with the North Atlantic sea level pressure field for the time periods 1904 - 1913 and 1951 - 1960. From the anomalies of the sea level pressure over the North Atlantic they estimated the rainfall over the Iberian Peninsula for the whole period from 1900 to 1980 which matches quite well the actual precipitation record measured at 30 stations (Universidad Complutense data set). Both records show a strong periodicity of approximately 30% amplitude with a period of about 20 years and, with the exclusion of the years 1900 and 1901, a superimposed increase of about 30% (or 10 - 15 mm per winter month) over the 80 years period. They then modelled the North Atlantic pressure anomaly under the assumption of a steady further increase of the carbon dioxide concentration of 1.3% per year and, with the canonical correlation obtained, estimated the future rainfall over the Iberian Peninsula. This resulted in a decrease of about -7 mm month^{-1} (100 years)$^{-1}$ with an amplitude of the interannual variability of the order of $\pm$ 10 mm month^{-1}. This result would suggest that the current trend reverses.

The observed increase of winter precipitation over the western part of the Iberian peninsula would be in accordance with an increase of the pressure gradient over the Atlantic Ocean which intensifies westerly winds between 40°N and 50°N in winter. North Africa and southern Europe become influenced by a strengthened subtropical high pressure belt during winter. It is not possible to decide yet, whether the observed increase of the subtropical high pressure belt in winter is accompanied by a northward shift of the belt. This would mean that southern Europe comes more under the influence of the high pressure cell also during winter, which would then explain the simulated reduction of future winter precipitation. It also can not be concluded yet, whether this change is part of a natural variability or a result of the increasing greenhouse effect. It must be kept in mind that the period from 1970 to 1990 is a warming period during which the system recovered from the cooling of the northern hemisphere, that occurred after the temperature optimum around 1940. It would therefore be of high interest, to know, how the pressure situation developed during the warming period from 1917 to 1940, which was not influenced by the increasing greenhouse effect.

Palutikof et al. (1999) summarized the results of data analysis and comes to the conclusion that with a few exceptions time series of the four sub-regions of the Mediterranean indicate a drying trend. In terms of the standardized anomaly index SAI this trend is of the order of 0.4 in 50 years.

The climate change assessment 1990 of the Intergovernmental Panel on Climate Change (IPCC) was mainly based upon model scenarios. It concluded, that for doubling of CO_2, which is expected to happen in year 2030, in southern Europe (35°-50°N, 10°W-40°E) "there is some indication of increased precipitation in winter, but summer precipitation decreases by 5 to 15%, and summer soil moisture by 15 to 25%" (Houghton et al., 1990). Dryer soils in summer result from enhanced evaporation due to the 2-3° higher temperatures. 1992, in its supplementary report

(Houghton et al., 1992), the IPCC compared the results of three climate models. The results show, that in the Mediterranean basin during the summer months a reduction of precipitation between zero and one mm/day is likely to occur and for the winter months a variation from an increase of up to one mm/day down to a decrease of one to two mm/day is possible. Wigley (1992) analysed the results of four independent climate simulations for the Mediterranean. According to his interpretation "the only common feature in which we can have any faith" is a large scale warming (of 1.2-3.5°C) in all seasons. The likely outcome of the precipitation analysis is a an increase, mainly in autumn, with a chance 1:4 for a decrease. In summer the western basin shows a tendency to become dryer, while the eastern part may become wetter. Similar results for the Mediterranean basin are obtained by the coupled ocean-atmosphere model ECHAM for the 100 years run, arriving at a threefold CO_2-increase.

Cubasch *et al.* (1996) compared the simulation results of five different transient coupled atmosphere-ocean models of low resolution and concluded that "the climate change predicted (for Southern Europe) by the 5 models is inconsistent and gives no clear result". The authors could, however, show that the quality of the simulation improves with increased model resolution. This is understandable in view of the importance of the complex topography of the area. In a model with higher resolution then also improved land cover information could be implemented. These should be effective land cover types representing the annual variability of albedo and latent heat flux.

The ACACIA project (Parry 2000) evaluated the scenarios underlying the estimate of the climate change impact on vegetation of five climate models of which finally one was selected for most of the future yield estimates. To give an idea of the obtained results some figures are extracted for Greece and Spain targeted at the year 2050. For Greece three of four versions of one model and two other independent models predict a decrease of winter precipitation with increasing temperature. For a 2 K temperature increase the winter precipitation should decrease by in the average 10%. For the summer precipitation all models predict a decrease of precipitation of in the average 18%. For Spain three out of five models suggest an increase of winter precipitation of 10% and all models show a decrease of summer precipitation of the order of 15% at 2 K temperature increase. There is a linear dependence between precipitation and temperature change with the effect that the precipitation change doubles if the temperature increase would be 4 K instead of 2 K.

The climate change signal for the persistence of droughts shows an upward trend of about ten percent for all seasons that is independent of resolution though its amplitude is resolution dependent. For precipitation a downward trend resulted for the spring, summer and autumn months with increasing amplitude as the resolution is improved. During the winter months, however, the trend is upwards. As an example, for the T42 ECHAM3 model of the DKRZ (German Climate Computer Center) the expected changes for southern Europe are of the order of + 0.2 mm/d in January/February, - 0.2 mm/d in May/June and -0.1mm/d throughout summer and autumn. In comparison the simulated present precipitation is of the order of 1.0 ±

0.5 mm/day throughout the year while measured amounts are twice as high. But also the measurement of local rainfall as well as the spatial integration of these punctual measurements can be in error by 15-30%. Climate model computations of the Hadley Centre confirm the trend towards a dryer climate in many areas for the past decennia and some models predict an even more dramatic decrease of (summer) precipitation from 2040 on. The temperature trend underlying these computations, however, is severely diminished if an increase of atmospheric sulphate aerosols would be taken into account. It also is not very clear to what degree regional effects of the sea are included in the models.

The question of the net water budget of the Mediterranean basin and its future development under global warming to date cannot be answered conclusively. The crucial question is to determine the net import of water into the area. Based upon the work of Grennon and Batisse (1988), Jeftic et al. (1990), Margat (1992), and Huber (1993), Wagner (2001) designed a map of the water influx to the Mediterranean sea. The net water influx from the North Atlantic Ocean and the Black Sea is $1.4 * 10^{12}$ m^3/a and the input from all rivers around the Mediterranean sea sums up to $0.439 * 10^{12}$ m^3/a. If in a rough estimate an average of 2.5 mm/day precipitation is assumed (see Table 1), the total annual rainfall over the sea would be of the order of $2.3 * 10^{12}$ m^3/a. The total water input of the order of $4.1 * 10^{12}$ m^3/a (equivalent to 4.5 mm/day) would have to evaporate to keep the sea level constant. This amount of water would be available for precipitation over land and export from the region. The air over the Mediterranean basin transports also water from the Atlantic Ocean. This water partly is included in the number of precipitation over the sea but a fraction may also be exported with the general circulation to the East. The moisture recycling in the Mediterranean basin is geographically most variable. Trenberth (1977) has considered the annual turnover of water at a scale of 1,000 km. Over North Africa only 2% of the local precipitation comes from the domain of 1,000 km, the rest is imported from more distant regions. For the Iberian Peninsula the number is about 18% while in central and eastern Mediterranean regions the values are more around 26% with "hot spots" of 34%.

To solve the question of the balance, a larger cage-type experiment, a GEWEX-like effort, would be necessary. The denser hydrological network that will include a number of new radiosonde stations, which the World Meteorological Organization plans to establish around the Mediterranean sea (MED-HYCOS project), will greatly contribute to answer this question (Abrate, 1999). On the basis of this knowledge then it would become possible to turn to the second part of the problem, how this budget would be changed if the world is warming and what impact this would make on its vegetational productivity.

Related to this problem is the question of the turnover of water within the Mediterranean Basin. Global temperatures seem to increase during the last ten years towards the highest values reached during this interglacial. If the Mediterranean Sea, which to a certain degree is decoupled from what happens in the world ocean, would warm at a considerable rate, then its hydrological cycle would be enhanced and more precipitation should also be expected over land. Global climate models in the average respond to a doubling of the CO_2 amount in the atmosphere with an acceleration of the hydrological cycle by in the average 15% which is equivalent to an average increase of precipitation of 0.42 mm/day (Randall et al., 1992). In the coupled atmosphere-ocean model ECHAM of the Max-Planck-Institut in Hamburg all components of the hydrological cycle are increased by about 5% if the CO_2-concentration is increased from 400 ppm to 1200 ppm within 100 years (IPCC scenario "A": rate of increase 1.3% as present). The already effective enhancement of the hydrological cycle manifests itself in the increase of the global recycling rate of water from 2.95 to 3.03 within the short period from 1988 to 1994 (Chahine, Hoskins, and Fetzer, 1997).

Over a warmer sea the air could take up more water vapour and if it is transported by the land-sea circulation systems over land, condensates and rains out, the land would stay cooler due to either enhanced cloudiness and consequently less net radiation flux or due to enhanced evaporation. One presupposition of the hypothesis that the hydrological cycle may be enhanced would be that the Mediterranean receives enough solar radiation to warm up the sea which is only the case if cloudiness remains low. Metaxas et al. (1991) investigated the variability of the sea surface temperature in the western, central and eastern Mediterranean during the last 120 years and showed that it reached a minimum in 1909 (19.18°C) and 1980 (19.3°C) with maxima 1940-1950 and 1965 (around 19.78°C). Since 1980 an steady increase was observed which did not reach the maximum of 1965 by the end of the study, 1990.

During the last few years in autumn the precipitation intensity seems to increase after hot and dry summers due to both thunderstorms and frontal activities. This resulted in disastrous floods in northern Italy, southern France, Spain and Algeria. Though the synoptic scale disturbances are an integral part of the general circulation they may have intensified by taking up, like local thunderstorms, additional latent heat from the Mediterranean Sea. One hypothesis for the generation of enhanced storm activities in autumn could be that the still warm sea leads to higher evaporation rates into the quicker cooling air which is labilized. The land surfaces are cooling not only because it is a later stage of the year but also because, with the first rains, the vegetation recovers and increases evaporation.

Worrying is that a combination of man's activities and an enhanced climate variability and trend may foster the process of desertification. This term does not imply a quick conversion from agricultural land into deserts but stands for the whole complex of possible land degradation. The increasing tendency of irrigation, overgrazing and land-surface conversion as well as the in parallel developing urbanization, tourism, and industrialization specifically in coastal areas will result in a continuation of the overexploitation of water resources. Wildfires will further denude large areas in which primarily low garrique-like vegetation will recover. This type of vegetation as well as dry agricultural crops generally will not have the cooling potential as forests and will reduce storage of water in the soil. Therefore summer temperatures may regionally increase and latent heat fluxes will only be strong for short times after rains by which most of the precipitated water returns to the atmosphere. During prolonged periods the flux of sensible heat will nearly equal the absorbed radiative energy and enhance dry convection over these areas. This reduces the probability of rain and the generation of clouds. The vegetation will suffer not only from extended droughts but also from enhanced temperatures and increased radiation which in addition reduces transpiration. For the soils the thread of salinisation and calcification increases. The regrowth of forests is retarded. Strong precipitation events, the frequency of which presently seem to increase, will not efficiently replenish natural water reservoirs because of the soil degradation. Additional artificial water reservoirs will be necessary to cope with the enhanced water requirements.

UNEP through its Oceans and Coastal Areas Programme Activity Centre (OCA/PAC) in 1987 launched the Mediterranean Action Plan and supported a number of studies in this region. The components of this Plan are: Pollution Monitoring and Research Programme (MED POL), Blue Plan, Priority Actions Programme, Specially Protected Areas and Regional Oil Combatting Centre. Already in early reports (Sestrini, Jeftic, Milliman 1989; Jeftic et al. 1992 and 1996) the impacts of climate change have been estimated based upon the Villach assessment of the role of carbon dioxide and of other greenhouse gases in climate variations and associated impacts (Bolin et al. 1986). *Inter alia* the relationship was emphasized between a reduction of rainfall respectively elongation of dry periods, the river flows, the water storage in reservoirs, wind erosion from the dry soils, sediment transports due to flash floods which reduce the efficiency of reservoirs, and salt accumulation in the top soils especially in coastal zones. A number of studies at different coastal sites (Rhodes, Kaštela Bay, Syrian coast, Malta, Cres/Lošinj archipelago) have been carried out which emphasize the sensibility of the Mediterranean system.

7 Conclusions

7.1 What Has Been Learned From Data Analysis?

Paleo-data show that the Mediterranean in its relatively short history as measured in the geological time scale went through major climate changes in the succession of which different types of vegetation covers developed. First the changes had geological reasons caused by plate movements. During the last million years the glaciations made their impacts. In the succession of these climate changes new forms of environment and life developed. Though climate provides the boundary condition for the development of the species, their actual distribution depends on many other factors as well. The changing vegetation cover and animal population is closely related to the changes of the hydrological regime.

In historical times the favourable climatic and environmental conditions which developed through the last climate optimum in the Mediterranean area stimulated mankind to extraordinary activities on the expense of the natural resources, which were rigorously exploited. This continued onto a time when climatic conditions became less favourable for a renewal of the natural resources at the pace of their consumption. In addition the pressure due to the expanding Sahara increased. This lead to devastation of large areas and an "exchange" of vegetation species due to the import of "useful" plants. Technical constructions (aqueducts) helped to maintain the living conditions for man and husbandry. The ecological-economical equilibrium began to become fragile. The natural land-surface processes were disturbed which had consequences for the regional climates. The decimation of the population due to epidemics from time to time restored equilibrium for some time periods.

Since about 350 years climate variability can be measured directly but for most locations comparable data sets are not much longer that 100 years. This period spans from the "Little Ice Age", which seems to have made not much impact on the Mediterranean area, to the present "global warming" era caused by the increase of the radiatively active gases in the atmosphere. In fact, in the average the climate was relatively balanced though large fluctuations occurred semi-periodically and interannually. The analysis of these data also shows locally large spatial gradients. An "antiphase" behaviour between the eastern and western Mediterranean and a periodicity in the climate records with frequencies around roughly 0.05, 0.1, 0.3 and 0.4 a^{-1} has been found. These frequencies vary slightly from investigator to investigator respectively from region to region. In the western Mediterranean the influence of the North Atlantic Oscillation seems to be strong. Sometimes weak correlations with the El Niño or Southern Oscillation are claimed. Generally it seems to be the case that the "antiphase" coupling between the western and the eastern Mediterranean is caused by external influences and is linked to the general circulation system. In coastal areas the influence of the land-sea circulation is superimposed to the variability caused by large scale influences and dominates it to the effect that its periodicities are not very marked here.

In the analysed data series a trend towards higher average temperatures is diagnosed. There are indications that this temperature increase mainly is due to increasing minimum temperatures. Insofar the Mediterranean seems to follow the general trend of global warming. Because of the large variability from year to year and an underlying quasi-periodicity of about 25 years it is difficult to say with final certainty that this is a trend which will continue in the future but the probability is high and supported by models. These models do also predict that the Mediterranean summers will become dryer. The data also show this tendency but in the case of precipitation the variability

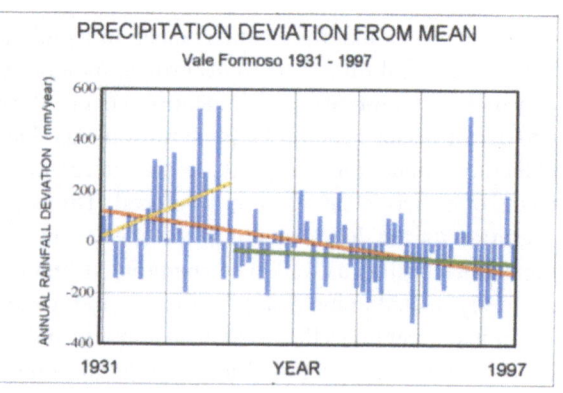

Fig. 16. Deviation of precipitation from the 1931 - 1997 mean with linear trends of different length after do Ó and Roxo (2001)

is even higher than for temperature. Sometimes such trends are deduced by linear trend analysis extended over hundred and more years and the trend is mainly determined by the higher precipitation in the early 20th century while presently such trends are not significant. Fig. 16 shows such an example taken from do Ó and Roxo (2001). The red line shows the impressive trend if the whole period from 1931 - 1997 is taken into account. Considering the large interannual variability the trend from 1952 - 1997 (green line) is very small and cannot be regarded as significant and if the years 1931 - 1951 are looked at the trend is very large and positive. Another example can be found in López-Bermúdez et al. (1999), where the 120 years negative rainfall trend of Murcia, Spain, is solely determined by high precipitation around 1890, while, if started in 1910, the trend would have a positive sign. These examples clearly indicate that linear regression is a dangerous tool to identify climate change, it depends very much on the interval chosen. The data for the last two years are not yet released for analysis and it is very likely that because of the high precipitation rates at least in some areas the trend of the annual precipitation of the last years may even be reversed. It needs an effort to merge these data sets and evaluate them under a superordinate point of view and with commensurable methods to obtain a consistent picture of climatological synergies in the Mediterranean area.

Climate change may manifest itself more in an increase of the amplitude and frequency of extreme events than in a smooth development of climate parameters. Research must concentrate on the causes of these extreme events and their impact on the annual totals. In this respect a distinction has to be made to what degree these extreme events are determined by large scale respectively regional processes. The flow of raw data to the analysts furthermore has to be accelerated to keep pace with the development of the global greenhouse effect.

7.2 To What Degree Are the Processes Understood Which Drive Climate Variability?

Through the atmosphere the Mediterranean basin is coupled to the North Atlantic Ocean and its oscillations (NAO), the Indian monsoon, the Sahara, the Hadley circulation and the polar front of the general circulation as schematically shown in Fig. 17. Substantial progress has been achieved by regional investigations about many details of the physical-biological system from groundwater respectively deep sea to the free troposphere. SVAT models have been constructed which simulate empirical results with great precision if initialized with correct data. Mesoscale models are capable to describe quite well regional circulation systems of both the atmosphere and the sea. The fact that notwithstanding these achievements present days climate models are not capable to provide reliable answers for Mediterranean regions is due to the lack of resolution and the accuracy of process descriptions in these large scale models to account for the complex topography of the Mediterranean basin, the internal variability of the Mediterranean Sea, the extraordinary manifold of land ecosystems, and the small scale processes like variability of the piezometric

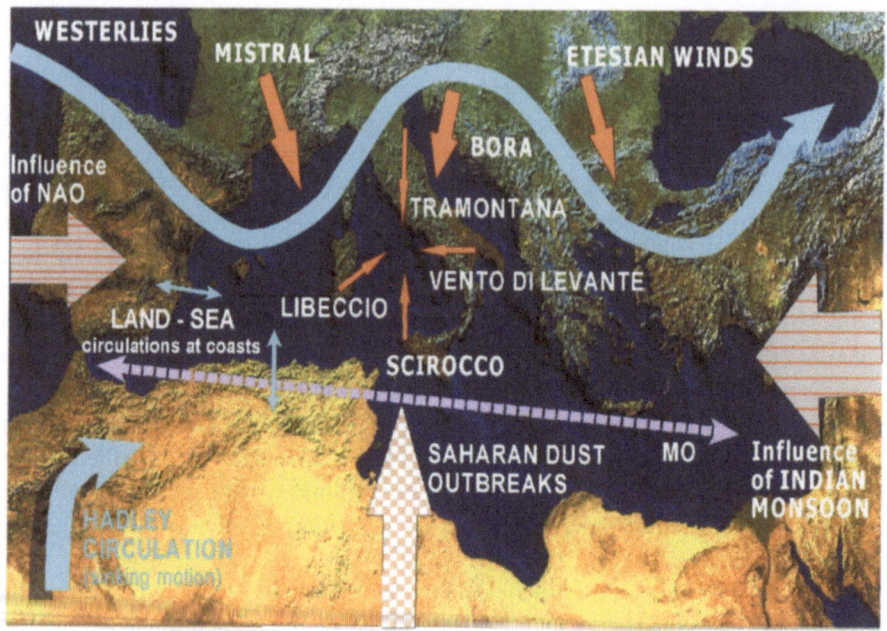

Fig. 17. The figure shows schematically some of the impacts on the Mediterranean basin due to large and regional scale wind systems and the Mediterranean Oscillation (MO). Regional wind systems as shown for the central region (red arrows) with Italian nomenclature exist under different names in other parts of the Mediterranean area as well.

level and its impact on evaporation, land-sea contrasts, and strong climatic as well as ecological gradients. For the formation of clouds and the related radiation budget the advection of aerosols from the Sahara plays an important role, especially in summer.

The combined action of all these components is not well represented in models and therefore the complete system is marginally understood. This in addition is a scaling problem, a matter of aggregation of empirical data into the grid widths of climate models in this complex landscape. To improve this situation it would be necessary to extend the so far in space and time very limited process studies to a number of representative areas and to extend the length of observation to full cycles. The validation of the output of models and of area covering measurements made from space furthermore requires supplemental information from "Anchor Stations" at which those more sophisticated measurements are performed which are needed for this task but are not included in the program of operational networks.

Regional water budgets are not very accurately known. Certain numbers are inherited from author to author without accounting for new evidence due to climate variability. This makes any extrapolation into the future ambiguous. The net water budget of the basin must be made a central question of future research.

7.3 Which Relationships Exist Between the Mediterranean Basin and the Earth System?

The outflow of heavy bottom water from the Mediterranean sea to the North Atlantic ocean is an important driver for the global conveyor belt and thus for North Atlantic SSTs and the NAO. The influx of Atlantic surface water not only compensates this outflow but in addition makes a large contribution to the Mediterranean water budget. During winter Mediterranean weather is strongly coupled with the large scale circulation system which is influenced by the state of the North Atlantic Ocean and the position and strength of the pressure systems which develop over the North Atlantic Ocean which in turn is coupled to the tropical system. The Mediterranean sea contributes to atmospheric humidity and labilization, but its influence decreases during winter. Extreme wet situations seem to be caused by an acceleration of the hydrological cycle which is a world wide phenomenon but no quantitative data are presently available for the Mediterranean. During summer the basin is topped by the sinking motion in the upper tropospheric high pressure belt and internal processes dominate underneath. Saharan dust outbreaks and regional emissions of industrial pollution, burning of vegetational debris, forest fires, and VOCs determine air quality, affect cloudiness, and net radiation.

Whether the Mediterranean basin is a sink or source for the global carbon

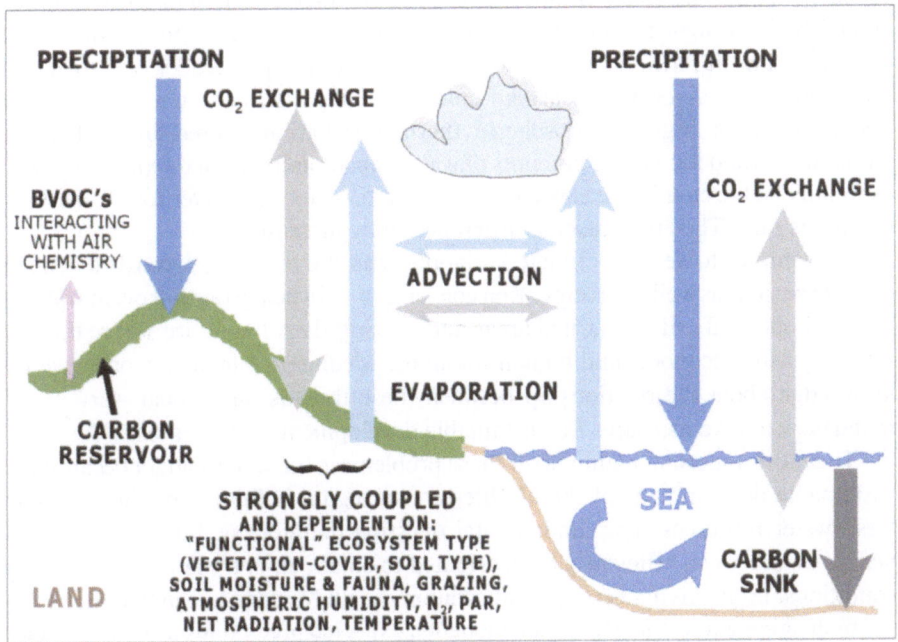

Fig. 18. Schematic presentation of the interconnections between of water vapour and carbon fluxes and of the coupling of sea and land by the exchange of these compounds through the atmosphere. The green line represents the vegetation.

dioxide pool cannot be concluded from a few local flux measurements. Careful integration over the ecosystems is necessary. Forest fires, forest or *macchia* regrowth, industrial emissions, and the carbon exchange through the sea surface and the fertilization of the sea with iron containing minerals have to be accounted for as well. The carbon and water cycles are intimately coupled to each other as shown in Fig. 18. A basin wide aggregation becomes urgent.

Whether the mostly man-made changes at the land surfaces with their impact on the regional climates reach out also to the larger scale via atmospheric exchange - are, so to say, "exported" to other regions - is difficult to decide. Some model studies seem to indicate it but the changes introduced into these model calculations have been very drastic and transient developments which have to include a manifold of interrelated parameters so far have not been completed.

7.4 What Are the Future Perspectives?

The answer to this question can only be given by climate models and therefor is related to the problem just referred to, that global climate models seem to be not very reliable in the case of the Mediterranean. Small shifts of atmospheric activity belts may cause large effects in the topographically highly structured landscape. At present it looks like as if the variability - the frequency and the amplitude of extreme events - is increasing, an observation which seems to be true for other regions of the world as well. If one effect of global change is the intensification of the Westerlies and of the hydrological cycle the hypothesis has much in favour that winters will become wetter. If further the Hadley circulation does not get weaker or retreat and the atmospheric aerosol content and cloudiness stay constant, summers may get hotter due to a higher atmospheric thermal radiation caused by higher air temperatures and increasing amounts of water vapour and carbon dioxide. This will especially be the case in areas where vegetation is scarce and water for evaporative cooling is rare. Thus the seasonal differences may increase.

Interannual to decadal climate variability and the tendency towards extended drought periods as well as precipitation concentrating in heavy rainfall events falling on dried-out hard soils or burned areas and causing flash floods are severe threads for ecosystems, economy and human life in the Mediterranean area. Consequently knowledge about the relationship between these threads and global warming is mandatory to take measures for sustainable development of the region.

The most important clamp of all these problems is the water budget because the sustainable development of the Mediterranean region is limited by the available fresh water resources. The fundamental question therefore is, how these limited water resources can efficiently be distributed to obtain an optimal result for living conditions, productivity, ecology and economy in all Mediterranean countries.

In the global map that shows the modelled expected precipitation changes under the increasing greenhouse effect in the recently published IPCC 2001 assessment report (Houghton et al., 2001, McCarthy, 2001) there is only one red spot: The Mediterranean. In the so-called scenario A2 - carbon dioxide increases by the year

2100 to 850 ppm and the global temperature increase is 3.8 K - models predict for 2100 a substantial decrease (-20%) of summer precipitation while the change will be only -1 to -2% in winter. The temperature increase will be a little bit higher than its global mean in winter and significantly higher in summer. It is astonishing that for a less strong rise of the carbon dioxide concentration (scenario B2, 600 ppm, +2.8 °C) the effect is much less than proportionally smaller. For the population that experiences present Mediterranean climate these results may sound somewhat curious because in regions where there is hardly any rain during the summer months it is very unlikely that there will be dramatically less in the future and in areas where during the last years extreme floods occurred in autumn or winter people ask themselves whether this is a first sign of climate change. Consequently there is need for interpretation and clarification. Firstly the scenarios of 2100 are compared with those of 1969-1990. Secondly the model output may need a different interpretation such as that the summer droughts may become longer. Or, thirdly, the Mediterranean climate gradients are so strong that a general statement for the whole basin is of not much relevance for its sub-regions. The conclusion would be that the Mediterranean system is not understood well enough to simulate regional climate development.

The problems of the water input into the region, its net water budget under the impact of changing global climate, the role of the vegetation which is changed by man and fires, and the role of the Mediterranean in the global carbon cycle must in the future be studied in an integrated way because these issues are so much interlaces that picking out one specific problem, as it has been done so much in the past, will not provide further deeper insight. One of the unknowns in this system still is the Mediterranean Sea the internal dynamics of which seems to undergo considerable changes. In other regions of the world where similar problems existed to understand the *interaction* of processes, scientists and governments found together and initiated a thorough study, a regional GEWEX experiment, to assess and integrate the available data, to improve the modelling of climate variability by combining regional with global climate models, and to study the less known processes and their interactions. The Mediterranean area because of its manifold of surface structures and its very critical geographical position with respect to climate change is predestinated to become in the future a focal point of such integrative research.

References

Aber JD (1998) Can we close the water/carbon/nitrogen budget for complex landscapes? In: Tenhunen JD and Kabat P (eds.): Integrating hydrology, ecosystem dynamics, and biogeochemistry in complex landscapes. Dahlem Workshop report. John Wiley & Sons, Chichester, pp 313-333

Abrate T (1999) Water availability and data availability - WMO policy and activity. International Workshop on Desertification Convention: Data and Information Requirements for

Interdisciplinary Research. Università di Sassari, Alghero, Italy, 9-10 October, 1999

Albaladejo J (1995) Soil rehabilitation and desertification control: Case study in Murcia. In: Fantechi R, Peter D, Balabanis P, Rubio JL: Desertification in a European context: Physical and socio-economic aspects, 213-226. European Commission, ISBN 92-827-4163-X, ECSC-EC-EAEC, Brussels, Luxembourg

Anderson OB, Knudsen P (2001) Long-Term Changes from ERS - Regional Changes in Sea level and Sea Surface Temperature. Earth Observation Quaterly. ESA-ESTEC, EOQ No. 69, Noordwijk, The Netherlands

Angelakis A, Kosmas C (1998) Water resources availability in relation to the threat for further degradation in the Mediterranean region: Need for quantitative and qualitative indicators. In: Enne G, D'Angelo M, Tanolla C: Indicators for assessing desertification in the Mediterranean. Proceedings of the International Seminar held in Porto Torres, Italy 18-20 September, 1998. Osservatorio Nazionale sulla Desertificazione, Nucleo Ricerca Desertificazione - University of Sassari (Italy)

Angelucci MG, Pinardi N, Castellari S (1998) Air-sea fluxes from operational analyses fields: Intercomparison between ECMWF and NCEP analyses ovber the Mediterranean area. Phys. Chem. Earth 23:569-574

Annual Bulletin on the Climate in WMO Region VI - Europe and Middle East - 1999, Deutscher Wetterdienst (ed.), Hamburg

Amanatidis GT, Paliatsos AG, Repapis CC, Bartzis JG (1993) Decreasing precipitation trend in the Marathon area, Greece. Intern. Journal of Climatology, 13:191-201

Artegiani A, Bregant D, Paschini E, Pinardi N, Raicich F, Russo A (1997) The Adriatic Sea General Circulation. Part I: Air-sea interactions and water mass structure. J. of Phys. Oceanography 27:1492-1514

Atalay I (2001) The ecological properties of the Turkish Mediterranean region. Report to RICAMARE

Baldocchi D, Valentini R, Running S, Oechel W, Dahlman R (1996) Strategies for measuring and modelling carbon dioxide and water vapour fluxes over terrestrial ecosystems. Glob. Change Biol. 2:159-168

Bethoux JP (1979) Budgets of the mediterranean sea: their dependence on the local climate and on the characteristics of the Atlantic waters. Ocean. Acta 2:157-163

Bethoux JP and Gentili B (1996) The Mediterranean Sea, coastal and deep-sea signatures of of climatic and environmental changes. J. Marine Systems, 7:383-394

Bindi M, Ferrini F, Miglietta F (1992) Climate change and shift in the cultivated area of olive trees. Journal of Agricultura Mediterranea, 22:41-44.

Bindi M, Fibbi L, Gozzini B, Orlandini S, Miglietta F (1996) Modelling the impact of future climate scenarios on yield and yield variability of grapevine. Climate Research 7:213-224

Bloch MR (1970) "Zur Entwicklung der vom Salz abhängigen Technologien. Auswirkungen von postglazialen Veränderungen der Ozeanküsten", Saeculum, XXI

Bolin B, Döös BR, Jäger J, Warrick RA (1986) The Greenhouse Effect, Climatic Change, and Ecosystems (SCOPE 29). John Wiley & Sons, Chichester, New York

Bolle H-J, André J-C, Arrue JL, Barth HK, Bessemoulin P, Brasa A, de Bruin HAR, Dugdale G, Engman ET, Evans DL, Fantechi R, Fiedler F, van de Griend A, Imeson AC, Jochum A, Kabat P, Kratzsch T, Lagouarde J-P, Langer I, Llamas R, Lopez-Baeza E, Melia Miralles J, Muniosguren LS, Nerry F, Noilhan J, Oliver HR, Roth R, Sanchez Diaz J, de Santa Olalla M, Shuttleworth WJ, Søgaard H, Stricker H, Thornes J, Vauclin M, Wickland D (1993) EFEDA: European Field Experiment in a Desertification-threatened Area. Ann. Geophysicae 11:173-189

Bolle H-J (1998) The EFEDA project. In: Mairota P, Thornes J B, Geeson N (eds.) Atlas of Mediterranean Environments in Europe. John Wiley & Sons, Chichester, 205 pp.

Bolle H-J (1999) Impact of climate variability on desertification processes. In: Balabanis P, Peter D, Ghazi A, Tsogas M (1999) Mediterranean Desertification Research results and policy implications. Proceedings of the International Conference 29 October to 1 November 1996, Crete, Greece, Volume 1, Plenary session - Keynote speakers, European Commission, Directorate-General Research, EUR 19303

Bouwman AF (1990) Global distribution of the major soils and land cover types. In: Bouwman AF, ed., Soils and the greenhouse effect. John Wiley & Sons Ltd., Chichester, UK

Bromley J (1995) EFEDA-2: Hydrology Group. Final Report Contract No. EV5V-CT93-0282.

Calvet J-C and 22 co-authors (1999) MUREX: A land-surface field experiment to study the annual cycle of the energy and water budgets. Ann. Geophysicae 17:838-854.

Calvet J-C (2000) personal communication

Castellari S, Pinardi N, Leaman K (1998) A model study of air-sea interactions in the Mediterranean Sea. J. Of Marine Systems 18:89-114

Chahine MT, Hoskins R, Fetzer E (1997) Observation of the recycling rate of moisture in the Atmosphere: 1988-1994. GEWEX-NEWS, WCRP, 7:1 and 3

Chbouki N, Stockton CW, Myers DE (1995) Spatio-temporal patterns of drought in Morocco. Intern. Journal of Climatology, 15:187-205

Chirici G, Maselli F, Lori C, Fibbi L, Bindi M (1999) Analysis of bio-geo-chemical fluxes in Mediterranean forest areas through the integration of remote sensing and GIS data by a modelling approach. RESYSMED Final Report.

Colacino M, Conte M (1993) Greenhouse effect and pressure patterns in the Mediterranean basin. Il Nuovo Cimento 16C:67-76

Colacino M, Diodato L, Malvestuto V (2000) "El Niño" and the Mediterranean climate. 3rd European Conference on Applied Climatology (ECAC2000), Pisa, Italy, October 16-20

Corte-Real J, Zhang X, Wang X (1995) Large-scale circulation regimes and surface climatic anomalies over the Mediterranean. Int. J.. of Climatology 15:1135-1150.

Conese C, Maselli F, Rodolfi A, Romanelli S, Bottai L (1999) In: Balabanis P, Peter D, Ghazi A, Tsogas M eds., Mediterranean Desertification. Research results and policy implications, Volume 2, Summary of project results. Proceedings of the International Conference 29 October to 1 November 1996, Crete, Greece. Directorate-General Research. EUR 19303, pp 549-558

Cubasch U, von Storch H, Waszkewitz J, Zorita E (1996) Estimates of climate change in Southern Europe derived from dynamical climate model output. Climate Research 7:129-149

D'Almeida G A (1987) On the variability of desert aerosol radiative characteristics. J. Geophys. Res. 92, D3:3017-3026

D'Almeida G A (1986) A model for Saharan dust transport. J. of Climate and Applied Meteorology 25:903-916

Da Silva AM, Young Ch, Levitus S (1994) Atlas of surface marine data 1994. NOAA Atlas NESDIS 7

Diaz JS, Hernandez RB, Ramirez AA, Cot CM, Colomer Marco JC (1991) Study of soil degradation, First Document. In: Bolle H-J, Streckenbach B (eds.) EFEDA First Annual Report, Contract EPOC-CT90-0030, Berlin

Diaz JS, Hernandez RB, Ramirez AA, Cot CM, Colomer Marco JC (1995) Metodologia para el estudio de la degradacion des suelo en el area Mediterranea (2ª aproximación). Documento preparado por el grupo UVAL.FF.DBV, EFEDA fase II.

Dickson R et al. (1996) Oceanography 38:241-295

do Ó A, Roxo MJ (2001) Driving forces of land use changes in Alentejo and its impact on souil and water. Paper presented at the workshop on "Land use changes & cover and water resources in the Mediterranean region", Medenine, Tunesia, 20-21 April

European Climate Support Network (1995) Climate of Europe. KNMI, de Bilt, NL

Faraco AM, Fernández F, Moreno JM (1993) Postfire vegetation dynamics of pine woodlands and shrublands in the Sierra de Gredos. In: Trabaud L, Prodron R (eds.) Fire in Mediterranean ecosystems. Ecosystems research report 5. Commission of the European Communities

Fehsenfeld F, Calvert J, Fall R, Goldan P, Guenther AB, Hewitt CN, Lamb B, Liu S, Trainer M, Westberg H, and Zimmermann P (1992) Emissions of volatile organic compounds from vegetation and the implications for atmospheric chemistry. Global Biogeochem. Cycles 6:389-430

Fiedler F (1994) private communication

Fiedler F, Adrian G, Baldauf M, Müller A (1996) Mesoscale and microscale circulations. Final Report Contract EV5V - CT93 - 0269

Flohn H (1993a) Klimaprobleme vor und nach der Rio-Konferenz (Juni 1992), preprint

Flohn H (1993b) Physical 3D-Climatology from Hann to the Satellite Era. Interactions Between Global Climate Subsystems, The Legacy of Hann. Geophys. Monograph 75, IUGG Volume 15

Font Tullot I (1988) Historia del clima de España - Cambios climáticos y sus causas. Instituto Nacional de Meteorología.

Fouquart Y, Bonnel B, Chaoui Roquai M, Santer R, Cerf A (1987a) Observation of Saharan rosols: Results of ECLATS field experiment. Part I: Optical thicknesses and aerosol size distributions. J. of Climate and Applied meteorology 26:28-37

Fouquart Y, Bonnel B, Brogniez G, Buriez J C, Smithg L, Mocrette J J (1987b) Observation of Saharan rosols: Results of ECLATS field experiment. Part II: Broadband characteristics of the aerosols and vertical radiative flux divergence. J. of Climate and Applied meteorology 26:38-52

Gabriel R, Schafer L, Gerlach C, Rausch T, Kesselmeier J (1999) Factors controlling the emissions of volatile organic acids from leaves of Quercus ilex L. (Holm oak) Atmos. Environment, Vol. 3 3, No. 9:1347-13 5 5.

Gaertner MA, Fernández C, Castro M (1993) A two-dimensional simulation of the Iberian summer thermal low. Monthly Weather Review 121:2740-2756

Georgelin M & 25 co-authors (2000) The second COMPARE exercise: A model intercomparison using the case of a typical mesoscale orographic flow, the PYREX IOP 3. Q.J.R.M.S. 126:991-1030

Gilman A and Thornes JB (1985) Land use and prehistory in south east Spain. Allen and Unwin, London

Giordani H, Noilhan J, Lacarrère P, Bessemoulin P, Mascart P (1996) Modelling the surface processes and the atmospheric boundary layer for semi-arid conditions. Agricultural and Forest Meteorology 80:263-287

Gleick P (ed.) (1993) Water in crisis - A guide to the world´s fresh water resources. Oxford University Press, Oxford, UK

Goossens C (1985) Principal component analysis of Mediterranean rainfall. J. of. Climatology 5:379-388

GRAPES (2000) European Commission Guidelines for the sustainable management of groundwater-fed catchments in Europe. Institite of Hydrology, Wallingford, UK.

Graßl H (1995) Klima und Mensch: Entsprechen sich Wissen und Handeln? Blick in die Wissenschaft. Forschungsmagazin der Universität Regensburg 4:22-30 and 69.

Grennon M, Batisse M, eds. (1988) Le Plan Bleu. Avenirs du Bassin Méditerranéen. Paris, France

Grove AT (1996) The historical context: Before 1850. In: Brandt CJ and Thornes JB (eds.) Mediterranean desertification and land use. John Wiley & Sons, Chichester, UK

Grove AT, Rackham O (1996) History of Mediterranean land use. In: Mairota P, Thornes JB, Geeson N, eds., Atlas of Mediterranean Environments in Europe. The Desertification Context. John Wiley & Sons, Chichester, pp. 76-78

Guzzi, R., 1981: Manuale di climatologia. Franco Muzzio & C., Padova.

GWP (2000) Towards Water Security: A Framework for Action (CD-ROM). Global Water Partnership. Stockholm, Sweden, and Wallingford, United Kingdom

Halpert M, Ropelewski C, Karl T, Angell J, Stowe L, Heim R, Miller A, and Rodenhuis D (1993) 1992 brings return to moderate global temperatures. EOS Trans. 74:433-437

Heintzenberg J (1980) Particle size distribution and optical properties of arctic haze. Tellus 32:251-260

Hibbard K (2000) EMDI update. Research Gaim 4:4-5

Houghton J T, Jenkins G J, Ephraums J J, eds. (1990) Climate Change: The IPCC scientific assessment. Cambridge University Press, Cambridge, UK

Houghton J T, Callander B A, Varney S K eds. (1992) Climate Change 1992: The supplementary report to the IPCC scientific assessment. Cambridge University Press, Cambridge, UK

Houghton JT, Ding Y, Griggs DJ, Noguer M, Van der Linden PJ, Xiaosu D (eds.) (2001) Climate Change 2001, The Scientific Basis. Cambridge University Press, Cambridge, UK

Hsü KJ (1987) The Mediterranean was a Desert: A Voyage of the Glomar Challenger. Princeton University Press, Princeton, NJ

Huber M (1993) Aktuelle Forschung zum Mittelmeerraum und ihre unterrichtliche Umsetzubng am Beispiel des Mittelmeeres. In: Struck (ed.) Aktuelle Strukturen und Entwicklungen im Mittelmeerraum. Passauer Kontaktstudium Erdkunde 3, Passau, pp. 33-48.

Ibáñez JJ, Benito G, García-Álvarez A, Saldaña A (1996) Mediterranean soils and landscapes. An overview. In: Rubiu JL, Calvo A, eds. Soil degradation and desertification in Mediterranean environments. Geoforma Ediciones, Logroño

Jeftic L et al. (1990) State of the marine environment in the Mediterranean region. UNEP Regional Seas Reports and Studies No. 132. Athens

Jeftic L, Milliman JD, Sestini G (1992) Climatic change and the Mediterranean. Edward Arnold, London

Jeftic L, Keckes S, Pernetta J (1996) climatic change and the Mediterranean . Vol. 2, Edward Arnold, London

Kesselmeier J, Staudt M. (1999) Biogenic volatile organic compounds (VOQ: an overview on emission, physiology and ecology. J. Atmos. Chemistry 33:23-88

Kesselmeier J, Bode K, Schafer L, Schebeske G, Wolf A, Brancaleoni E, Cecinato A, Ciccioli P, Frattoni M, Dutaur L, Fugit JL, Simon V, Torres L (1998) Simultaneous field medsurements of terpene and isoprene emissions from two don-,driant Mediterranean oak species in relation to a North American species. Atmos. Envirorim. 32 (11):1947-1953

Kesselmeier J, Schäfer L, Ciccioli P, Brancaleoni E, Cecinato A, Frattoni M, Foster P, Jacob V, Denis J, Fugit JL, Dutaur L, Torres L (1996) Emission of monoterpenes and isoprene from a Mediterranean oak species Quercus ilex L. measured within the BEMA (Biogenic Emissions in the Mediterranean Area) project. Atmos. Environment 30:1841-1850

Kesselmeier J, Bode K, Gerlach C, Jork E-M (1998) Exchange of atmospheric formic and acetic acid with trees and crop plants under controlled chamber and purified air conditions. Atmos. Environm.32 (10):1765-1775

Kesselmeier J, Bode K, Hofmann U, Müller H, Schafer L, Wolf A, Ciccioli P, Brancaleoni E, Cecinato A, Frattoni M, Foster P, Ferrari C, Jacob V, Fugit JL, Dutaur L, Simon V, Torres L (1997) Emission of short chained organic acids, aldehydes and monoterpenes from Quercus ilex L. and Pinus pinea L. in relation to physiological activities, carbon budget and emission algorithms. Atmospheric Environment, 3 1 (SI):119-134

Kömüşçü AÜ (1998) An analysis of the fluctuations in the long-term annual mean air temperature data of Turkey. Int. J. of Climatology 18:199-213

Korres G, Pinardi N, Lascaratos A (2000a) The ocean response to low frequency interannual atmospheric variabilityin the Mediterraean Sea. Part I: Sensitivity experiments and energy analysis. Journal of Climate 13 (4):705-731

Korres G, Pinardi N, Lascaratos A (2000b) The ocean response to low frequency interannual atmospheric variabilityin the Mediterraean Sea. Part II: Empirical orthogonal functions. Journal of Climate 13 (4):732-745

Koslowsky D (1999) In: Synthesis of change detection parameters into a land-surface change indicator for long term desertification studies in the Mediterranean area (RESYSMED). Final Report EC Project ENV4-CT97-0684.

Kossman M, Corsmeier U, de Wekker SFJ, Fiedler F, Vögtlin R, Kalthoff N, Güsten H, Neininger B (1999) Observations of handover processes between the Atmospheric Boundary Layer and the Free Troposphere over mountainous terrain. Contr. Atmos. Phys. :329-350

Kutiel H, Maheras P (1998) Variations in the temperature regime across the Mediterranean during the last century and their relationship with circulation indices. Theoretical and Applied Climatology 61:39-53

Lamb H H (1982) Climate, history and the modern world. University Press, Cambridge, UK

Lamb H H (1972/7) Climate: Present. past and future. Vol. I-II. Methuen. London

Latif M, Grötzner A. (2000) The equatorial Atlantic oscillation and its response to ENSO. Climate Dynamics Abstract Volume 16, pp 213-218

Legates DR and Willmott CJ (1990a) Mean seasonal and spatialvariability in global surface air temperature. Theor. Appl. Climatol. 41:11-21

Legates DR and Willmott CJ (1990b) Mean seasonal and spatial variability in gauge-corrected global precipitation. Int. J. Climatology 10:111-127

Lerdau M (1991) Plant function and biogenic terpene emission. In: Sharkey Th D, Holland EA, and Mooney HA, Trace gas emissions from plants. Academic Press, San Diego, pp. 121-134

López-Bermúdez F, Barberá GG, Alonso-Sarría F, Romero-Díaz MA (1999) In: Balabanis P, Peter D, Ghazi A, Tsogas M eds., Mediterranean Desertification. Research results and policy implications, Volume 1, Plenary session - Keynote speakers. Proceedings of the International Conference 29 October to 1 November 1996, Crete, Greece. Directorate-General Research. EUR 19303, pp 399-423

Lovett R F (1978) Quantitative measurement of airborne sea salt in the North Atlantic. Tellus 30:358-364

Luterbacher J, Schmutz J, Gyalistras D, Xoplaki E, Wanner H (1999) Reconstruction of monthly NAO and EU indices back to AD 1675

Macdonald A, Candela J, Bryden HL (1994) An estimate of the net heat transport throught he Strait of Gibraltar. In: Violette PE, ed., Coastal Estuarine Stud. 46, AGU, Washington D.C., USA

Maggiore A, Zavatarelli M, Angelucci MG, Pinardi N (1998) Surface heat and water fluxes in the Adriatic Sea: Seasonal and interannual variability. Phys. Chem. Earth 23:561-567

Maheras P (1988) Changes in precipitation conditions in the western Mediterranean over the last century. Journal of Climatology 8:179-189

Maheras P, Xoplaki E, and H Kutiel (1999) Wet and Dry Monthly Anomalies Across the Mediterranean Basin and their Relationship with Circulation, 1860-1990. Theoretical and Applied Climatology 64:189-199

Malberg H and Bökens G (1993) Changes in pressure-, geopotential-, and temperature fields between the subtropics and the subpolar region over the Atlantic in the period 1960-1990. Met. Zeitschrift, N.F. 2:131-137

Malberg H and Frattesi G (1995) Changes of the North Atlantic sea surface temperature related to the atmospheric circulation in the period 1973 tp 1992. Meteorol. Zeitschrift, N. F. 4:37-42

Manes F, Seufert G, and Vitale M (1997) The BEMA project: Ecophysiological studies of Mediterranean plant species at the Castelporziano estate. Atmospheric Environment 31, S1***

Manes F, Grignetti A, Tinelli A, Lenz R, Cicciolo P (1997) The BEMA project: General features of the Castelporziano test site. Atmospheric Environment 31, S1 ***

Maracchi G, Crisci A, Grifoni D, Gozzini B, Meneguzzo F, Zipoli G (2000) Some features of climate variation in central Italy: Need for a Mediterranean cooperation. 3rd European Conference on Applied Climatology (ECAC2000), Pisa, Italy, October 16-20

Margat J (1992) L'Eau dans le bassin Méditerranéen. Plan d'action pour la Méditerranée. Les Fascicules du Plan Bleu Vol. 6, Paris

Margat J, Vallée D (1999) Water Resources and Uses in the mediterranean Countries: Figures and facts. Blue Plan - facts on water for the Mediterranean (CD-ROM)

Marsh R (2000) Recent variability of the North Atlantic Thermohaline Circulation inferred from surface heat and freshwater fluxes. J. of Climate 13:3239-3260

McCarthy JJ, Canziani OF, Leary NA, Dokken DJ, White KS (2001) Climate Change 2001: Impacts, Adaptation, and Vulnerability. IPCC, Cambridge University Press, Cambridge, UK

McGlade J, van der Leeuv SE (1998) Environmental dynamics in the Vera basin. Summary. In: Van der Leeuw SE (ed.). Understanding the natural and anthropogenic causes of land degradation and desertification in the Mediterranean basin. The ARCHAEOMEDES Project. Synthesis Volume. Directorate General XII of the European Commission, Belgium, EUR 18181 EN

MED-HYCOS/IRD (1999) Mediterranean Hydrological Cycle system (CD-ROM Version 2.0). Institut de recherche pour le développement, Orstom, Montpellier, France

Meixner FX (1994) Surface exchange of odd nitrogene oxides. Nova Acta Leopoldina NF70:299-348

Metaxas, DA, Bartzokas A, Vitsas A (1991) Temperature fluctuations in the Mediterranean area during the last 120 years. Int. J. of Climatology 11:897-908.

Miglietta F, Peressotti A (1999) MEDEFLU: Summer drought reduces carbon fluxes in Mediterranean forest. Global Change News Letter 39:15-16

Moreno JM (1999) Forest fires: Trends and implications in desertification prone areas of southern Europe. In: Balabanis P, Peter D, Ghazi A, Tsogas M eds., Mediterranean Desertification. Research results and policy implications, Volume 1, Plenary session - Keynote speakers. Proceedings of the International Conference 29 October to 1 November 1996, Crete, Greece. Directorate-General Research. EUR 19303

Myers P, Haines K, Josey A (1998) On the importance of the choice of wind stress forcing to the modelling of the Med-sea circulation. Journal of geophysics Research 103:15729-15749

Nunes Correia F (1999) Water resources under the threat od desertification. In:Balabanis P, Peter D,

Ghazi A, Tsogas M eds., Mediterranean Desertification. Research results and policy implications, Volume 1, Plenary session - Keynote speakers. Proceedings of the International Conference 29 October to 1 November 1996, Crete, Greece. Directorate-General Research. EUR 19303

Otterman J, Manes A, Rubin S, Alpert P, O'C Starr D (1990) An increase of early rains in southern Israel following land-use change? Boundary-Layer Meteorology 53:333-351

Otterman J, Starr D O´C (1995): Alternative regimes of surface and climate conditions in sandy arid regions: Possible relevance to Mesopotamian drought 2200-1900 B.C.. Journal of Arid Environments 30

Özsoy E (2000) Sensitivity to Global Change in temperate Euro-Asian seas (the Mediterranean, Black Sea and Caspian Sea): A Review. In press.

Paccalet Y (1981) Mittelmeerflora. Belser A G, Stuttgart and Zürich

Palutikof JP, Trigo RM, Adcock ST (1999) Scenarios of future rainfall over the Mediterranean: Is the region drying? In: Balabanis P, Peter D, Ghazi A, Tsogas M eds., Mediterranean Desertification. Research results and policy implications, Volume 1, Plenary session - Keynote speakers. Proceedings of the International Conference 29 October to 1 November 1996, Crete, Greece. Directorate-General Research. EUR 19303

Papanastasis V P (2000) Land degradation caused by overgrazing and wildfires and management strategies to prevent and mitigate their effects. In: Enne G, Zanolla Ch, Peter D (eds.): Desertification in Europe: Mitigation strategies, land-use planning. Office for Official Publications of the European Communities, Luxembourg

Parry ML, Editor (2000) Assessment of potential effects and adaptations for climate change in Europe: The Europe ACACIA Project. Jackson Environment Institute, University of East Anglia, Norwich, UK

Piervitali E, Conte M, Colacino M (1999) Rainfall over the Central-Western Mediterranean basin in the period 1951-1995. Part II: Precipitation scenarios. Il Nuovo Cimento 22 C:649-662

Piervitali E, Colacino M, Conte M (1998) Rainfall over the Central-Western Mediterranean basin in the period 1951-1995. Part I: Precipitation trends. Il Nuovo Cimento 21 C:331-344

Pinardi N, Masetti E (2000) Variability of the large scale general circulation of the Mediterranean Sea from observations and modelling: A review. Paleageography, Palaeoclimatology, Palaeoecology 158:153-174.

Pongácz R, Bartholy J, Tar K (2000) El Niño forcing on macrocirculation and regional wind structure. 3rd European Conference on Applied Climatology (ECAC2000), Pisa, Italy, October 16-20.

Portrella A, Castro M (1996) Summer thermal low in the Iberian peninsula: A three-dimensional simulation. Quart. J. Roy. Met. Soc. Part A, 122:1-22.

Raich JW and Schlesinger WH (1992) The global carbon dioxide flux in soil respiration and its relationship to vegetation and climate. Tellus, 44B:81-99

Randall, D.A., R.D. Cess, J.P. Bölanchet, et al. (thirty co-authors) (1992) Intercomarison and interpretation of surface energy fluxes in atmospheric General Circulation Models. J. of Geophys. Res., 97, D4:3711-3724

Reynolds RW (1988) Areal-time global sea surface temperature analysis. J. Clim., 1:75-86

Reale O, Dirmeyer P (2000) Modeling the effects of vegetation on Mediterranean climate during the Roman classical period. Part I: History and model sensitivity. Global and Planetary Change (in press).

Reale O, Shukla J (2000) Modeling the effects of vegetation on Mediterranean climate during the Roman classical period. Part II: Model simulation. Global and Planetary Change (in press)

RESRAPS (1994) Remote Sensing and Radiometric Properties of the Surface: Assessment of Desertification from Space FINAL REPORT EFEDA Phase II, Group V Contract No.: EV5V-CT93-0284

Rodrigo FS, Esteban-Parra MJ, Pozo-Vázquez D, and Castro-Díez Y (2000) Rainfall variability in southern Spain on decadal to centennial time scales. International J. of Climatology 20:721-732

Roether W, Manca B, Klein B, Bregant D, Georgopopoulos B, Beizel V, Kovacevic V, and Lucetta A (1996) Recent changes in the eastern Mediterranean deep waters. Science, 271:2119-2122

Roldán Fernández A (1988) Notas para una climatología de Albacete, Cuenca, Toledo (respectively). Ministerio de Tranportes, Turismo y Comunicaciones, Instituto Nacional de Meteorologia,

Madrid, Publicacíon Serie K no. 28, 33, 58 and I.S.B.N. No. 84-505-7925-2, 84-7837-003-X, 84-7837-013-7 respectively.

Runnels CN (1995) Environmental degradation in ancient Greece. Scientific American, March:72-75

Running SW, Coughlan JC (1988) A general model of forest ecosystem processes for regional applications. I. Hydrologic balance, canopy gas exchange and primary production processes. Ecological Modelling 42:125-154

Running SW, Nemani RR, Peterson DL, Band LE Potts DF, Pierce LL, Spanner MA (1989) Mapping regional forest evapotranspiration and photosynthesis by coupling satellite data with ecosystem simulation. Ecology, 70:1090-1101.

Saaroni H, Ziv B (2000) Summer rain episodes in a Mediterranean climate, the case of Israel: climatological dynamical analysis. Int. Journal of Climatology 20:191-209

Sarachik ES and Alverson K (2000) Opportunities for CLIVASR/PAGES NAO studies. PAGES Newsletter 8, No. 1:14-16

Sarnthein, M., 1978: Sand deserts during glacial maximum and climatic optimum. Nature, 272, No. 5648:43-46

Schönwiese C-D, Rapp J, Fuchs T, Denhard M 1994: Observed climate trends in Europe 1891-1990. Meteorol. Zeitschrift, N.F. 3:22-28.

Schultze-Westrum T (1994) Ecological impact of goat pasturage in semi- and low-productivity regions especially in Greece. Intern. Symposium on the optimal exploitation of marginal Mediterranean areas by extensive ruminant production systems. Thessaloniki, Greece

Schultze-Westrum T (1999) private communication

Sestini G, Jeftic L, Milliman JD (1989) Implications of the expected climate changes in the Mediterranean region. UNEP, MAP Technical Reports Series No. 27, Athens

Seufert and 17 co-authors (1997) An overview of the Castelporziano experiments. Atmospheric Environment 31:5-17

Singh HB and Zimmermann PB (1992) Atmospheric distribution and sourcesof nonmerthane hydrocarbons. In: Nriagu JO (ed.) Gaseous pollutants: Characterization and cycling. John Wlley

Smith, E.A., 1986a: The structure of the Arabian heat low. Part I: Surface energy budget. Mon. Wea. Rev. 114:1067-1083

Staudt, M., Wolf, A. and Kesselmeier, J. (2000) Influence of Environmental Factors on the Emissions of gaseous formic and acetic acids from orange (Citrus sinensis L.) foliage. Biogeochemistry 48 (2):199-216

Strahler AN (1975) Physical Geography. John Wiley & Sons, USA

Trenberth KE (1997) Atmospheric moisture residence times and cycling: Implications for how precipitation may change an climate changes. GEWEX NEWS, WCRP, 7:1 and 4.

Trigo RM and da Camera CC (2000) Circulation weather types and their impact on the precipitation regime in Portugal. Internationale Journal of Climatology, in press.

Türkeş M, Sümer UM, Kiliç G (1995) Variations and trends in annual mean air temperatures in Turkey with respect to climate variability. Int. J. of Climatology 15:557-569

Tzedakis PC (1993) Long term tree population in northwestern Greece through multiple climatic cycles. Nature 364: 437-440

Ünal Y, Karaca M, Dalfes N (2000) Regional climate change due to the southeastern Anatolian irrigation project in Turkey. 3rd European Conference on Applied Climatology (ECAC2000), Pisa, Italy, October 16-20

Valentini R, Baldocchi DD, Running S (1997) The IGBP-BAHC global flux network initiative (FLUXNET): Current status and perspectives. Global Change newsletter 28:14-16

Valentini R, Baldocchi DD, Tenhunen JD (1998) Ecological controls on land-surface atmospheric interactions. In: Tenhunen JD and Kabat P (eds.): Integrating hydrology, ecosystem dynamucs, and biogeochemistry in complex landscapes. Dahlem Workshop Report. John Wiley & Sons, Chichester

Valentini R, Baldocchi D, Olson R (1999) FLUXNET: A challenge that is becoming a reality. IGBP Global Change NewsLetter 37:15-17

Valentini R, G. Matteucci, A.J. Dolman, E.-D. Schulze, C. Rebmann, E.J., Moors, A. Granier, P. Gross, N.O. Jensen, K. Pilegaard, A. Lindroth, A., Grelle, Ch. Bernhofer, T. Grünwald, M. Aubinet, R. Ceulemans, A.S. Kowalski, T. Vesala, Ü. Rannik, P. Berbigier, D. Loustau, J.

Gudmundsson, H., Thorgeirsson, A. Ibrom, K. Morgenstern, R. Clement, J. Moncrieff, L., Montagnani, S. Minerbi, P.G. Jarvis (2000) Respiration as the main determinant of carbon balance in European forests. Nature 404:861-865

Van der Leeuw SE ed. (1998) Understanding the natural and anthropogenic causes of land degradation and desertification in the Mediterranean basin. The ARCHAEOMEDES Project. Synthesis Volume. Directorate General XII of the European Commission, Belgium, EUR 18181 EN

Vélez R (1990) Los incendios forestales en España. Ecología 1:213-222

Verheye WH (1991) The role and impact of biophysical determinants on present and future land use patterns in Europe. In: Brouwer FB, Thomas AJ, and Chadwick MJ (eds.) Land use changes in Europe. Processes of change, environmental transformations and future patterns. Kluwer Academic Publishers, Dordrecht, The Netherlands

Von Storch, H., E. Zorita, and U. Cubasch, 1993: Downscaling of Global Change estimates to regional scales: An application to Iberian rainfall in wintertime. J. of Climate, 6:1161-1171

Wagner H-G (2001) Mitelmeerraum. Wissenschaftliche Buchgesellschaft, Darmstadt

Wanner H, Brönnimann S, Casty C, Gyalistras D, Luterbacher J, Schmutz C, Stephenson DB, Xoplaki E (2001) North Atlantic Oscillation - concepts and studies. Surveys in Geophysics 22:321-382.

Wallace JM (2000) North Atlantic Oscillation/annual mode: Two paradigms - one phenomenon. Q.J.R.M.S. 126:791-806

Wigley, T.M.L., 1992: Future climate of the Mediterranean Basin with particular emphasis on changes in precipitation. In: Jeftic, L., J.D. Milliman and G. Sestini, eds.: Climatic change and the Mediterranean. Edward Arnold, London

WMO (1999) MED-HYCOS: A water observing and information system

Wu P, Haines K, Pinardi N (2000) Toward an Understanding of Deep-Water Renewal in the Eastern Mediterranean. Journal of Physical Ovceanography 30:443-458

Wüst G (1961) On the vertical circulation of the Mediterranean Sea. J. Geophys. Res., 66:3261-3271

Yassoglou NJ (2000) History and development of desertification in the Mediterranean and its contemporary reality. In: Enne G, Zanolla Ch, Peter D (eds.): Desertification in Europe: Mitigation strategies, land-use planning. Office for Official Publications of the European Communities, Luxembourg

Yassoglou NJ (1998) History of desertification in the European Mediterranean. In: Enne G, D'Angelo M, Zanolla C (eds.): Indicators for assessing desertification in the Mediterranean. Proceedings of the International Seminar held in Porto Torres, Italy 18-20 September, 1998. Osservatorio

Appendix 1
The ECHIVAL Field Experiment in Desertification Threatened Areas (EFEDA)

EFEDA, the "ECHIVAL Field Experiment in Desertification-threatened Areas" was initiated by the former CEC in response to the international activity on land-surface processes started by the International Satellite Land-Surface Climatology Project (ISLSCP) and the Hydrologic Atmospheric Pilot Experiment (HAPEX) initiative of the World Climate Research Programme. The basis for the EFEDA Experiment plan was the *European International Project on Climatic and Hydrological Interactions between Vegetation, Atmosphere, and Land-surfaces* (ECHIVAL) developed by Andre, Bolle and Shuttleworth (1989). A related field experiment was carried out in Crete (Messara Valley Project). The experiments were organised according to the strategy developed by ISLSCP (Becker, Bolle, and Rowntree, 1988) and subsequently adopted by IGBP-BAHC (1993).

EFEDA was designed as a multidisciplinary activity to improve our understanding of climate related land-surface processes in semi-arid areas which suffer from desertification. Its primary goal was the investigation of the transfer of water between soils, vegetation, the atmospheric boundary layer and the free troposphere as well as the assessment of the importance of this chain of processes for land-degradation and desertification. The experiment the first time combined the forces of European geologists, soil scientists, hydrologists, geographers, botanist, ecologists, atmospheric boundary layer researchers, aeronomists, modellers, remote sensing specialists, and socio-economists including a NASA and a German instrumented research aircraft and German motorgliders. The first time in the Mediterranean region the full range of water transfer processes from sub-surface transports up into the free troposphere have been studied within one experiment. In addition socio-economic aspects such as the overexploitation of water resources due to water abstraction for other regions and irrigation had to be taken into account, because the water cycle in these regions cannot be investigated without considering the impact of man´s actions. To be able to compare the results with the output of wide meshed climate models and to contribute to an improvement of these models the experiment had to span over an area compatible with the grid width of climate models. The first short term pilot field experiment was carried out in June 1991 in the plains and adjacent smooth hills of Castilla-La Mancha, Spain, that offered an existing infrastructure and little complication by topographic effects (Bolle et al., 1993). The area was revisited by part of the experimental team in summer 1994. Part of the experiment was a remote sensing activity (a) to assist in the aggregation over the whole experimental area of parameters needed to model land-surface processes and energy fluxes which at the ground were measured only at individual points, and (b) to explore the capability of presently available remote sensing data to quantify changes at the land surfaces, specifically land degradation and desertification. These remote sensing activities found a continuation in the EC Projects RESRAPS (1994), RESMEDES (1998), and RESYSMED (1999).

The overall goal of the experiment was to determine the interactions between soil, vegetation, and atmosphere at a scale of approximately 10^4 km^2. Within this area at three representative experimental sites - one dry farming, one with mixed dry farming and irrigated fields, and one with mainly matorral vegetation - fixed instrumented stations were installed. In each of these areas was a nucleus of heavily instrumented masts surrounded by some additional stations up to a few kilometre distant to investigate spatial inhomogeneities. With portable equipment and remote sensing measurements from aircraft and satellites the areas in between these densely instrumented sites were bridged over and a synoptic view of the whole area was generated. Due to a number of instrument intercomparisons that have been made during EFEDA, the error range in the determination of energy budget components (for which no absolute standard exists) could be quantified.

The detailed experimental data of the surface energy budget and water transfer within the soils and through the plants into the atmosphere lead to an improvement and adaption for the semi-arid conditions of models that can now be used with much more confidence. For this purpose also an extensive intercomparison of soil-vegetation-atmosphere transfer (SVAT) schemes was conducted. The structure of the atmospheric boundary layer that is responsible for the escape into the free troposphere of the water vapour evaporated at the surface was studied with the dense radiosonde network that could be established in the area, sodar, and flux measurements from aircraft that provided detailed information on fluxes and water vapour as well as aerosol distributions. The availability of these aeronomical data allowed to initiate and to test mesoscale models which simulated the atmospheric circulation system in the area and the effect of the land-sea circulation that develops at the Mediterranean coast (Bolle, 1998). The evaluation of satellite data gained immensely from the possibility to validate the inference of primary quantities such as reflectance, temperature and vegetation indices by intercomparisons with ground measurements. In addition, more complex evaluation schemes such as needed to derive surface energy fluxes from remote sensing data could be tested and applied for the whole area. Based upon these experiences completely new satellite data processing and correction schemes have been developed during EFEDA.

To embed the hydro-meteorological process studies into the general ecological situation of the area, intensive studies of the hydro-geology of the area, of the structure and the degree of degradation of the soils, of vegetation parameters including the CO_2-exchange, and of the land use as well as its recent changes have been carried out.

The EFEDA project lead to

1 a general improvement of experimental and model techniques to study complex land-surface processes,
2 a thorough understanding of the importance of and interactions between the land-surface processes during a drying period and thus for aridification which is one precurser of desertification, and
3 the conclusion that the sustainable development of the investigated area is in a very critical phase, if
   a)   the water resources are not recharged (e.g. due to climatic change

respectively variability) to a degree that they are able to support intensive irrigation,

b) the available water cannot be spent more efficiently (e.g. by application of modern irrigation techniques and/or change to different species), and#

c) the cultivation of species with roots reaching moist soil layer in some depth does not receive continued economic support.

The intensive measuring phase of EFEDA was a pilot experiment during which only processes could be investigated with short characteristic time scales and under dry conditions. But the occurrence of such dry conditions depends on long term climate variability respectively climate change and the impact of dry spells in alternation with wet events on the condition of the land-surfaces including its vegetation *inter alia* is a function of the time constants of sub-surface hydrological processes. These are much longer than could readily be studied within EFEDA. To receive full return from the achieved insights into the investigated land-surface processes it therefore is necessary to lead the developed methodology into an operational mode that relies upon very few but well equipped ground stations (so called "anchor stations") and to extend especially the hydrological studies over an number of years and into hilly terrain.

One important task during the EFEDA pilot experiment was the re-calibration of AVHRR satellite data and the validation of information derived from these data (Koslowsky, 1997). The continuation of the evaluation of satellite data and some scattered ground truth operations throughout the Mediterranean area (Castilla-La Mancha, Tuscany, Messara Valley on Crete, and Israel) during the years 1994 - 1999 nevertheless allow to embed the measurements made 1991 and 1994 in Spain into a larger space and time scale. Ten years of satellite data time series make it possible to analyse the interannual variability of the Castilla-La Mancha area in the context of changes all over the Iberian Peninsula. Moreover for few years carefully calibrated AVHRR data are now available for the whole Mediterranean area which enables first studies of synergisms throughout the basin.

The draw back of such regional to large scale EFEDA-type experiments so far was that for each single parameter which enters the process studies the measuring equipment had to be implemented or the relevant data had to be assessed by labourious efforts from various offices. To embed such regional short-term investigations into the larger scales of time and space variability a long term data set of at least a reduced spatial coverage must be available. Fortunately in the case of the Castilla-La Mancha experiment a long term hydrological data set was available at the University Complutense in Madrid. Also the meteorological support was excellent, weather maps and general meteorological data were made available for the time of the experiment. But for comparison with the larger scale satellite data so far no surface data backup exist. For the correction of the satellite data a study was undertaken within the relevant EC projects to determine the water vapour field over the Iberian peninsula based upon a ECMWF re-analysis. Such data must in the future be available in nearly real time to correct the satellite data.

EFEDA-type experiments gained a lot from already existing data and served as a vehicle to bring these data together, relate them to each other and to aggregate them into a new level of understanding and monitoring of processes. This aggregation often can be accomplished by models. EFEDA-type experiments per se, mainly because of their short duration, do not solve environmental problems, sometimes they indicate where problems may arise, but their main benefit is that they develop and prove new approaches and methodologies how to *attribute* changes to the "Global Change" problem. The problem is not to provide experimenters with additional data but to gather and analyse those data which have been identified by EFEDA-type experiments in a quasi-operational manner for time scales which are recognized as essential to isolate trends from interannual variability.

In an continuation of the efforts started with EFEDA and especially if it is desired to transform this approach into a large scale GEWEX-type study it would be necessary to gather in addition to meteorological and hydrological routine measurements at a certain number of "anchor stations" placed into representative areas around the Mediterranean and to merge the data of the operational system with this scientific network. The emphasis of the "anchor stations" must lie on those quantities which are not measured by the operational network. In the past there have been often a misunderstanding in this respect as standard climatological equipment was offered as "anchor station". But it is now feasible to infer from satellite measurements in addition to the "primary" quantities like surface reflectance, surface temperature, and various vegetation indices also information such as on vegetation quantities, fluxes between surface and atmosphere, and soil top moisture. These "secondary level" data, which are in the first instance Land-surface Change Indicators need to be "clamped" to reality by continuous observations at anchor stations - and from time to time field experiments to remove uncertainties - to enfold the whole observing potential of measurements made from space.

Considering the use of satellites it is mandatory that
- atmospheric data (three dimensional temperature, pressure, water vapor and aerosol fields) become available in quasi real time which allow in combination with topographic models to correct the measurements made in space for atmospheric effects, and
- that anchor stations be established around the Mediterranean Sea to continuously control at representative points the validity of the information deduced from the satellite measurements.

To embed short term EFEDA-type experiments into long term developments it is essential to obtain in addition to operationally available data also those data which are not operationally archived and are needed especially to expand local measurements into area averages. It is not as important if a few hectare of unmanaged forest burns down in an area where the environmental conditions allow a fast regrowth as it is if a large agricultural area slowly gets into an irreversible permanent water shortage situation because of extended droughts during the summers and not enough winter rain to compensate for the water evaporating during summer and extracted for households, irrigation, industry and tourism.

References

Andre JC, Bolle H-J, and Shuttleworth WJ (1989) Echival: Proposal for an European International Project on Climatic and Hydrological Interactions Between Vegetation, Atmosphere and Land-surfaces.

Becker F, Bolle H-J, Rowntree PR (1988) The International Satellite Land-surface Climatology Project. ISLSCP Secretariat, FU Berlin, 100 pp.

Bolle H-J, ed. (1993) Biospheric Aspects of the Hydrological Cycle (BAHC): The operational plan, IGBP Report No 27, Stockholm

Bolle H-J, André J-C, Arrue JL, Barth HK, Bessemoulin P, Brasa A, de Bruin HAR, Dugdale G, Engman ET, Evans DL, Fantechi R, Fiedler F, van de Griend A, Imeson AC, Jochum A, Kabat P, Kratzsch T, Lagouarde JP, Langer I, Llamas R, Lopez-Baeza E, Melia Miralles J, Muniosguren LS, Nerry F, Noilhan J, Oliver HR, Roth R, Sanchez Diaz J, de Santa Olalla M, Shuttleworth WJ, Søgaard H, Stricker H, Thornes J, Vauclin M, Wickland D (1993) EFEDA: European Field Experiment in a Desertification-threatened Area. Ann. Geophysicae 11:173-189

Bolle H-J (1998) The EFEDA Project. In: Mairota P, Thornes JB and Geeson N (eds.): Atlas of the Mediterranean Environments in Europe. John Wiley & Sons, Chichester, ISBN 0-471-96092-6, pp 12-15.

Koslowsky D (1997a) Signal degradation of the AVHRR shortwave channels of NOAA 11 and NOAA 14 by daily monitoring of desert targets. Adv. Space Res. 19(1):1355-1358.

RESRAPS (1994) Remote Sensing and Radiometric Properties of the Surface: Assessment of Desertification from Space FINAL REPORT EFEDA Phase II, Group V Contract No.: EV5V-CT93-0284

RESMEDES (1998) Final Report of EC Contract Contract No.: ENV4-CT95-0094, European Commission, Directorate-General, Science, Research and Development

RESYSMED (1999) Synthesis of Change Detection Parameters into a Land-surface Change Indicator for Long Term Desertification Studies in the Mediterranean Area. Final Report Contract No.: ENV4-CT97-0683, Firenze and München

Chapter 3
Challenges for Climate Data Analysis

Indicators and Information Requirements for Combatting Desertification

G. Enne, C. Zucca, C. Zanolla

1 Introduction

The development of indicators and the availability and circulation of information are two correlated and relevant issues in the fight against desertification as highlighted by the United Nations Convention to Combat Desertification (UN-CCD). Article 16 of the UN-CCD notifies that the ratifying countries have accepted integrating and co-ordinating the collection, analysis and exchange of relevant short and long term data and information to ensure systematic observation of land degradation in the affected areas and to better understand and assess the processes and effects of desertification and drought.

Desertification integrates a complex system of dimensions, parameters and variables involving the environment and society. In order to facilitate global understanding of the status of desertification and be able to assess what is happening to soil, vegetation, water resources and the quality of life of populations in the affected areas, indicators must be developed to represent precisely and unambiguously quantifiable qualities or properties, symptoms or parameters of a phenomenon relating to a feature of the environment or of society, a change in which over time or space is related to this process.

Affected countries are invited to develop Desertification Monitoring Systems (DMS). DMS should be able to generate diagnoses integrated in space and time with respect to the status of natural resources and populations in the regions affected. They must likewise act as a support to the decision making process by supplying information on the biophysical and socio-economic environmental problems. DMS should aid actors at different levels to assess the progress achieved, examine their priorities and improve their situation as against the degradation process. However the identification of indicators would be a useless effort unless they are supported by the necessary data. It is therefore essential:

- to identify indicators from a compromise between knowledge of processes (objective need for information) and real and cost effective data availability;
- to make data supply and demand to meet so that the agencies involved in monitoring and relevant data production are stimulated to concentrate all efforts on the topical issues and to adopt protocols that better meet the needs of the DSM;

- to set up mechanisms and terms of reference to favour an increasingly better circulation and availability of data.

In this context the Nucleo di Ricerca sulla Desertificazione (NRD) of the University of Sassari has organised several initiatives that, although mainly addressed to the Annex IV Countries, have had wider impact particularly from a methodological standpoint:

▶ International Seminar on *"Indicators for Assessing Desertification in the Mediterranean"*, held in Porto Torres, Italy, from 18th to 20th September 1998[1].

▶ International Workshop on *"Desertification Convention: Data and Information Requirements for Interdisciplinary Research"*, held in Alghero, Italy, 9-11 October 1999[2]

▶ Elaboration of a study on *"Desertification indicators for the European Mediterranean Region. State of the Art and Possible Methodological Approaches"*[3]

The major results of these three activities are summarized in the following sections.

2 International Seminar on Indicators for Assessing Desertification in the Mediterranean

2.1 Objectives and Assumptions

This Seminar was the first initiative promoted by the constituting Italian National Observatory on Desertification, among whose main objectives is the implementation of UN-CCD in Italy and in the Mediterranean basin.

The main aim of this seminar was to allow a general overview on the indicators of desertification and their applicability to the Mediterranean basin, with particular reference to the UN-CCD Annex IV Countries.

This event constituted an important opportunity for researchers from European Mediterranean Countries to examine jointly the outputs of years of experimental activities in the field of land degradation and desertification, and a rare chance of discussion among scientists and decision makers involved in land management. It

---

[1]   Organised in collaboration with the Italian National Observatory on Desertification - Italian Ministry of the Environment, the Town of Porto Torres and the Universities of Sassari and Cagliari.

[2]   Enabled by DG Research and DG VIII of the European Commission (EC) and organised in collaboration with the Italian National Observatory on Desertification - Italian Ministry of the Environment.

[3]   Supported by ANPA and the Italian National Observatory on Desertification - Italian Ministry of the Environment.

contributed to highlight the main issues on which much work must still be carried out.

Discussion was based on the following general assumptions:
(i) Definitions and terms as defined by the UN-CCD must be the only term of reference.
(ii) Indicators are the objective answer to the necessity for a direct knowledge on desertification, its state and its evolution. This means that an indicator must be able to describe key issues on desertification in a specific area or in a particular country.

2.2 Results and Conclusions

The general discussion at the end of the event allowed to better define:
- The importance of the identification and utilisation of indicators;
- The limits of the current knowledge on and approaches to the research on indicators and the advantages that would derive by adopting the criteria suggested by the UN-CCD.
- The necessity to identify useful indicators given by the widespread need to simplify in a limited number of variables a complex phenomenon like desertification, particularly when actions must be taken to halt and/or mitigate its effects.

One of the main problems encountered by policy-makers, end users and all institutions involved in land planning is connected to the application of scientific theories to real situations. It is very difficult for them to translate the scientific abstractions into actions, and the use of indicators would really constitute an important step towards the implementation of actions to combat desertification. Indicators would in fact make it possible:
▸ to provide information to policy makers and the wider public on the current state and changes in the conditions of the environment;
▸ to allow policy makers to better understand the linkages between causes and effects of the impact of environmental policies; and
▸ to contribute to monitor and evaluate the effectiveness of policies in promoting sustainable development.

From a scientific perspective, indicators would:
▸ enable the assessment of the present status of the environment and the prediction of future trends;
▸ ease the identification of the main causes of desertification and therefore suggest proper management of natural resources; and
▸ enable an identification of detrimental changes and therefore the adoption of the necessary measures to prevent irreversible damage.

In the light of all the examples, case studies and methodological contributions presented, it appeared clear that there is a urgent need to carry out specific studies on the identification of indicators for the Mediterranean area.

Most research on the subject of desertification was undertaken before that the recent guidelines on research and application of useful indicators were issued by the CST-UN-CCD, the Committee on Science and Technology of the United Nations Convention to Combat Desertification. The concept of Desertification itself in the context of our regions is still quite new for most of the scientific community. Consequently, it is not surprising that international bibliography on the subject often seems to lack uniformity and is frequently conditioned by the specificities of individual disciplines, generally ill-suited to the extrapolation of elements that could be applied to more general and complex situations.

The task of developing these indicators is a multidisciplinary one, therefore involving natural, social and economic sciences. Moreover, since among the causes of Desertification the Convention mentions political and cultural factors, the contribution of political science and anthropology is absolutely necessary to complete the frame of knowledge.

Hitherto scientific work on the subject has not only been deficient in terms of interdisciplinarity, often restricted to the natural sciences, but even within the disciplines themselves where comparisons have not been carried out to overcome the major differences in approach prevailing to this day.

Finally, the task of identifying a limited set of essential and efficient indicators does not only require the contribution of all the disciplines involved, but also calls for an awareness of the problem and a thorough grasp of objectives within the individual groups of disciplines. With regard to this aspect the international scientific community has understood the impossibility to determine a set of universal indicators and, on the other hand, the importance to devise a method for the identification of context driven indicators able to effectively represent the different aspects of Desertification. This requires a higher level of perception of the desertification processes and of the real priorities, which can only be attained by an adequate social participation.

The results of the Porto Torres Seminar are proof of the existence of these problems. Consequently, there still remains much work to be done with particular reference to harmonisation and comparison, which hopefully could be done by interdisciplinary groups specially set up for the purpose and through international concerted actions.

In the light of the above and of the experiences matured in this field, the following priorities were identified at the Porto Torres Seminar:
- Need for methodologies that, starting from the experiences already matured in national and international research projects, integrate the evaluation of the physical, biological and socio-economic processes contributing to the land degradation and desertification so as to identify indicators useful in land planning and information dissemination at all levels.
- Need for supporting the testing of indicators already available.
- Need for setting up a network to monitor desertification both at national and at Mediterranean basin scale and to improve availability and accessibility of data sets.

– Need for a greater participation of local populations by means of activities aiming at stimulating local administrations and NGOs to contribute to identify problem-driven indicators.

3 International Workshop on Desertification Convention: Data and Information Requirements for Interdisciplinary Research

3.1 Objectives

The main objectives of the international workshop held in Alghero were
• to assess data and information systems with relevance to UN-CCD, and
• to agree on requirements for data and meta-data needed to monitor causes and effects of man induced desertification.

These objectives were based on the assumption that the identification of the available data for the Convention can help to propose suitable instruments methodologies and information systems for planning needs in the Mediterranean countries, and to evaluate the costs for the implementation and running-up of easy information systems.

Introductory statements set the scene relating to research approach and methodology, and the requirements of scientific support for implementing the UN-CCD. Data problems related to soil degradation, water availability, economy, policy, and population were then presented, as well as interdisciplinary approaches for regional focal areas. Data availability in international institutions was also outlined, including FAO, the European Soil Bureau, WMO, Joint Research Centres of the EU and NGOs.

Working groups specially set up on the themes climate/remote sensing (WG1 co-ordinated by Prof. Bolle), soil/water/vegetation (WG2 co-ordinated by Prof. Imeson), and socio-economy/policy (WG3 co-ordinated by Profs. Homewood and Steen) looked into needs, existence and availability of data and information, as well as into the requirements for better exchange of information and data. The open discussions were strongly finalised to the achievement of concrete solutions. Their specific objectives were:
• to identify the core data sets and information (soil, water and vegetation) available at different scales (local, national and continental) in Mediterranean Europe and Africa;
• to document the experience of interdisciplinary desertification projects and data providers with respect to the current situation and future possibilities;
• to identify the data sets and information needed to meet future requirements;
• to consider how data and information should be organised and made available;

- to propose the way forwards and formulate appropriate actions.

In particular, WG1 stressed that the analysis of data availability and requirements for climate/remote sensing related research aims at assessing the role of climate variability and global climate change in land degradation and desertification in the Mediterranean - North African region, to achieve an improved disentanglement and quantification of climatic versus human influences.

WG3 underlined the importance of the assessment of data requirements by focusing on needs of different users and on improving data flows from "micro to macro" (local communities to sectorial programmes), also highlighting the different perceptions of researchers and managers.

3.2 Results

The Workshop highlighted that many institutions in Europe have a long-standing scientific experience in research and information exchange relating to desertification, both in Europe's own affected territory, as well as through research collaboration with developing countries, particularly in Africa. A great number of national and international institutions located in Europe own established databases and information systems with high relevance to monitoring, modelling and managing desertification processes and mitigation strategies.

In view of emerging UN-CCD activities, European institutions bear the potential to become major partners in research and development co-operation towards finding and implementing viable solutions to the syndrome of desertification, both in Europe and its partner countries. Improvements relating to research strategies, data accessibility and information sharing can be addressed with the help of the Second International Co-operation Programme (INCO II) of the European Commission. This will add further value to European current research for combating desertification.

Data availability

An overview of data which exist at the level of the Joint Research Centres (JRC), the international institutions, and the individual research institutions revealed the wealth of information available for Mediterranean and African regions. Of particular interest for desertification research are specific case studies in desertified regions. Existing information is primarily concerned with describing the status and dynamics of symptoms of desertification, and to a lesser extent their causes and effects. Models which allow predictions and the analysis of trends are available for key natural resources. Indicators of pressure and societal responses have been assessed in case studies and in an exemplary manner.

A major challenge in data and information exchange, however, is to make them better available for research and implementation. *There exist many more data than are actually accessible*. This was a major concern of all thematic groups during the workshop. In addition, the need for better integration of information, for better definition of vulnerability indicators, and for more information directed towards the support of mitigation strategies was expressed.

The working groups added some specific considerations.

Working Group 1: Climate and remote sensing related
Relevant existing data are archived at international data centres as well as at national meteorological and hydrological services. These data basis should be broadened by data which exist in regional and "private" archives (individual researchers, research institutes, universities, industries such as energy producers, etc); existing data should be made easier available for this research and, as far as possible, be brought physically together; reliability and accuracy control should be mandatory.

Working Group 2: Soil, water, and vegetation related
Both Mediterranean and African research teams have relevant interdisciplinary data and information that could be of benefit to the National Action Plans.

With respect to *hydrological data*, organisations responsible for collecting and storing hydrological data have long been established, but there is a threat to existing data, where institutions formerly responsible for running and documenting gauging stations etc., are being disbanded or privatised. The following needs were identified: a) a data and information retrieval or preservation initiative, comparable to that described for meteorological data should be considered, and b) meta data from sets from completed desertification projects should be archived at one of the identified centres.

With reference to *soil data*, although the availability of large scale surveys may give the impression that much is known about the distribution of degraded soils or those threatened by desertification, in practise, the data and information used for compilation and validation of the maps is limited. At more detailed scales, important data and information is often available outside the national catalogues and archives.

As with the hydrological data, efforts should be made to retrieve, compile and incorporate this information into existing archives. Furthermore, there has been a contraction in the number of organisations compiling detailed information on soils and there is urgency in rescuing archives before they are lost. Finally, soil map legendas and profile descriptions were usually made for purposes other than desertification or soil degradation.

In relation to *vegetation*, much information is available with respect to "land cover types" and data from remote sensing. However, it was the experience of the working group that the data available, although in itself an achievement, often missed critical qualities and did not provide what was needed for monitoring and understanding change.

Working Group 3: Socio-economy and policy related
Despite differences in aims, scope, coverage and depth, managers and researchers
are largely seeking data from the same wide array of variables and indicators.
Although there is great availability of these data, they are for the most part
aggregated at national level, and have therefore little relevance to the real situation
of an area affected by land degradation. The needs for these data must therefore be
defined in a case specific way.

Data needs

Defining data need was considered to be a difficult task in view of the complexity
of desertification problems observed in the different bio-physical, social and
economic settings. It appeared necessary to identify the needs for each specific
research field of the different working groups. Among the general needs the
followings can be emphasised:
- better aggregation of data sets at multiple scales;
- improved compilation of information;
- better integration of different case studies through improved networking among
  all institutions dealing with desertification research;
- more efficient up- and downscaling of information for different user groups, be
  it research teams, or policy and other decision makers at different levels of
  intervention, from local to community, regional to national, and the international
  levels.

Future research needs

It was concluded that such needs be based on a research strategy, that would have
to be defined beforehand. A general consensus was reached on the following:
- research focusing on monitoring and understanding the processes and causes of
  desertification would have to be ongoing;
- new research initiatives focusing on aspects of mitigation, restoration, and
  prevention of desertification should receive equal attention. Here major gaps
  were identified. Research in support of pro-active interventions would have to be
  multi-level, multi-scale, and multi-temporal, and include different
  methodologies, from disciplinary to multi-, inter- and transdisciplinary
  approaches, the latter being the most participatory type of research;
- a key requirement for such types of research would be the identification of joint
  case study areas, where more demand-driven, participatory activities could lead
  to better decision-support systems serving different users of the information. A
  better understanding of institutions, policies and desertification systems, and the
  development of tools for mitigation should be at the centre of future research.
  The working groups introduced the following more specific considerations.

Working Group 1: Climate and remote sensing related
The problem which has to be solved is the scaling between global modelling and local impact. The downscaling from global models requires additional techniques to include topography and vegetation and small scale circulation systems which are available from different studies but not used in a systematic way. Similarly, the studies of regional and local effects need regional expertise which may be available but is not yet organized enough.

A Mediterranean/North African GEWEX (Global Energy and Water-cycle Experiment) is proposed for which main requirements are:
- Measurements in representative areas necessary to provide input data for models and to validate model results.
- Satellite data bridge from isolated measuring sites to larger land-surface units.

A GEWEX has to be build upon a networked scientific community able to provide the logistic support for experiments as well as regional expertise. Some ground stations should in any case be selected as *"anchor stations"* for satellite data. At these stations surface properties should be measured such as albedo, net radiation, carbon dioxide and water exchange between surface and atmosphere, state and density of vegetation, as well as subsurface quantities such as the piezometric level, soil moisture profiles and root depth precipitation, evaporation, river run-off, maps of "functional land-cover types". Time series of Satellite data must be extended to 20 years minimum.

Working Group 2: Soil, water, and vegetation related
Concerning water resources, there are major gaps with respect to data and information relating to the monitoring of water use, the use of water for irrigation and for public water supply.

Data should be collected at a variety of scales in catchments where all the collateral data relating to land use change, vegetation, soil and climate are available, as well as information on socio-economic and policy drivers. Too often critical data sets required to interpret results are missing.

With particular reference to soil, particularly valuable are observations related to the dynamic features of soils such as various aspects of soil structure (including surface crusts, bulk density and roughness), soil chemistry and nutrient dynamics and various soil biological indicators. Soil moisture amounts, fluctuations and distributions were also considered to be important.

It was stressed that these data and the indicators that will be derived from them should be specifically targeted at :
- the reconstruction or rehabilitation of soils that are already degraded
- the prevention of degradation in areas that are likely to be threatened.
- deriving indicators that could provide early warning of degradation.

Concerning methods, it was thought that low-cost simple field tests or measurements should be given priority. Detailed DEMs are very important for erosion modelling.

Concerning vegetation, information is specially needed with respect to a) the vegetation cover and land use and their dynamics, b) data and information on vegetation cover, biomass, vegetation growth and plant phenology.

Understanding how different techniques of land use and management affect vegetation was mentioned as a gap.

A general conclusion of the working group 2 was that *trans-disciplinary pilot areas* would be of extreme benefit to:

- Provide locations within which the complete trans-disciplinary data and information requirements, identified by the working groups as being needed for combating desertification, would be obtained. The pilot areas would constitute a desertification research and mitigation facility.
- Provide areas, selected by the National Committees, within which the scientific knowledge built up during previous "Environment" and INCO programmes projects would be efficiently used to support and underpin the action programmes proposed by the National Committees.
- Provide areas within which restoration and prevention strategies would be monitored and tested and best practices be demonstrated. The impact of policies in different sectors would be evaluated.

Working Group 3: Socio-economy and policy related

There exist a urgent need for indicators of response, particularly those which are local and site specific. Understanding responses means understanding local knowledge and perceptions, the ways in which broad scale incentives translate into local opportunities and constraints, understanding locally successful initiatives that form and inform local behaviour, and understanding how the impact of interventions are seen at the local scale.

This is the type of information that managers and researchers commonly lack. It can only be gained by in-depth field work using participatory approaches. It cannot readily be extrapolated because of the complexity of interacting variables, some operating from afar and some intimately local, producing the behaviour observed in any given site.

In synthesis, WG3 highlighted the following:

- data requirements can only be defined in relation to specific problems, but local land users commonly have a different perception of 'the problem', and radically different priorities, from researchers and managers at regional, national and international levels;
- there is a strong flow of information from macro to micro, but all levels sense a lack of bottom-up, micro to macro flow.
- it is only possible to compile inventories and identify gaps of existing data sets with respect to specific sites. However there is a general need for basic data at a disaggregated (potentially household) level on demography, health, education, economy and land use trends for:
    - populations living in marginal areas – by definition hard to document (mobile people, minorities, remote area dwellers, refugees)

– local level information (on a socio-economic and political/ cultural "catchment" scale) for sites of special concern on local knowledge, local level priorities/ perceptions of environmental issues, decision making on land use/ resources, response to change in climate, policy, institutions and interventions.

3.3 Conclusions

The scientific contribution of the European Union to UN-CCD is a valuable asset and an added value to help combating desertification processes both within and outside Europe. Many of the existing research programmes and databases have a documented potential which can be used at multiple levels of intervention.
A major homework for the European Commission should be:
• to stimulate improved access, quality, user friendliness, documentation, and integration of the various sources of data and information;
• to co-ordinate its activities with other international and national institutions;
• to provide a conceptual framework for mitigation research.

The research community, on the other hand, should make better use of the instruments offered by the EC, such as INCO II, particularly in relation to joint research actions, concerted actions, and thematic networks related to combating desertification.

It was suggested to propose a *thematic network dealing with 'mitigation of desertification'*, to be organised among research and database institutions working in this field inside and outside Europe. This would allow to enhance methodological approaches, from disciplinary to transdisciplinary, and to develop a research strategy focusing on desertification mitigation, including the monitoring of key problems and the finding of pathways for mitigation. In particular, tools could be developed for monitoring, appraisals, predictions, and mitigation scenarios, including the identification of bio-physical and socio-economic indicator systems.

4 Study on Desertification Indicators for the European Mediterranean Region: State of the Art and Possible Methodological Approaches

4.1 Objectives

This study, that the Nucelo di Ricerca sulla Desertificazione has presented at the UN-CCD Secretariat at the Fourth Conference of the Parties, constitutes the natural prosecution of the Seminar held in Porto Torres and of the Workshop held in Alghero, providing a first answer to the questions emerged from them. It also constitutes the NRD fulfilment of the engagement taken on at the International Seminar on the CCD Impact indicators, organised by OSS-CILSS at the UNESCO in Paris in the summer 1999. On that occasion a first definition of common terms of reference was achieved. In particular, this study is an attempt of the NRD to provide the international scientific community with a comprehensive framework of desertification indicators as well as guidelines deriving from an accurate methodological research.

The report is the result of a critical appraisal of the state of the art in the field of Desertification, starting from the Rio de Janeiro conference up to the present day, setting out a methodological approach in line with the CCD mandate and inspired from the recommendations and decisions of the CST-UN-CCD and of the first three COP held in Rome, Dakar and Recife. The COP-3 in particular, held in Recife in November 1999, focused on the urgent need to draw up a synthesis of work accomplished in the sphere of indicators in order to proceed as soon as possible to the next phase of testing these indicators.

This report is thought to provide both a comprehensive vision of the approach adopted by international organisations to the problem of desertification, and also and above all, important methodological indications for the identification and selection of the different types of indicators and for their use within the NAPs and RAPs. Indications as to the method and the course to follow, overriding the hierarchical and relational confusion that often exists when different disciplinary fields are at play, to lay the ground, without bias, for an unequivocal and objective interpretation of indicators and their application.

4.2 Contents

The report is divided into three parts, as follows:

The first part deals with the background of the Desertification indicator concept, with a summary of scientific progress made from Rio onwards by the Conference of Parties technical and scientific bodies on the subject of indicators and a brief analysis is made of the environmental and socio-economic specificities of the Northern Mediterranean Annex. This section also clarifies the position and role of all the indicators that are part of the large family of indicators for use in the implementation of the Convention (Fig. 1).

In part two, a synthesis of the most recent achievements in the field of Desertification indicators for the European countries of the Mediterranean is made and some suggested methods are described with reference to the identification and management of indicators.

The third and last part contains useful information for those wishing to undertake further research on the subject of indicators to assess Desertification (bibliographical references, organizations and research and experimental programmes under way, addresses and web sites devoted to indicators).

The major methodological contributions are summarised below.
a) Operational definition of desertification indicators.
b) Characterisation system for the desertification indicators.
c) Classification system for the desertification indicators.
d) Definition of the indicators development procedure .
e) Proposal of an implementation indicators matrix for the Italian National Action Programme.

The first 4 methodological contributions, in particular, are worth further clarification.

a) Operational definition of the indicator.
What identifies a parameter or an index as an indicator (for a specific, well defined use) is a set of characteristics, amongst which are the following:

Objective: An indicator is such if it serves tocharacterise/measure/monitor a state or process in a determined context, for a specific purpose.

Method: The method of measurement and/or calculation of the value of a parameter/index. Required is the level of accuracy, number of repetitions, periodicity, statistical processing, fields and contexts of application, spatial scale at which it is significant.

Benchmarks of the parameter/index: Threshold values, reference intervals.

Type and quality of the final information: Site-specific or distributed spatially; spatial density of the sampling; specific techniques for space/time interpolation of point data (data from a single site are often interpolated with undiscerning use of geostatistical algorithms, whereas the choice of a suitable method is an important aspect).

To conclude, it can be said that an environmental indicator is not a parameter/index, but a set defined by at least five elements: parameter/index; objective; method of measurement; benchmarks; type and quality of information output. The following formula borrowed from vector algebra may be applied:

indicator = {parameter, objective, method, benchmark, spatial scale/type of spatial extension}

b) Framework for characterising desertification indicators
It was obtained by adopting a conceptual frame of reference (the well-known DPSIR also adopted by the European Environment Agency) and by elaborating a technical-methodological chart that lists all the necessary information that should accompany a ready-for-use indicator. In particular, it highlights the

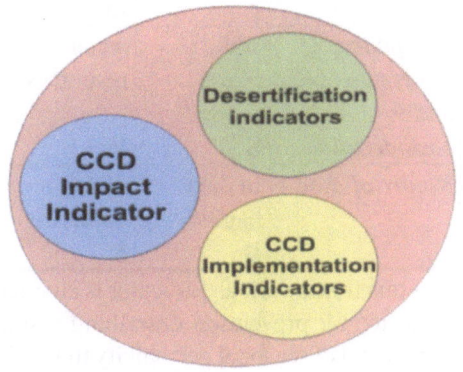

Fig. 1. The family of indicators related to CCD implementation

relevance of the criteria summarised below (for more detail, please refer to the volume):

- Accurate definition of the indicator;
- Position within the logical framework (DPSIR);
- Target and political pertinence (position within the Agenda 21, importance with respect to sustainable development, to the Conventions and to the international agreements);
- Methodological description and basic definitions, with particular reference to the methods of measurement and to the characteristics and operational limits of the indicator;
- Necessary data to calculate the indicator and evaluation of data availability;
- Additional information, authors and bibliographical references.

Table 1. Classification framework for indicators[1] in five hierarchical levels

Criteria	Classes and relative codes				
Operational Objective	prevention **P**	monitoring **Mo**	mitigation **Mi**		
Position in the logical framework	driving force **D**	pressure **P**	state **S**	impact **I**	response **R**
S c a l e	*Space*	punctual **P**	local **L**	sub-region **Sr**	region **R**	European Mediterranean region **M**
	Time	daily or more **g**	monthly or seasonal **m**	annual **a**	less than annual **b**	single event **s**
Component of the system under consideration	soil **S**	water resources **W**	vegetation **V**	climate **C**	socio-economic aspects **SE**
Nature of data	in data banks **B**	direct gathering **F**	remote sensing **RS**		

[1] In this framework an indicator is characterized by five letters. As an example an indicator for prevention describing a state of a spatial scale from sub-region to region and of seasonal periodicity that refers to vegetation cover and is assessed by remote sensing is classified by he following sequence of letters:

$$P \quad S \quad Sr/R - m \quad V \quad RS$$

Table 2. Schematic procedure for the production of impact indicators (simplified from Enne and Zucca, 2000)

Step	Synthetic description
1	Identification of the general and specific <u>objectives</u> and their classification according to the specific field of application and users
2	Establishment of a mechanism for consultation amongst all the potential <u>users</u> and the potential providers of data.
3	Integrated analysis of the objectives and <u>key issues</u> to which they pertain, referring the DPSIR logical framework.
4	Identification of <u>indicators</u> able to best describe, at each level, the key issues identified above.
5	Identification and characterisation of <u>necessary data</u> to measure and/or derive the adopted indicators and benchmarks.
6	Analysis of national and/or local situations relative to production and/or <u>availability of data</u> on the different key issues identified.
7	<u>Calculation/measurement</u> and analysis of indicators
8	Preparation of an <u>action plan</u> to provide for the production of necessary but as yet unavailable data.
9	Dissemination of <u>results</u>
10	Testing <u>perception</u> (feed-back from users)

c) *Classification framework for desertification indicators*
The NRD has proposed a classification framework for desertification indicators structured on five hierarchical levels corresponding to the following criteria:
1. Operational objectives of the Convention (Prevention, Monitoring, Mitigation).
2. Position in the DPSIR logical framework (Driving force, Pressure, State, Impact, Response).
3. Operational spatial scale (punctual, local, sub-region, regional, European Mediterranean region) and operational time scale (high frequency, daily or more; monthly or seasonally; annual; less than annual; single measure).
4. Component of the environmental or socio-economic system (climate, soil, water resources, vegetation; socio-economic aspects).
5. Type of data and acquisition platform (pre-existing data banks, direct gathering, remote sensed).
Table 1 describes schematically the proposed framework.

d) *Definition of the procedure for the production of indicators*
This procedure described in Table 2 was elaborated with the aim of facilitating the establishment of a selection mechanism for the indicators as described by the COP.

In addition to the fundamental principles outlined by the COP, indications given result from the analysis of contributions from various international bodies (in particular OSS-CILSS, 1999).

5 Conclusions and Perspectives

On the basis of the experience acquired during the above mentioned activities, the following conclusions can be drawn:
- Studies and interventions related to indicators should be qualified and finalised to respond to the criteria and methodologies proposed by the UN-CCD.
- The guidelines proposed in the volume "Desertification Indicators for the European Mediterranean region. State of the Art and Possible Methodological Approaches" can effectively contribute the necessary harmonisation of the efforts towards this direction.
- It is necessary to promote a better integration, both horizontal and vertical, of the frameworks to facilitate the exchange of data and experiences.

With the aim of further contributing to the considerable efforts already made, the Nucleo Ricerca Desertificazione wishes to promote a fully participative type of process for developing indicators based on the methodology proposed by the Conference of Parties and revised by NRD, and in which organisations in possession of relevant knowledge, from the scientific to knowledge at all levels, will undertake to organise their experience systematically in order to come up with simple indicators so that they may be used by a wide range of users and be clearly and unequivocally interpreted.

This process should lead to the creation of a data bank for indicators that might be organised according to the classification framework and the descriptive schemes proposed. The steps leading to the production of impact indicators are specified in Table 2.

The future user of this data bank, possibly available on the web, must have the possibility to find the desired indicators by means of keys constituted by the five classification criteria (that provide an answer, among the others, to the questions "what is it for", "at what scale does it apply", "what kind of data is it based on"). Once the user has found the indicators, he must also have the possibility to reach the related methodological charts.

The NRD is now co-ordinating the MEDRAP project, a three- year Concerted Action funded by EC, whose main objective is to support the elaboration of the Regional Action Programme of the Annex IV Countries "and which include a workshop on desertification indicators". The National Focal Points are also involved in this project based on the participation of a wide telematic network of stakeholders. The CA will surely also provide relevant information to the other UN-CCD regional annexes.

References

Enne G, d'Angelo M., Zanolla C (1999) Proceedings of the International Seminar on Indicators for Assessing Desertification in the Mediterranean, Porto Torres (Italy) 18-20 September. Pp 333. ANPA. Rome.

Enne G, Zucca C (2000) Desertification indicators for the European Mediterranean region. State of the art and possible methodological approaches. Pp. 261. ANPA. Rome.

Enne G, Peter D, Pottier D (2001). Proceedings of the International Workshop on the Desertification Convention: Data and Information Requirements for Interdisciplinary Research. Alghero, Sardinia, Italy, 9-11 October 1999. European Commission EUR 19496, pp. 374

OSS-CILSS (1999) Les indicateurs d'impact de la CCD. Compte rendu et communications de l'Atelier international de Paris, 29 juin – 2 juillet.

References

Climate Changes in the Mediterranean Region: Physical Aspects and Effects on Agriculture

A. Iglesias

1 Background

While agriculture is a complex sector, the system is still dependent on climate, because heat, light, and water are the main drivers of crop growth. Plant diseases and pest infestations, as well as the supply of and demand for irrigation water are also dependent on climate.

There is now concern that the effects of climate variability on food production and costs will be exacerbated due to global warming with its potential for affecting the climatic regimes of entire regions (IPCC, 2001). Furthermore, such shifts in climate in different nations may have different effects on agricultural productivity and costs.

The effects of climate change on agriculture are thus likely to vary between different regions and different scales (global, regional and local). As a result, it is most important that impact assessments be undertaken for as many different locations as possible and for different sizes of study region, focusing not only in final production, but in other indicators of vulnerability of the agricultural sector. This will only be useful, however, if the methods of assessment are broadly compatible, enabling the generation of sets of results that can be compared and integrated into a wider picture. The purpose of this chapter is to provide guidance toward a set of approaches that will enable progress toward this objective.

The Third Assessment Report of the IPCC concludes that projected adverse impacts of climate change include a general reduction in potential crop yields in most tropical, subtropical and mid-latitude regions, and decreases in the water availability for agriculture and population in those regions. In addition, the assessment concludes that potential change in climate extremes could have major consequences, and those impacts are expected to fall disproportionately on the poor.

2 Why is Climate Change of Concern in Agriculture?

World food production varies by several percent from year to year, largely as a result of weather conditions such as the inter-annual climatic variability in the Mediterranean and Sahel regions. But agriculture in some regions is more sensitive than in others. Typically, sensitivity to weather is greatest firstly in developing countries, where technological buffering to droughts and floods is less advanced, and secondly in those regions where the main physical factors affecting production (soils, terrain and climate) are less suited to farming. A key task facing those concerned with conducting climate impact assessments is to identify those regions likely to be most vulnerable to climate change, so that impacts can be avoided (or at least reduced) through implementation of appropriate measures of adaptation.

3 Understanding the Vulnerability of Mediterranean Agriculture

Current crop production in the Mediterranean region often fails to meet planned internal demand and/or export targets, thus altering the economic balance of nations in the region. Mediterranean crop production is extremely sensitive to large year-to-year weather fluctuations, especially in southern Mediterranean countries, where technological buffering to droughts is less advanced, and in those areas where other physical factors affecting production (e.g., soils, terrain) are less suited to farming. Crop diseases and pest infestations are also weather-dependent, and tend to cause more damage in countries with lower technological levels.

There is emerging evidence that some areas in the Mediterranean region have been affected by droughts as a result of the regional temperature increases observed in the 20th century (IPCC TAR, 2001). Projected adverse impacts of climate change in the Mediterranean region include general reduction of crop yields and decreased water availability for agriculture and populations, exacerbated by the projected changes in extreme weather events (IPCC TAR, 2001). Adaptation of the food production systems and water resource management is an urgent necessary strategy in the Southern Mediterranean region at all scales.

4 Which Impacts are Likely to be Important?

Even without climate change, Mediterranean agriculture faces some serious challenges in the coming decades. The most striking of these are increasing domestic demand, loss of comparative advantage vis-à-vis international growers, land deterioration (including salinization), and rising costs due to environmental protection policies. Competition for international markets will intensify.

Costs of production are likely to rise in a changing climate, as producers adjust crop varieties and species, scheduling of operations, and land and water management. Successful adaptations to climate change may imply significant changes to current agricultural systems, and some of the required changes may be costly. There is likely to be a need for investment in new technologies and infrastructure. New irrigation systems may be required where aridity or instability of precipitation ensues. Damages from flooding may increase, as well. Costs may include greater applications of and/or development of new agricultural chemicals, particularly herbicides and pesticides.

If climate change is taken into account on a global basis, the role of some Mediterranean countries as providers of grains for export may be affected.

Because of the growing interdependence of the world food system, the impact of climate change on agriculture in each country depends more and more on what happens elsewhere. International trade policy issues, especially the movement to lower agricultural trade barriers, will be crucial in climate change response strategies.

National farm policy can be a critical determinant in the adaptation of the farming sector to changing conditions. In the Mediterranean countries farm subsidies may either help or hinder necessary adaptation to the eventuality of a changing climate. An important policy consideration is the assessment of risk due to weather anomalies. If drought frequency increase, the need for emergency allocations will also increase. Anticipating the probability and the potential magnitude of such anomalies can help make timely adjustments that may reduce social costs.

5 Need for Changes in Policy

Mediterranean policy-makers recognize the urgent need for investment in new technologies, infrastructure, and knowledge related to agriculture in response to major problems in the region:

Increasing domestic demand, urbanization, high unemployment level, and low rent of the traditional rural populationLoss of comparative advantage vis-à-vis international growersWater resources competition and land deterioration (erosion, desertification, salinization, pollution, and deforestation)Rising costs due to environmental protection policiesAs regional planners are meeting these challenges they will benefit from an improved understanding of the climate-agriculture-economics interactions in the region.

The three regional application studies will look at the benefits of climate information and projections for designing adaptation options for different farm types and cropping systems. The work focuses on designing agricultural land and water management options that will improve food production and efficiency of irrigation water use. The regional studies focus on optimising traditional production systems since they are the current basis of agricultural production in the Mediterranean, but the research also has benefits for large-scale commercial systems.

6 A Survey of Previous Studies

Several hundred impact studies have now been completed of impacts of climate change on agriculture, and these can provide an indication both of the types and magnitude of climate change likely to be most important. A survey of such studies can provide an approximate and initial indication of the types of impact to expect and, thus, the likely methods of analysis that will be most effective. The survey is important because different methods of impact assessment will yield information on some, but probably not all, types of impact. For example, analysis of large-area shifts of cropping zones will require broad-scale use of simple agro-climatic indices, whilst analysis of yields can best be achieved through use of process-based crop growth models. Effects on income and employment can only be assessed using economic and social forms of analysis.

Climate variability is emerging as an important issue to be considered in the impact studies, specially when evaluating associated risks of spatial agricultural production. Recent studies consider explicitly the impact of climatic variability in addition to climate change in the evaluations of crop responses.

In regions where current inter-annual climate variability is a mayor factor determining agricultural output, there is an additional challenge for projecting climate change impacts on crop patterns.

7 Assessment of Global Impacts

Potential impacts of climate change on world food supply are estimated for three climate change scenarios developed from global climate models. Results show that some regions may improve production, while others suffer yield losses. This could lead to shifts of agricultural production zones around the nation. Furthermore, different crops will be affected differently, leading to the need for adaptation of supporting industries and markets. Climate change may alter the competitive position of countries with respect, for example, to exports of agricultural products. This may result from yields increasing as a result of altered climate in one country, whilst being reduced in another. The altered competitive position may not only affect exports, but also regional and farm-level income, rural employment and, of course, the type of crops grown in a region. While most studies are unlikely to include an analysis of competitiveness itself, it is possible to evaluate the relative position of a country by studying the few analyses of climate change effects on global food trade. Indeed, some data on country-level output are available as part of the global studies.

8 Case Study: Egypt

Background: Agriculture in Egypt is restricted to the fertile lands of the narrow Nile valley from Aswan to Cairo and the flat Nile Delta north of Cairo. Together this comprises only 3 per cent of the country's land area. Egypt's entire agricultural water supply comes from irrigation, solely from the Nile River. In 1990, agriculture (crops and livestock) accounted for 17 per cent of Egypt's gross domestic product.

Problem: The study sought to assess the potential impact of a change in climate and sea level on Egypt's agricultural sector, accounting for changes in land area, water resources, crop production and world agricultural trade. The aim was not to predict Egypt's future under a changed climate, but rather to examine the combined effects on agriculture of different natural factors and the adaptability of the economic system. This Box summarizes only the agricultural component of the study.

Methods: A physically-based water balance model of the Nile Basin was used to evaluate river runoff and thus enable inferences to be drawn concerning water supply for agriculture. Process-based agronomic models were used to estimate crop yields and crop water requirements.

External factors such as world food prices were introduced from a study of climate change which used a global food model to assess climate-change impacts on world food supply and demand. Supply and demand at the national level were then input to a national agricultural sector model to determine effects on land use, water use, agricultural employment, etc.

Testing of methods: Each of the sub-models used in the study was validated against local data.

Scenarios: The current baseline adopted for the socio-economic projections was 1990 and the climatological baseline, 1951-1980. The time horizon of the study, 1990-2060, was largely dictated by the climate change projections.

Impacts: An agricultural water productivity index was used to measure impacts on agriculture: total agricultural production (tonnes) divided by total agricultural water use (cubic meters). Under these 2 x CO2 GCM-derived scenarios the index declined between 13 and 45 percent.

Adaptive responses: Adaptations in water resources (major river diversion schemes), irrigation (improved water delivery systems), agriculture (altered crop varieties and crop management) and coastal protection against sea-level rise were all tested. They achieve a modest 7-8 per cent increase in agricultural sector performance compared to no adaptation, but together would be extremely expensive to implement. However, investment in improving irrigation efficiency appears to be a robust, 'no regrets' policy that would be beneficial whether or not the climate changes.

Chapter 4
Large Scale Aspects

Development of Priority Climate Indices for Africa: A CCl/CLIVAR Workshop of the World Meteorological Organization

A. Mokssit

Preamble

In the framework of the Regional Rapporteur on Climate Change Detection and Attribution Studies (CCDAS) activity in Africa, difficulties were met in term of inviting Africans countries to report on their activities in this area of CCDAS due to a number of reasons:

- Absence of such CCDAS,
- lack of the necessary daily data to perform such kind of CCDAS,
- difficulties of countries to exchange their raw data.

In this situation it was necessary to think about an idea to make countries perform Climate Indices by themselves and exchange these indices "rather than raw data" through a Capacity Building process; So the role of Rapporteur was hence converted in Capacity Building Implementation.

This idea was introduced and endorsed at the CCl/CLIVAR working group on climate change detection meeting, during November 1999 in Geneva. Hence the working group recommended the organization of a number of workshops on regional climate indices. The current workshop is intended to help African countries making a great progress in this area. Many climate indices will be developed using the longest available and reliable station data. These indices will be shared with the international research community interested in studying the African climate such as the IPCC.

This Capacity building process concerned 28 participants from 23 African countries (including Morocco) and Spain (listed below) and was conducted by a resource team composed by Lisa Alexander (Hadley Centre, UK), David R. Easterling (NCDC/NOAA, USA), Abdalah Mokssit & Rachid Sebbari (National Met. Research Centre/ DMN, Morocco) and managed by Valery Detemermann from WMO which sponsored the workshop.

1 Introduction

Interest in the climate system of Africa has been constantly increasing in recent years. The scientific interest and studies has yield many discoveries such as teleconnections between the African climate and other centres of the tropical variability. Both statistical techniques applied to raw data and the use of General Circulation Models (GCM) allowed to make such important progress in studying the African climate. The social and economic characteristics of most African countries depend on the seasonal and interannual variability of temperature and rainfall patterns. While anomalous variations in these shorter time scale patterns can have serious effects on society, the vulnerability of society to longer-term climate change, occurring over periods of decades to centuries, will depend on its ability to understand and respond to this change.

The rising interest in Climate Change and its impacts is the consequence of human action on his environment through mainly release of green-house gases. Many studies in Europe and North America show an increasing trend of regional air and surface temperatures. In Africa, it has been difficult for scientists to carry out studies in climate change detection due to lack of data.

Through the initiative of the regional Rapporteur on climate change detection and attribution studies in Africa, a CCI/CLIVAR workshop to develop climate indices for Africa has been held in Morocco from 18 to 23 February, 2001. The aim of this workshop was to bring together scientists from a number of African countries to enable them make inventories of their data bank, quality control the existing data and produce climate change related indices for each station using daily rainfall and temperature data. The derived indices will be used for climate change assessment at the regional level. The particularity of this workshop was related to its agenda which was divided in three parts: 1- tele-forum: an intensive exchange of emails, fax and letters took place to inform, to send examples and to explain the work to be done during the workshop; 2- pre-forum which consisted in training the participants (capacity building action); 3- the forum where each participant demonstrates his ability to perform national and local study, write a report for their countries and make a presentation.

At the end of the workshop, the participants have been provided with the statistical software used to produce climate indices during the workshop and have been encouraged to update the indices regularly and compute a full set for all the daily records back at their institutions.

2 African Climate Overview

Africa is a vast continent and consequently experiences a wide variety of climate regimes. It has the largest tropical land mass with its east-west extent in the northern Hemisphere being particularly impressive. While the poleward extremes of the continent experience winter rainfall associated with the passage of midlatitude airmasses, a majority of the continent is strongly influenced by circulations which also extend across large parts of the Atlantic and Indian Oceans. These direct circulations have a pronounced annual cycle and associated variations in rainfall, often described in terms of the movement of the Inter-Tropical-Convergence-Zone (ITCZ).

3 Data and Indices

To monitor changes in climate and changes in climate extremes, a set of key indices has been computed. A good index is expected to have a clear meaning, be highly relevant to people, provide insights into climate change, be homogeneous, be easy to interpret, be relevant to the practical concerns of policy makers and do not smooth out potentially important changes. Other desirable characteristics of an index for monitoring climate extremes include: a good signal-to-noise ratio for the detection of a trend from one period to another, relevance to economic activity and other aspects of human society, sensitivity to likely anthropogenically-induced or natural variations in climate and it should be calculable from available observational and model data.

Many types of indices can be drawn up from daily data. Table 1 is a list of the climate indices that the software compute. Some indices are based on thresholds, some on varying extremes, some are just normalize indices, others are combined indices. The indices are expressed in various ways to facilitate spatial and temporal trend detection and impact analysis.

Daily rainfall and temperature data is needed to compute the indices above. A selection of stations from the Region I GSN station list has been done. Most participants brought daily data for the selected stations existing in their countries.

The period covered by the daily data depend from one country to another and from one station, in the same country, to another.

4 Software and Analysis

One of the most important step in treating daily data is to make a quality control and look to inhomogeneities in these data. An inhomogeneity in a time series is defined as any change in this time series that is not due to change in weather or climate. Among the causes that may lead to such inhomogeneities we find : changes in instruments, changes in processing, changes in the environment around the shelter, changes in observing practices and changes in location of stations.

The software used to control the data, to control the homogeneity of the data and to compute the indices is called ClimDex software. It runs under Excell and was developed by Byron Gleason from NCDC/NOAA, USA. ClimDex provides users with a way to detect temperature (TMAX and TMIN) inhomogeneities. These values

Table 1. Climate Indices

INDEX NO.	ABBRE-VIATION	TITLE	Units
125	FD	Number of Days with Frost (Tmin < 0 deg C)	days
141	ETR	Intra-Annual Extreme Temperature Range (Th-Tl)	days
143	GSL	Growing Season Length (when T>5 deg C for >5 days and: T<5 deg C for >5 days)	days
144	HWDI	Heat Wave Duration Index	days
191	Tx10	Percent of Time Tmax < 10th Percentile of Daily Maximum Temperature	% of time
192	Tx90	Percent of Time Tmax > 90th Percentile of Daily Maximum Temperature	% of time
193	Tn10	Percent of Time Tmin < 10th Percentile of Daily Minimum Temperature	% of time
194	Tn90	Percent of Time Tmin > 90th Percentile of Daily Minimum Temperature	% of time
606	R10	No. of days with Precipitation >= 10.0 mm/day	days
641	CDD	Maximum Number of Consecutive Dry Days (Rday < 1 mm)	days
644	R5d	Largest 5-day Rainfall Total	mm
646	SDII	Simple Daily Intensity Index	mm/day
695	R95T	Fraction of Annual Total Rainfall due to Events Above the 95th Percentile	-
001	TxGE	Number of Days Tmax >= user defined threshold	days
002	TxLE	Number of Days Tmax <= user defined threshold	days
003	TnGE	Number of Days Tmin >= user defined threshold	days
004	TnLE	Number of Days Tmin <= user defined threshold	days
005	Prcp	Number of Days Prcp >= user defined threshold	days

can be thought of as discontinuities or shifts in the data record or time series of maximum or minimum temperature. These abrupt or sometimes gradual changes can be traced to both natural and artificial (human induced) changes. User's are generally more interested in eliminating or mitigating the effects of the latter (artificial) and trying to detect and/or explain the former (natural). To examine a temperature time series for inhomogeneities, ClimDex utilizes both visual inspection of a temperature time series and statistical test (t-test) to test the difference between two adjacent period mean values In the first step, ClimDex simply provides the user with a time series of annual mean (temperature) and accumulated values (precipitation). These time series can then be examined in conjunction with any existing metadata to identify potential inhomogeneities. The second step involves the user defining a "window" size in years. This window size is then split into two adjacent periods and then the difference between the two mean values are tested for any differences from 0 (e.g. two-sided t-test, for Beta not equal to 0). The resultant probabilities from this statistical test are plotted for the user.

5 Results

Each participant has undertaken a Climate Indices Study for his country and elaborate a report which will be available to the scientific community. Hereafter, only global African maps of selected indices are reported. Six selected indices are shown:

- ▸ The Percent of Time the Minimum Temperature is less or equal to the 10th Percentile. This can show us the trend of the lowest minimum temperatures.
- ▸ The percent of Time TMIN is greater or equal to 90th Percentile to see also the trend of the warm night temperature.
- ▸ The Percent of Time the Maximum Temperature is less or equal to the 10th Percentile.
- ▸ The percent of Time TAMX is greater or equal to 90th Percentile to see also the trend of the Hot night temperature.
- ▸ The number of days that precipitation is greater or equal to a chosen threshold like 10 mm.
- ▸ The percent of annual precipitation due to events exceeding 95th percentile.

The first two figures concerning minimum temperature show a positive trend of the warm night temperature and a negative trend of the lowest minimum temperatures over most countries, including Madagascar and Seychelles, except in the extreme eastern coast of Tanzania. Similar results are found for maximum temperature but this time the extreme eastern coast of Tanzania show similar results to those obtained over all countries. There is evidence of warming in almost all the countries, this warming is found to be statistically significant in many cases.

The last two figures show the last two precipitation indices listed above. Unlike the temperature, the trend in these cases are very different from one location to another even in the same country making it very difficult to make a regional analysis of the results. Extreme rainfall events seem to increase in some locations but to decrease in other locations sometimes very close from the former locations, e.g. case of Tunisia. There is a need to consider other kind of indices to be able to make conclusion at regional level like indices which characterize regional precipitation features instead of local precipitation climate.

6 Conclusion

This workshop was an opportunity to reach all the initial objectives:
- an inventory of 23 national daily data in Africa.
- 23 national daily data homogenized and quality controlled.
- 23 national detection studies based on a set of indices listed in appendix A were performed and papers produced and presented.
- A regional set of cumulative indices over Africa.
- A validated software ready to be circulated to the remaining African countries

Taking into account the short period (5 days) and achievement of all objectives the concept of tri-steps workshop is demonstrated to be efficient in Africa .

Most of the selected indices seem to be very relevant for temperature and give a regional overview of climate change however for precipitation the same indices seem to be highly dependent on location and though for regional studies there is a need to consider other indices more relevant to regional level.

The workshop was also an occasion to validate a new concept of exchange of data through "Elaborated data" rather than "Raw data". Indeed, all participants show their agreement to exchange indices rather than raw data due to many constraints, mainly: national sovereignty and commercialization. It was also a good opportunity to strengthen the collaboration between the African national meteorological services. The usefulness of such kind of workshops have been expressed by all the participants and we therefore recommend that the ClimDex software, used in the workshop for computing some climate indices, and its manual should be sent to the others African countries that were enable to attend this workshop.

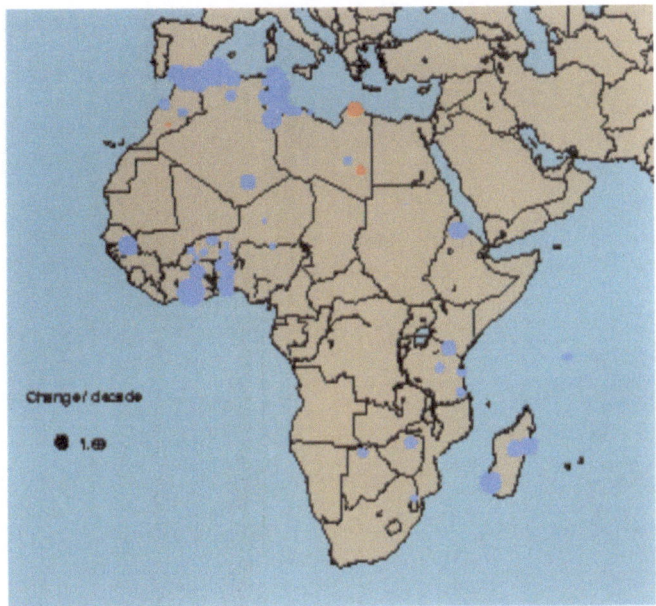

Fig. 1. Percentage of the time the maximum temperature is less or equal to the 10th percentile

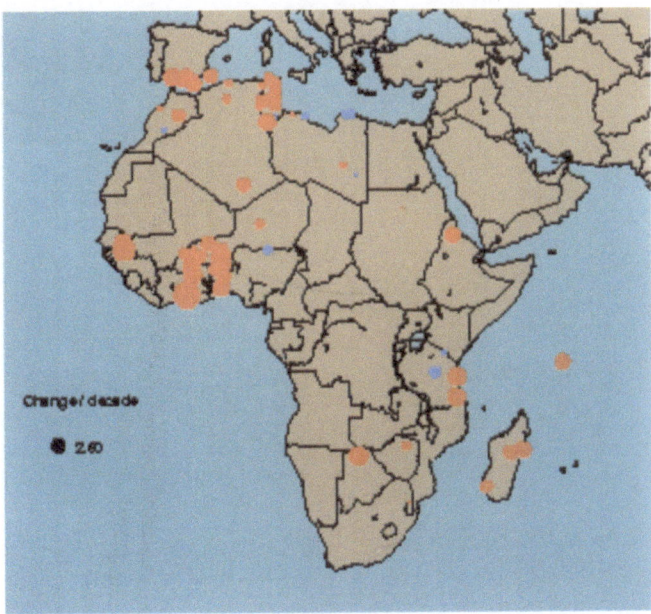

Fig. 2. Percentage of the time the maximum temperature is larger or equal to the 90th percentile indicating the trend of the hot night temperature

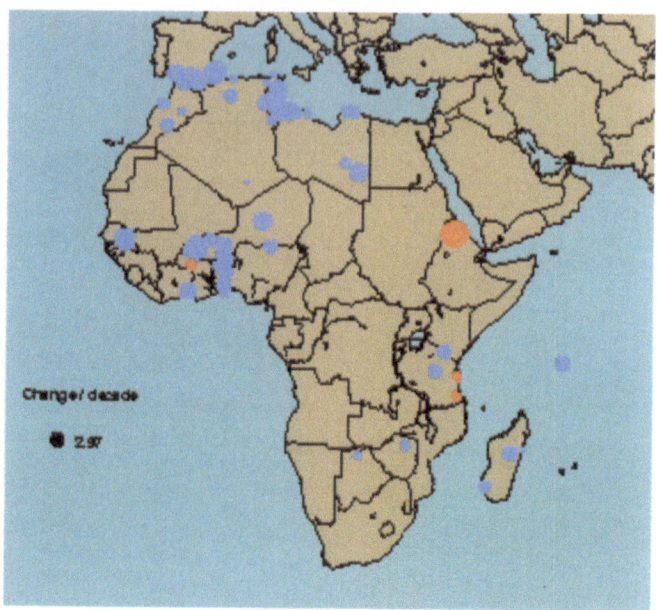

Fig. 3. The percent of time the minimum temperature is less or equal to the 10th percentile. This can show us the trend of the lowest minimum temperatures

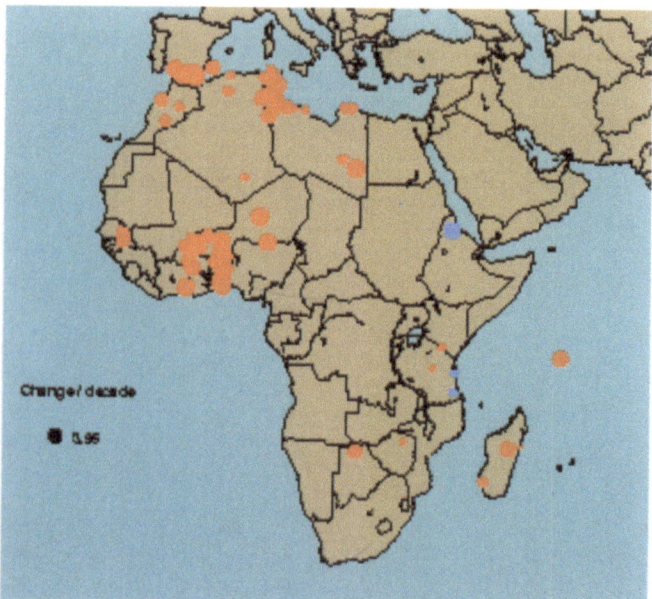

Fig. 4. The percent of time the minimum temperature is larger or equal to 90th percentile which indicates the trend of the warm night temperature.

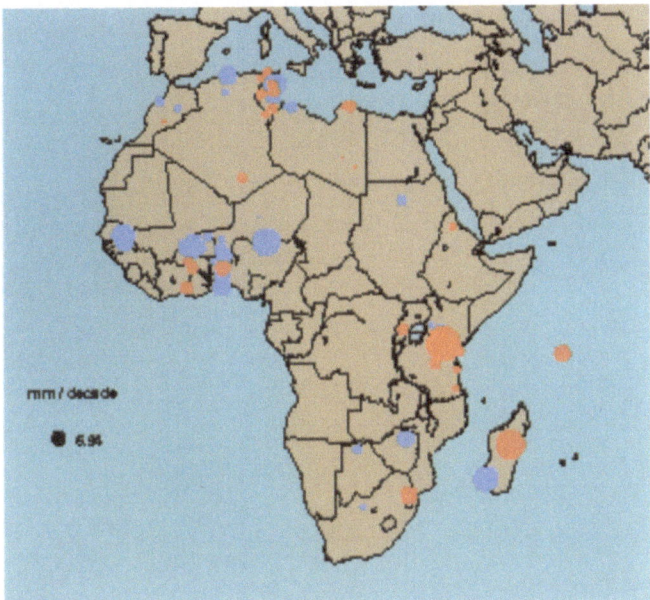

Fig. 5. The number of days that precipitation is greater or equal to a chosen threshold like 10 mm. Maximum 5-day precipitation total

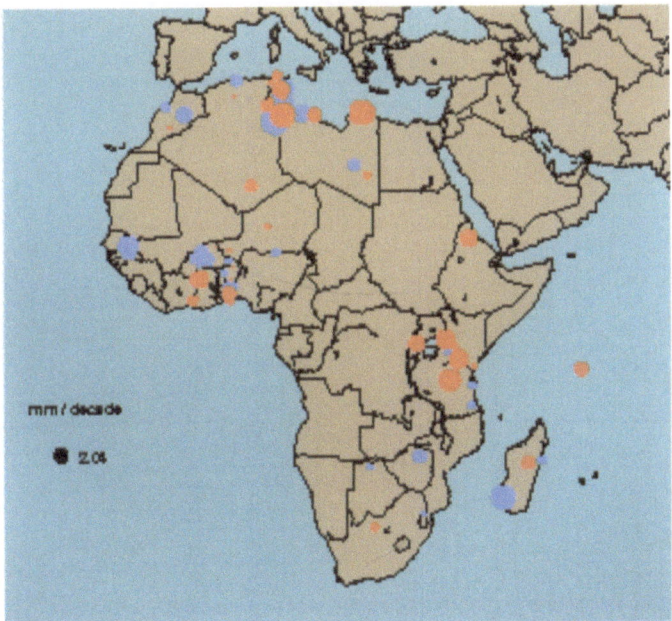

Fig. 6. Fraction of Annual Total Rainfall due to Events Above the 95th Percentile

Analysis of Mediterranean Climate Data: Measured and Modelled

J. Palutikof

1 Introduction

There are four principal sources of meteorological information:
1. Surface observations of present-day conditions from 'traditional' instrument arrays:
   - raw station values
   - gridded observations
2. Remotely-sensed observations, principally from satellites,
3. Proxy data sets from tree rings, sediments etc., important for the study of past climates,
4. Modelled data:
   - climate model output, to provide information on potential future conditions, and
   - reanalyses, from forecasting models and observations, to provide information on the recent past.

This paper concentrates on just two of these: instrumental observations (1) and output from models (4).

Climate data is normally used to try and answer four fundamental questions:
- To understand trends in observations
- To understand the underlying causes of the observed trends
- To understand what may happen in the future
- To understand the relationship between what is happening now, and what may happen in the future.

2 Understanding Trends in Observations

By compositing data from a number of stations, it is often possible to obtain a clear signal of trend in the data, which would not be apparent by examining time series from individual stations. Because the mean and standard deviations of individual records differ, it is necessary to standardize prior to compositing. The formula for

Figure 1 Selected standardized anomaly indices for Mediterranean rainfall

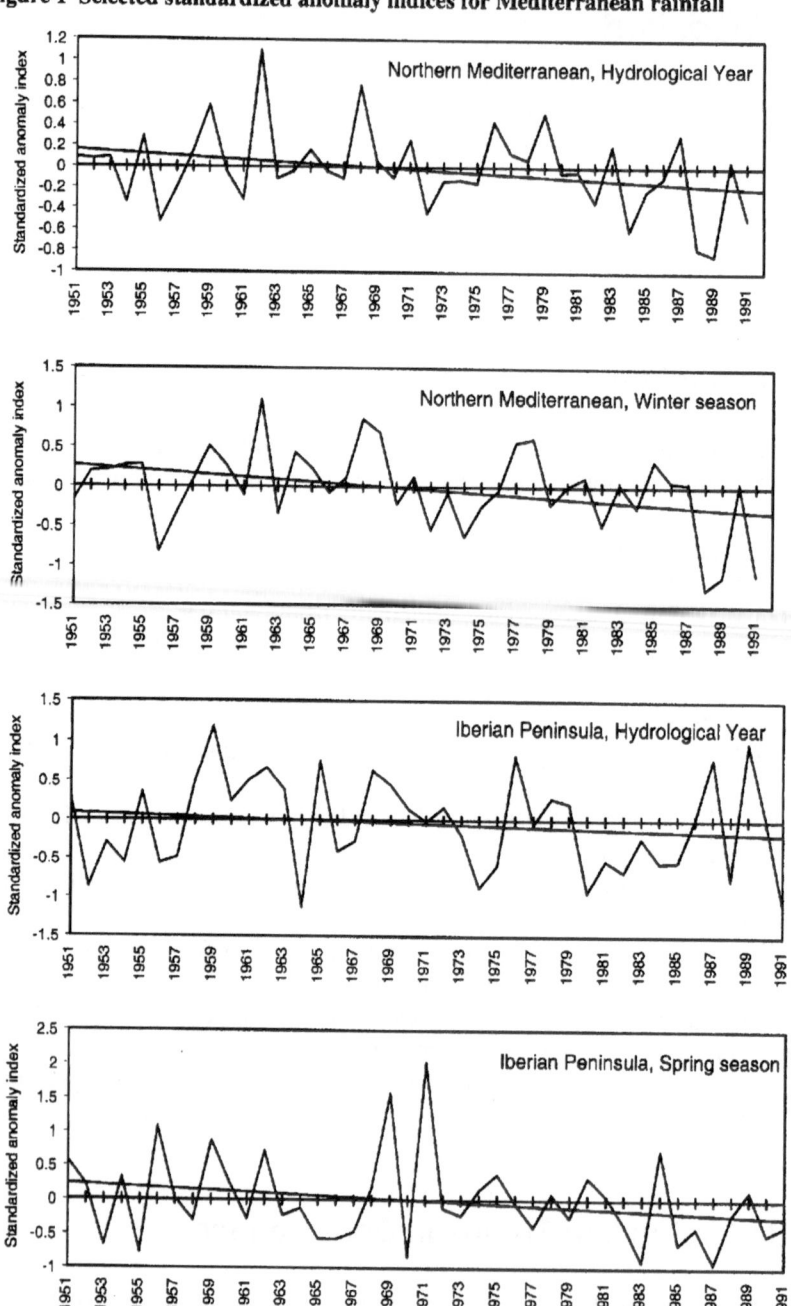

Fig. 1. Selected standardized anomaly indices for Mediterranean rainfall

the calculation of standardized anomaly indices (SAIs) is:

$$SAI = \frac{1}{n} \sum_{i=1}^{i=n} \left[\left(x_{ij} - \bar{x}_i \right) \sigma_i \right] \tag{1}$$

where n is the number of stations, x is the precipitation at station i for period j, $\bar{x}$ is the mean and σ is the standard deviation.

Composite precipitation series have been produced for four regions in the Mediterranean. Some selected results are shown in Figure 1. What is immediately apparent is that, in general, the trend is towards decreasing precipitation, with a turning point in the early 1960s. Although these graphs have been selected to show the trend most clearly, inspection of the full results shows a decreasing trend for almost all regions in all seasons. The only clear positive trend is in the eastern Mediterranean in autumn.

These graphs extend to the early 1990s. Extension of the time series to the end of the 1990s, using the NCEP re-analyses, shows that these downward trends appear to level off.

3 Understanding the Underlying Causes of the Observed Trends

The SAIs were correlated with two measures of spatial sea level pressure (SLP) difference. The first is the North Atlantic Oscillation (NAO), calculated from the pressure difference between Ponta Delgada in the Azores and Stykkisholmur in Iceland. It is an indicator of activity in the North Atlantic. The second measure is the difference between SLP at Gibraltar (Northern Frontier) and Israel (Lod Airport). The existence of a Mediterranean pressure oscillation has been suggested by a number of authors, based on the similarity between the east-west distance across the basin and the length of atmospheric long waves. Some results from the correlation analyses are shown in Table 1.

In the winter season, the NAO is significantly correlated with the western Mediterranean SAI, as predicted in the literature. However, the relationship appears to be more extensive than indicated, since significant relationships are also found with the northern and eastern Mediterranean SAIs. The MPI is significantly correlated with a number of the SAIs in winter, but since the sign is the same as that for the NAO, we cannot assume at this stage that any additional variance is explained - this may simply be a function of a strong relationship between the NAO and the MPI. However, significant relationships are found between the MPI and the western and northern Mediterranean SAIs in autumn, and the central and northern Mediterranean SAIs in summer, when the NAO/SAI relationships are weak.

Further analyses were performed to investigate the validity of the MPI. It is necessary to show (a) that the MPI is a true indicator of a pressure oscillation, rather

Table 1 Correlation coefficients between rainfall SAIs and two pressure indices, the NAO and a Mediterranean pressure index (MPI). Coefficients in bold are significant at the 99% level

Season	Winter	Spring	Summer	Autumn
Western Mediterranean				
NAO	**-0.64**	-0.26	-0.32	0.02
MPI	**-0.74**	0	-0.17	**-0.62**
Central Mediterranean				
NAO	-0.35	0.02	0.05	0.02
MPI	**-0.5**	-0.17	**-0.41**	-0.3
Eastern Mediterranean				
NAO	**-0.45**	0.1	0.05	-0.13
MPI	-0.22	0.18	-0.37	-0.13
Northern Mediterranean				
NAO	**-0.66**	-0.22	0.03	-0.29
MPI	**-0.62**	-0.24	**-0.56**	**-0.61**

than simply a measure of a dominant pressure variation at either the western or eastern end of the Basin, and (b) that it is independent from the NAO. First, the SAIs were correlated with the observations of SLP from the individual stations used to calculate the NAO and MPI. If the MPI is a true measure of an oscillation, then we would expect the sign of the relationships between the two individual SLP time series and the SAIs to be opposed, and for the correlations between the SAIs and the MPI to be higher than those between the SAIs and the individual SLP observations. The results are shown in Table 2. The results from Table 1 are repeated in order to aid understanding. The following conclusions can be drawn.

i. For the pressure observations which make up the NAO, in winter the sign of the correlation between the SAIs and Ponta Delgado SLP is always negative, and the sign of the correlation between the SAIs and Stykkisholmur SLP is always positive, indicating a true oscillation. However, the size of the correlation between the NAO and the SAIs is not always greater than it is between the individual SLP time series and the SAIs. In fact, it is only for the eastern Mediterranean that the NAO has a higher correlation with the SAI than it does with the individual SLP time series.

ii. For the pressure observations which make up the MPI, it is only in summer that the sign of the correlation between the SAIs and the SLP observations is always opposed, with Gibraltar having a negative correlation and Lod Airport a positive sign. In this season, and for all four SAI regions, the correlation with the MPI is always higher than it is with the individual SLP time series.

iii. In winter, spring and autumn, there is a negative relationship between all the SAIs and the pressure at both Lod Airport and Gibraltar. The only exception is for the western Mediterranean in spring. The highest correlations are with Gibraltar, with the exception of the eastern Mediterranean.

We conclude from this analysis that the MPI is only a true oscillation, having a real relationship with rainfall amounts over the Mediterranean region, in the summer months.

Finally, we correlated the NAO and the MPI values to establish whether the MPI is an independent entity from the NAO. The results for the four seasons are as follows:

Winter	0.72
Spring	0.37
Summer	-0.13
Autumn	0.30

The correlation in winter is significant at the 99% level, and the spring and autumn results are significant at the 95% level. This result suggests that in summer (but only in this season) the MPI is a phenomenon truly independent of the NAO.

4 Understanding What May Happen in Future

Information is available from a whole range of climate models, both coarse-resolution General Circulation Models and, increasingly, finer-resolution Regional Climate Models. As an example for the Mediterranean, Table 3 shows the predicted composite changes, based on four recent climate model simulations. Between both decades the greatest change in temperature occurs in summer and the smallest change is in spring. For precipitation, the greatest change occurs in spring and the smallest in winter. In the latter season, between 2030-1990 there is a decrease and between 2090-1990 there is an increase.

5 Understanding the Relationship Between Present and Future Climates

It is important to understand the relationships between present-day observational data and the future as predicted by climate models. This allows us to address such

questions as:
- Can the climate models accurately simulate the climates of the present day? The answer to this question helps us to assess the reliability of model estimates of the future.
- Do we see the same trends in climate model output as we do in present-day observations? Detection is an important area of climate research at present. In a single region, a similarity between present-day and future trends is most likely to be a coincidence. However, by comparing trends in many regions, we can begin to address the detection issue meaningfully. What is the relationship between natural variability and the climate perturbation produced by changes in greenhouse gas emissions? In many areas, it is likely that the signal from greenhouse warming will be small in the context of natural interannual and decadal variability. This can have important implications for impacts analyses.

6 Conclusions

The major lack of climate data sets is with respect to observed data sets which are:
- quality controlled,
- readily accessible, and
- regularly up-dated.

Achieving these three goals is neither straightforward nor cheap. A major function of this meeting must be to discuss how the on-going compilation of such data sets can be achieved.

Table 2. Correlation coefficients between rainfall SAIs and the SLP observations used to construct the NAO and the MPI. Bold denotes significance at the 99% level

Season	Winter	Spring	Summer	Autumn
Western Mediterranean				
Ponta Delgada	**-0.71**	-0.07	-0.35	-0.20
Stykkisholmur	**0.52**	0.32	0.23	-0.10
NAO	**-0.64**	-0.26	-0.32	0.02
Gibraltar	**-0.75**	0.14	-0.04	**-0.66**
Lod Airport	-0.36	0.20	0.15	-0.21
MP1	**-0.74**	-0.00	-0.17	**-0.62**
Central Mediterranean				
Ponta Delgada	-0.16	0.13	-0.28	0.12
Stykkisholmur	**0.43**	**0.43**	-0.18	-0.02
NAO	-0.35	0.02	0.05	0.02
Gibraltar	**-0.53**	**-0.53**	-0.29	-0.37
Lod Airport	-0.30	-0.30	0.20	-0.18
MP1	**-0.50**	-0.17	**-0.41**	-0.30
Eastern Mediterranean				
Ponta Delgada	-0.21	-0.02	-0.03	-0.14
Stykkisholmur	0.38	-0.02	-0.02	-0.14
NAO	**-0.45**	0.10	0.05	-0.13
Gibraltar	-0.28	-0.20	-0.15	-0.04
Lod Airport	**-0.59**	-0.37	0.33	-0.16
MP1	-0.22	0.18	-0.37	-0.13
Northern Mediterranean				
Ponta Delgada	**-0.47**	-0.11	-0.18	-0.37
Stykkisholmur	**0.70**	0.25	-0.11	0.22
NAO	**-0.66**	-0.22	0.03	-0.29
Gibraltar	**-0.74**	**-0.58**	-0.37	**-0.63**
Lod Airport	**-0.58**	**-0.50**	0.30	-0.17
MP1	**-0.62**	-0.24	**-0.56**	**-0.61**

Table 3. Predicted 'composite' change in mean temperature and precipitation between
decades, averaged over observational sites

Season	Temperature 2030/39- 1990/99	Temperature 2090/99- 1990/99	Precipitation 2030/39- 1990/99	Precipitation 2090/99- 1990/99
Annual	1.2	1.21	-3.8	-2.17
DJF	1.23	1.12	-0.95	1.5
MAM	1.07	1.11	-6.78	-4.43
JJA	1.41	1.43	-2.66	-1.56
SON	1.19	1.27	-1.21	-3.02

500-year Winter Temperature and Precipitation Variability over the Mediterranean Area and its Connection to the Large-scale Atmospheric Circulation

J. Luterbacher and E. Xoplaki

Summary

Spatio-temporal highly resolved estimates of past natural climate variability over the Mediterranean are important to assess recent significant climate trends. Principal component regression analysis has been performed to reconstruct monthly (AD 1659-1900) and seasonal (AD 1500-1658) temperature and precipitation fields over the land areas including Europe, North Africa and the Near East (Luterbacher et al. 2001c). The reconstructions are based on the combination of early instrumental station series (temperature, pressure and precipitation) and proxy reconstructions from Eurasian sites. The statistical relationships were derived over the 1901-1990 period and were applied to the available data prior to 1900 in order to derive the temperature and precipitation reconstructions back to AD 1500.

We derived a winter mean Mediterranean precipitation and temperature time series from AD 1500-1995 through averaging 2159 gridpoints over the larger Mediterranean area (10°W-40°E; 30°N-47°N). Several cold relapses and warm intervals as well as dry and wet periods on the decadal timescale, on which shorter-period quasi-oscillatory behaviour was superimposed, could be shown.

1891 was the coldest, 1772 the warmest, 1609 the driest and 1684 the wettest Mediterranean winter over the last few centuries. The corresponding connected sea level pressure (SLP) fields are analysed and compared with the long-term 20th century mean. The potential and the limitations of these reconstructions are presented.

A negative highly significant correlation has been found between the winter North Atlantic Oscillation Index (NAOI) and winter Mediterranean precipitation, but only a weak connection between the NAOI and Mediterranean temperature.

1 Introduction

The Mediterranean region lies in an area of great climatic interest. It is influenced by some of the most relevant mechanisms influencing the global climate system: it marks a transitional zone between the deserts of North Africa, which are situated within the arid zone of the subtropical high, and central and northern Europe, which is influenced by the westerly flow during the whole year. In addition, the Mediterranean climate is exposed to the South Asian Monsoon (SAM), the Siberian High Pressure System, the Southern Oscillation (SO) and the North Atlantic Oscillation (NAO). There is recent evidence that the teleconnection of ENSO has extended its reach into parts of the Mediterranean in recent decades (Rodó et al. 1997; Prize et al. 1998; Moron and Ward 1998; Kadioglu et al. 1999).

The main physical and physico-geographical factors controlling the spatial distribution of the climatic conditions over the Mediterranean are the atmospheric circulation, the latitude, the altitude and, generally, the orography, the Atlantic and Mediterranean sea surface temperature (SST) distribution, the land-sea interactions (distance from the sea) and smaller-scale processes (Xoplaki et al. 2000). The Mediterranean area is a climate-sensitive region which is climatically stressed by limited water resources and extremes of heat which help to create or exacerbate existing socio-political tensions (Mann 2001). Especially high-frequency (monthly, seasonal, annual and interannual) as well as low-frequency (interdecadal) variations of precipitation plays a crucial role in the management of regional agriculture, ecosystems, environment and socio-economics and water resources (Xoplaki et al. 2000).

The current climate debate remains centered around questions related to the reasons and the detection of spatial and temporal patterns of climate change over the last few centuries. Better understanding of natural climate variability is of importance to assist in the detection of any 'anthropogenic signal'. Reliable, spatially and temporally highly resolved estimates of past natural climate variability before the instrumental period are necessary to assess recent significant climate trends. Over the last half millennium, Europe inclusive the whole Mediterranean area experienced complex climatic change, which is believed to have been connected with a marked change of the atmospheric circulation. Therefore, understanding of natural climate variability is of much importance but it is difficult to diagnose from the relatively short instrumental record.

For the western, central and eastern parts of the Mediterranean area there are numerous studies dealing with the description of past climate variability and its interactions with the tropical and extratropical modes of the global climate system, climatic change or climate reconstructions (mainly temperature, precipitation and atmospheric circulation over parts of the Mediterranean and for limited time periods) prior to the instrumental period (Serre-Bachet and Guiot 1987; Guiot 1991; Serre-Bachet et al. 1992; Camuffo and Enzi 1991, 1992, 1994, 1995; Grove and Conterio 1994, 1995; Martin-Vide and Barriendos 1995; Barriendos 1997; Glaser et al. 1999; Camuffo et al. 2000; Alcoforado et al. 2000; Felis et al. 2000; Xoplaki et al. 2001a, Grove 2001; Rodrigo et al. 2001; Mann 2001; D'Arrigo and Cullen 2001; Briffa et al. 2001ab and many others). These climate reconstructions include 'multiproxy' networks consisting of diverse proxy indicators. They can be divided in documentary proxy data and proxies derived from natural archives. The documentary proxy evidence consists of chronicles and historiographies, narratives, annals, records of public administration and government, scientific writings, monastery records, etc. (Xoplaki et al. 2001a) where meteorological indications such as number of rainy days, direction of cloud movement, wind direction and related events such as warm and cold spells, freezing of water bodies, droughts, floods, advanced or delayed vegetation and other phenological and biological observations can be found (Pfister et al. 1994; Martin-Vide and Barriendos 1995; Barriendos 1997; Glaser et al. 1999; Luterbacher et al., 2000; Alcoforado et al. 2000; Rodrigo et al. 2001; Xoplaki et al., 2001a). The natural proxy climatic indicators are from direct natural archives such as tree-ring data, corals, lake and marine warves and ice cores which are all limited to specific regions. All these proxies contain direct or indirect information about the course of meteorological phenomena or climate or they describe natural phenomena and social events related to weather and climate (Pfister et al. 1998; Xoplaki et al. 2001a). Besides their obvious seasonal and spatial limitations, different proxies are also potentially limited in their ability to represent climatic variations over a range of different time scales (Jones et al. 1998).

To our knowledge, there are not yet continuous spatio-temporal highly resolved (monthly, seasonal) temperature and precipitation reconstructions for the whole Mediterranean area for the last few centuries. New et al. (1999, 2000) published gridded temperature and precipitation data of global land areas for the period 1901-1996 based on station data which were optimally interpolated on a very dense grid with 0.5° resolution. We extend this data set (0.5° longitude-latitude resolution) and present seasonal gridded reconstructions of temperature and precipitation over land including the whole of Europe, the North African coast and the Near East region back to AD 1500 by using early instrumental station series (pressure, temperature and precipitation) combined with documentary proxy data. We first describe the available data back to AD 1500 and present the statistical model to reconstruct gridded temperature and precipitation fields back to AD.

We then define a time series of winter precipitation and temperature variations by averaging the 2159 gridpoints in the larger Mediterranean region (30°N-47°N latitude and 10°W-40°E) and study its variability from AD 1500 to 1995.

From these long-term time series of temperature and precipitation we derive the coldest, warmest, driest and wettest winters and plot the reconstructed temperature and precipitation fields, show the departure from the 1901-1995 long-term mean and present the quality of these reconstructions. For these four anomalous winters the corresponding large-scale reconstructed sea level pressure (SLP) distribution (Luterbacher et al. 2001a), the departures from the long-term mean and the quality of the reconstructions are given and discussed.

The temperature and precipitation time series for the larger Mediterranean area are then correlated with recent independent North Atlantic Oscillation Index

(NAOI) reconstructions (Cook et al. 2001) in order to see whether there is any connection between Mediterranean climate and the atmospheric circulation over the eastern North Atlantic over the last 5 centuries. The discussions and conclusions are presented in the last sections.

2 Data and Methods

2.1   Predictor Data (Instrumental and Documentary Proxy Data for the Last 500 Years)

The available time series of documentary proxy and early instrumental data and the domain of the reconstructed area are presented in Figure 1a (Luterbacher et al. 2001a). Circles mark time-varying instrumental data series (pressure, temperature and precipitation) provided by different sources, triangles mark data series estimated from high resolution documentary evidence (see above), thus not directly measured values. Prior to AD 1659 only reconstructed climatic indices mostly on a seasonal resolution are available which are given in red triangles (Figure 1a). These seasonal predictors are continuously available or for parts of the AD 1500 to 1658 period. For the Mediterranean, relevant time series are the documentary proxy based reconstructed seasonal precipitation of Andalusia (Southern Spain, Rodrigo et al. 2001), the tree-ring based reconstructions of precipitation over Central Turkey (D'Arrigo and Cullen 2001) and Southern Jordan (Touchan et al. 1999).

Thus, from AD 1500 to 1658, seasonal temperature and precipitation fields were reconstructed. From AD 1659 the reconstructions are monthly based (Luterbacher et al. 2001c). The same data have recently been used to successively reconstruct monthly (seasonally) atmospheric circulation indices (NAOI) and entire SLP fields back to AD 1659 (AD 1500-1658) (Luterbacher et al. 2001ab).

The dependent variables (predictands) consist of gridded monthly temperature and precipitation (0.5°x0.5° resolution) data on the land areas from 30°N-70°N and 30°W-40°E. The gridded monthly temperature and precipitation data for the period 1901-1990 were taken from New et al. (2000).

2.2   Statistical Reconstruction of Large-scale 500 Years Temperature and Precipitation Fields

To reconstruct the gridded temperature and precipitation fields we follow the description of Luterbacher et al. (2001c). We first identified all predictors for which data were available for that particular month (season) and all other months from the same climatological season (i.e. winter is December, January and February). Since the data base varies over the whole 500 years (Fig. 1b), this resulted in a total of 320 different cases for the monthly and seasonal temperature and precipitation estimations back to AD 1500. Second, we selected the period 1901-1960 for fitting the statistical models. Both, the gridded temperature and precipitation and the instrumental data together with the proxy based reconstructions for each climatological season were transformed into Empirical Orthogonal Functions (EOFs). The leading predictor data EOFs sharing 95% of total variance and the leading temperature and precipitation EOFs explaining 90% of the total variability

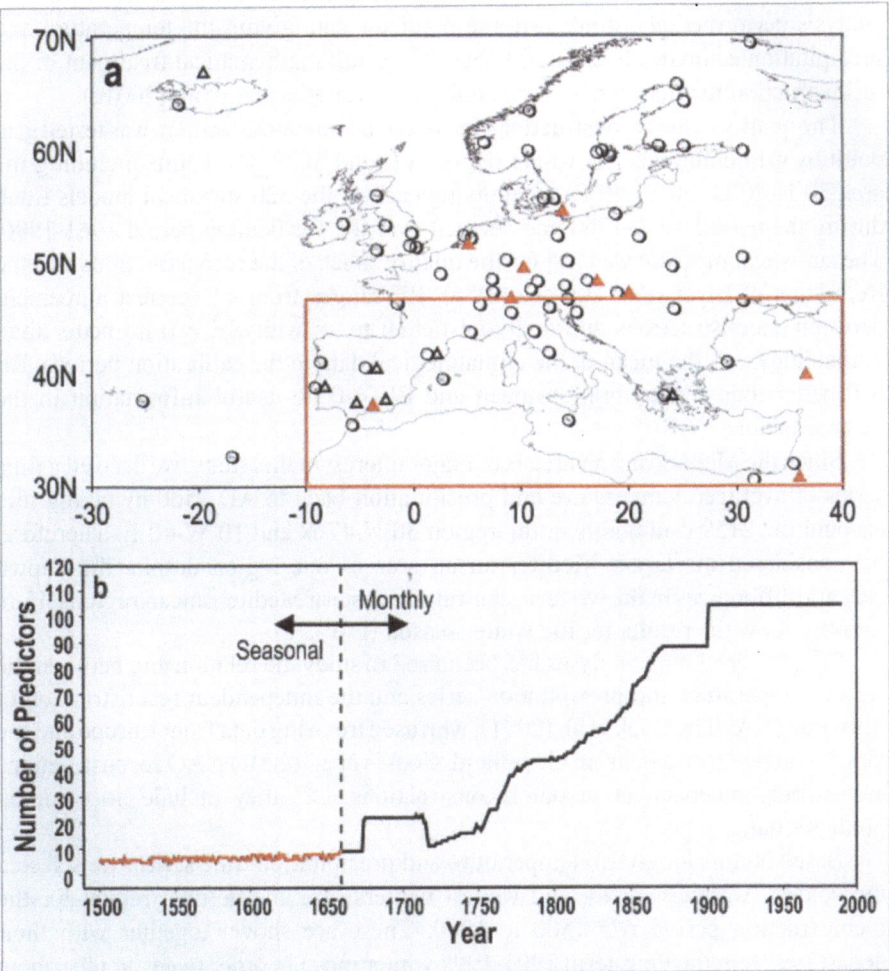

Fig. 1 (a) Locations of the available data (predictors) for the temperature and precipitation reconstructions and domain of the area to be reconstructed (30°N-70°N; 30°W-40°E). The gridded temperature and precipitation (0.5°x0.5°) over the land mass include 5829 grid-points. The extended Mediterranean land area which is studied in more detail is enclosed by the red rectangular box. Circles mark instrumental data series (pressure, temperature and precipitation), triangles mark data series estimated from high resolution documentary evidence (see text for details). Red triangles indicate the predictors available for parts or for the whole pre-1659 period. (b) Temporal development of the number of predictors used for the temperature and precipitation reconstructions (from Luterbacher et al. 2001a).

were calculated and the higher order EOFs have been discarded. The subsequent multivariate regression was performed linking the predictors to the predictands, i.e. the determination of the linear transfer functions. Then, the quality of our reconstructions was tested by applying each of the 320 multivariate regression models to independent data from the verification period 1961-1990.

Finally, the EOFs for the predictors and predictands were derived over the entire 90 years from 1901-1990 and the subsequent principal components regression analysis performed and applied to the predictor data giving the temperature and precipitation estimates back to AD 1500. For a full mathematical treatment of the reconstruction method, the reader is referred to Luterbacher et al. (2001c).

The quality of the reconstructions for selected anomalous winters was tested grid point by grid point over the whole region (in total 5829 grid points including the area 30°N-70°N; 30°W-40°E) by applying each of the 320 statistical models fitted during the period 1901-1960 and verified over the verification period 1961-1990. The statistical measure we used for the quality check of the reconstructions was the Reduction of Error (RE) (Lorenz 1956). RE ranges from +1 (perfect agreement between reconstructions and analysed fields) to -∞ with RE = 0 no better than climatology (i.e. the mean of the climatological data in the calibration period), RE > 0 better than the calibration mean and RE < 0 no useful information in the reconstructions.

Since the Mediterranean area is of major interest in this study, we derived a time series of averaged temperature and precipitation back to AD 1500 by taking into account the 2159 grid points in the region 30°N-47°N and 10°W-40°E. Therefore, we considered the larger Mediterranean area as one region despite the known climatic differences in the western, central and eastern Mediterranean regions. Here we only show the results for the winter season (DJF).

Simple correlation analyses has been used to study the relationship between the winter temperature and precipitation series and the independent reconstruction of the winter NAOI by Cook et al. (2001), who used tree-ring data from Europe and the U.S.A. and ice core data from Greenland. Cook's et al. (2001) NAO reconstructions are entirely independent of our reconstructions, i.e., they include no common predictor data.

Based on this long-term temperature and precipitation time series, we selected the coldest, warmest, driest and wettest winters over the defined region for the reconstruction period AD 1500 to 1900. These are shown together with their departures from the long-term 1901-1995 winter mean in order to get an idea about the manifestation of the anomalies over the whole of Europe, North Africa and the Near East. In addition, the reconstruction quality (RE) (see description above) for these four extreme winters will be presented to have an indication where meaningful reconstructions can be expected. The corresponding reconstructed SLP fields, their anomalies and their reconstruction quality are given as well in order to interpret the temperature and precipitation anomalies in terms of the large-scale atmospheric circulation.

3 Results

3.1 Mediterranean Winter Temperature and Precipitation Variability AD 1500-1995

Figure 2 shows the winter mean Mediterranean precipitation time series from AD 1500-1995 derived through averaging all the 2159 grid points over the defined region (10°W-40°E; 30°N-47°N). The mean winter values for the period AD 1500 to AD 1900 are our reconstructions, the values from the 20th century are taken from the New et al. (2000) data set. The long-term (AD 1500-1995) winter precipitation mean is around 60 mm. The standard deviation is 5.7 mm. The time series clearly exhibit strong decadal to interdecadal precipitation variations and sub-periods of enhanced and reduced variability can be discerned. Wet periods can be found in parts of the 16th century, during the so-called Maunder Minimum (AD 1645-1715), a period with reduced solar activity, several explosive volcanoes, dry and cold

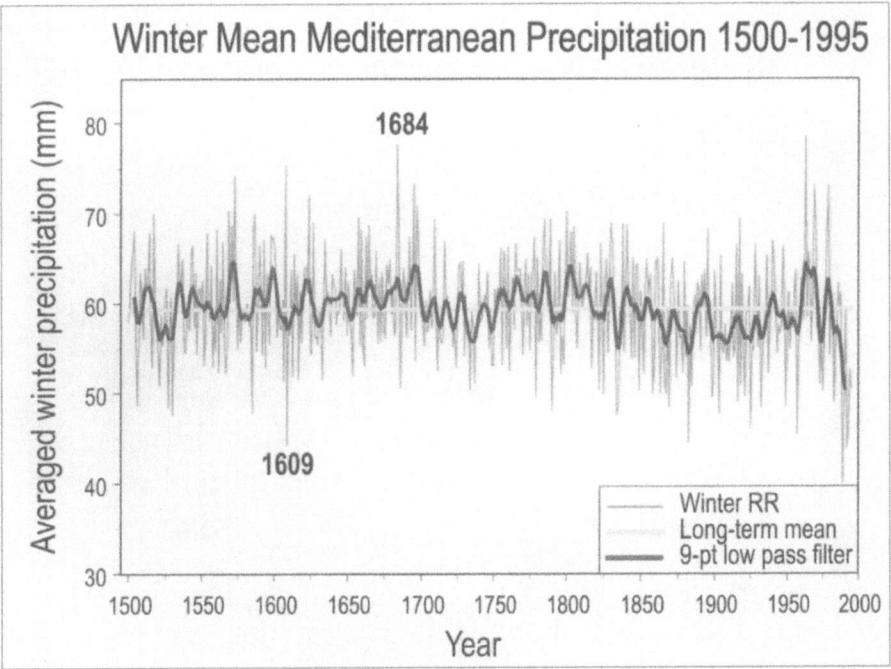

Fig. 2 Winter (DJF) mean Mediterranean precipitation time series from AD 1500-1995 defined as the average over 2159 gridpoints (10°W-40°E; 30°N-47°N) (thin green line). The mean winter values for the period AD 1500 to AD 1900 are the reconstructions, the values from the 20th century are the ones derived from the New et al. (2000) data set. The thick dark green line is the 9 point low pass filtered time series. The long-term (AD 1500-1995) winter precipitation mean is given in yellow. The wettest and the driest winters of the reconstruction period are denoted by black letters

conditions over wide parts of Europe and even the Northern Hemisphere. It was also rather wet at the end of the 18th century and the beginning of the 19th century. However, from the 1810s the winter precipitation over the Mediterranean decreased significantly. After this, the values increased again with a maximum in the early 1970s. The time series further shows the unprecedented trend towards winter dryness over the last few decades of the 20th century. Several droughts in many parts of the Mediterranean during these periods were connected with problems related to water supply.

The winter of 1608/1609 was the driest from 1500 to 1900 and has been chosen for further analysis and to show its reconstruction quality (Fig. 4). During the 20th century, the two winters 1989 and 1992 were even drier than 1609 with a reduction of precipitation of around 35% compared to the long-term mean (AD 1500-1995).

Figure 3 shows the winter mean Mediterranean temperature time series from AD 1500-1995 derived through averaging all the 2159 gridpoints over the defined region (10°W-40°E; 30°N-47°N). As for the precipitation series the mean winter values for the period AD 1500 to AD 1900 stem from our reconstructions, the values from the 20th century are the ones from the New et al. (2000) data set. The long-term

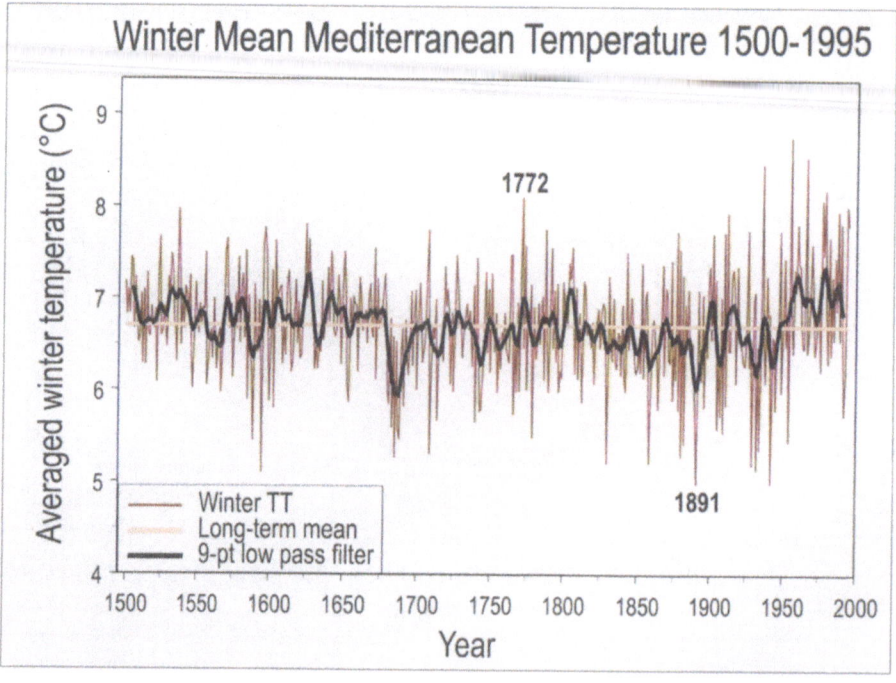

Fig. 3. Winter (DJF) mean Mediterranean temperature time series from AD 1500-1995 defined as the average over 2159 gridpoints (10°W-40°E; 30°N-47°N) (thin red line). The mean winter values for the period AD 1500 to AD 1900 are the reconstructions, the values from the 20th century are the ones derived from the New et al. (2000) data set. The thick black line is the 9 point low pass filtered time series. The long-term (AD 1500-1995) winter temperature mean is given in light red. The warmest and the coldest winters of the reconstruction period are denoted by black letters

(AD 1500-1995) winter temperature mean is around 6.7°C. The standard deviation is0.58°C. However, the time series clearly shows periods with extended coldness and warm intervals on which shorter-period quasi-oscillatory behaviour was superimposed. The 16th and most of the 17th century were rather warm. An abrupt drop of mean winter Mediterranean temperature of the order of 1°C followed at the end of the 17th century (Late Maunder Minimum). The 18th century shows winter temperatures fluctuating around the long-term mean without extended cold or warm periods. The winter of 1772 was the warmest winter for the reconstructed period with around 1.5°C higher reconstructed temperatures than the long-term 1500-1995 mean. This winter temperature distribution will be presented in Figure 4. In the second part of the 20th century, however, there were a few warmer winters. After a warm beginning of the 19th century, the temperature decreased successively till the 1890s with the coldest winter being 1891 (~1.7°C reduced winter mean temperature, see also Figure 4). The 20th century was characterised by strong fluctuations around the long-term mean and temperature increase from the 1950s.

3.2   Mediterranean Winter Climate Extremes Over the Last Five Centuries

Figure 4 (top) presents the reconstructed fields of the coldest (1891), warmest (1772), driest (1609) and wettest (1684) Mediterranean winter over the last centuries according to Figs. 2 and 3, respectively. The charts in the middle of Figure 4 show the anomalies of these extremes from the long-term 1901-1995 means. The bottom charts present the model performance (RE) for the four anomalous winter temperature and precipitation reconstructions, therefore they give rise where the reconstructions are trustworthy and where not.

The largest anomalies for the coldest winter (1891) can be found in a zonally elongated band between around 44°N to 52°N (Figure 4 middle). These anomalies are of the order of 3 to 5°C. The Mediterranean area south of around 40°N shows up to 3°C colder conditions than the long-term mean. A warmer winter was experienced over Scandinavia with maximum positive anomalies of 4°C. The corresponding quality of this reconstructed winter (Figure 4 bottom) clearly indicates reliable reconstructions all over the grid. The lowest RE values (but still positive) are found over the western Mediterranean area and the Northern African coast.

The warmest winter over the larger Mediterranean area between AD 1500 and 1900 was 1772. It reveals large positive anomalies up to 3°C over the Balkans and Turkey (Figure 4 middle). The majority of the grid points south of 50°N show departures of the order of 0 to 2°C from the 20th century mean. Negative anomalies are prevalent over Northern Europe. The quality of the reconstruction of this winter is worse than 1891. This is due to less predictors, mainly over southern Europe during this time (Luterbacher et al. 2001a). For the Mediterranean area only the station temperature and precipitation of Milan (Italy) and Marseille (France) were available, therefore, the reconstruction quality over the Mediterranean region is reduced. However, even without predictors from the Mediterranean there are extended regions that can be partly meaningful reconstructed (Figure 4 bottom). Areas with negative RE, therefore no skill in the reconstructions, are found over parts of Africa, the centre of the Iberian Peninsula, Southern Italy, Central Turkey and Syria. Therefore, one should be careful in interpreting these areas.

The driest reconstructed winter over the Mediterranean from AD 1500-1900 was 1609. The reconstructed precipitation field is given on top of Figure 4. The anomaly fields indicate below normal precipitation generally south of around 50°N. The largest negative precipitation anomalies can be found over Portugal, western Greece and western Turkey. Higher then normal precipitation is visible mainly over Iceland, the western part of the British Isles and Scandinavia. A striking feature is the above normal precipitation values along the Northern African coast east of the prime Meridian. The corresponding reconstruction quality clearly indicates the more heterogeneous pattern compared to the temperature pattern. This is mainly due to the fact that precipitation data reflect much greater variability, both in space and time, compared to temperature. The majority of the grid points shows positive RE values, thus there is an indication of potential to reconstruct precipitation from just a few temperature and/or precipitation series (mainly from Switzerland, Germany, Hungary, Southern Spain and Southern Jordan). Especially for the areas close to these predictors, the quality is high but decreases fast further away from these regions. Though some Mediterranean areas do provide skill in the precipitation reconstruction for 1609, the general potential for trustworthy precipitation reconstructions is rather small.

The wettest winter over the entire 496 years was 1684, within the Late Maunder Minimum period. The reconstructed precipitation field and the departures from the long-term 20th century mean are given in Figure 4. It shows negative anomalies north of around 48°N and wetter than normal conditions over the larger Mediterranean area. Parts of the Mediterranean such as Portugal, Italy and the western Balkans obtained 30 to 90 mm more rainfall than usual. However, there were not everywhere wetter conditions over the Mediterranean region. The Northern African coast east of 10°E for instance was even drier than usual. The quality of the reconstruction (Figure 4 bottom) shows a similar pattern as for 1609. For 1684 there were a few more predictors for the Mediterranean available, such as documentary based reconstructed precipitation from Greece (Xoplaki et al. 2001a), from different places of Spain (Martin-Vide and Barriendos, 1995; Barriendos 1997) and Portugal (Alcoforado et al. 2000). As for 1609, there is skill for some areas, though for the majority trustworthy reconstructions is restricted.

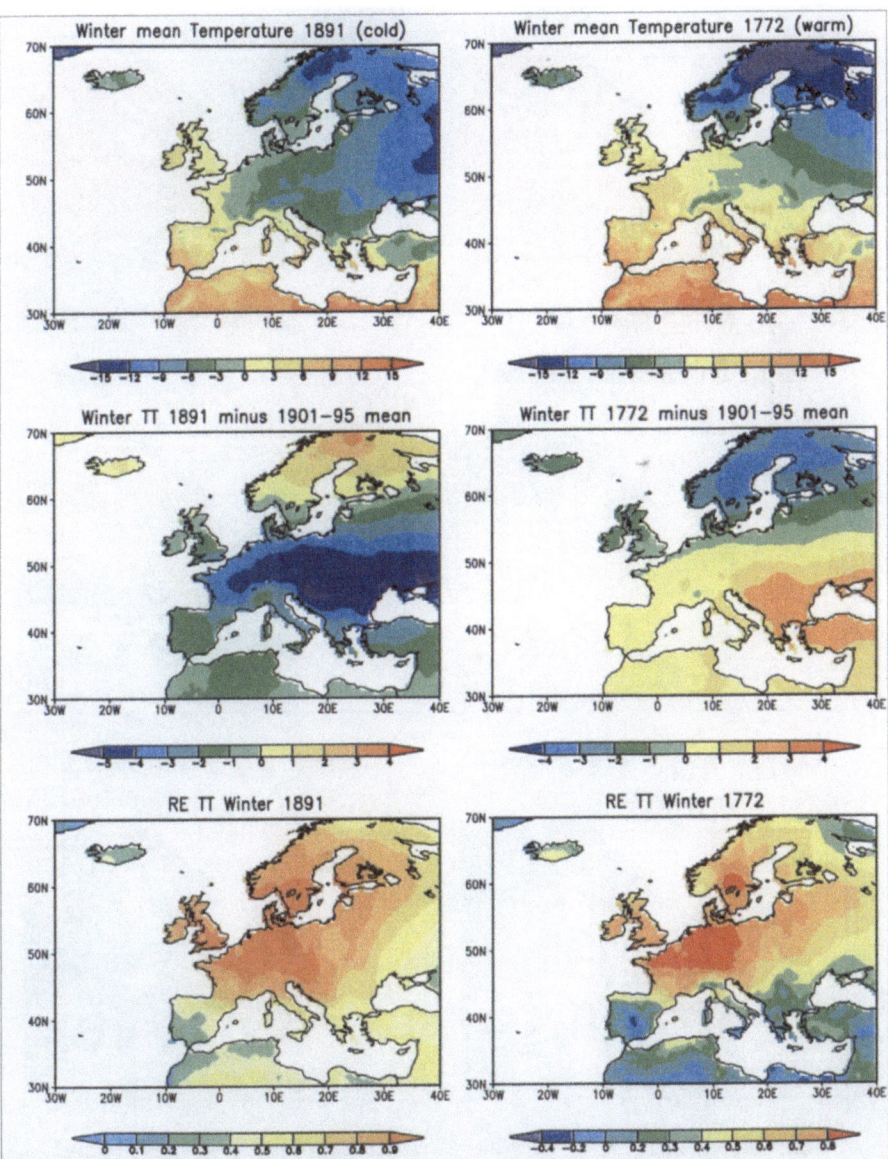

Fig. 4a. Top: Mean reconstructed temperature fields for the coldest (1891) and the warmest (1772) winter (DJF) over the larger Mediterranean area. Middle: Seasonal temperature difference patterns of the extreme winters (Winter minus long-term seasonal temperature mean of 1901-1995). Bottom: Model performance (RE) for the anomalous winter temperature reconstructions. For areas with RE > 0 meaningful reconstructions can be expected.

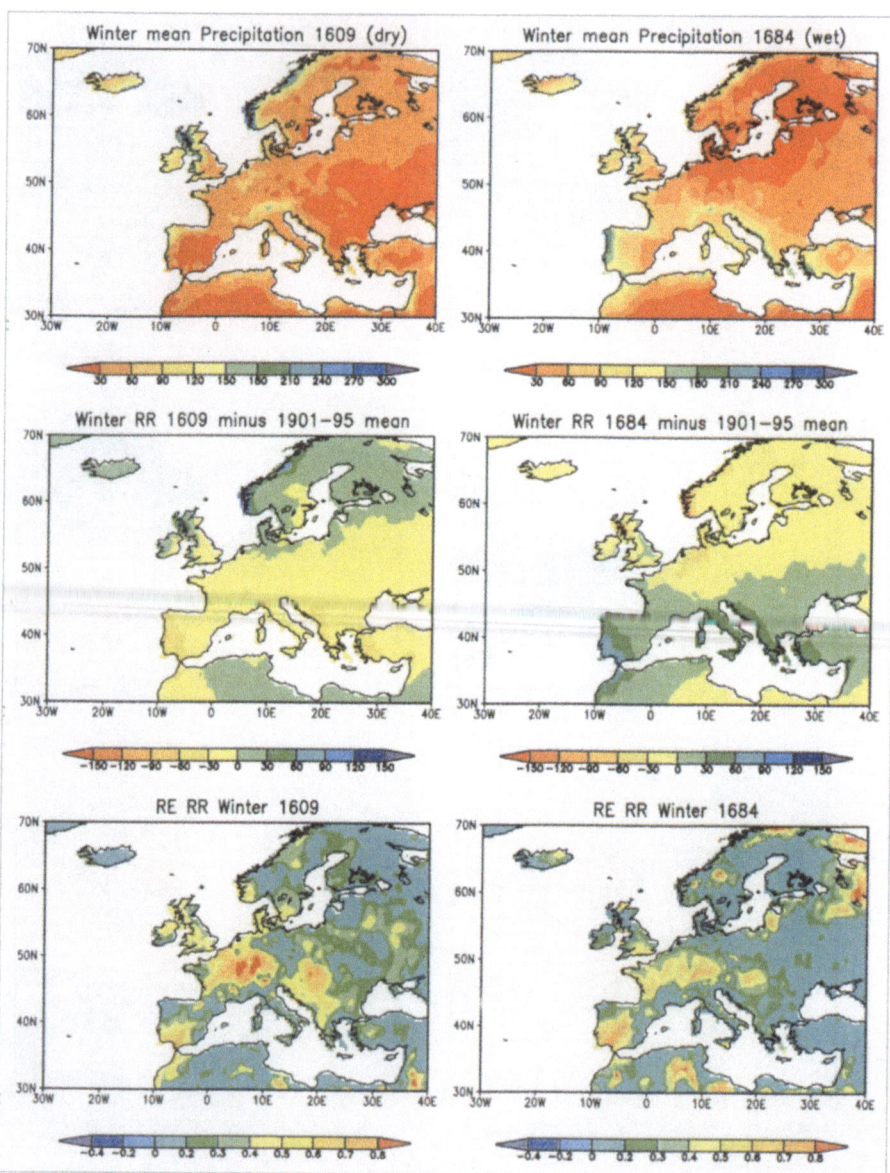

Fig. 4b. Top: Mean reconstructed precipitation fields for the driest (1609) and wettest (1684) winter (DJF) over the larger Mediterranean area. Middle: Seasonal precipitation difference patterns of the extreme winters (Winter minus long-term seasonal precipitation mean of 1901-1995). Bottom: Model performance (RE) for the anomalous winter precipitation reconstructions. For areas with RE > 0 meaningful reconstructions can be expected

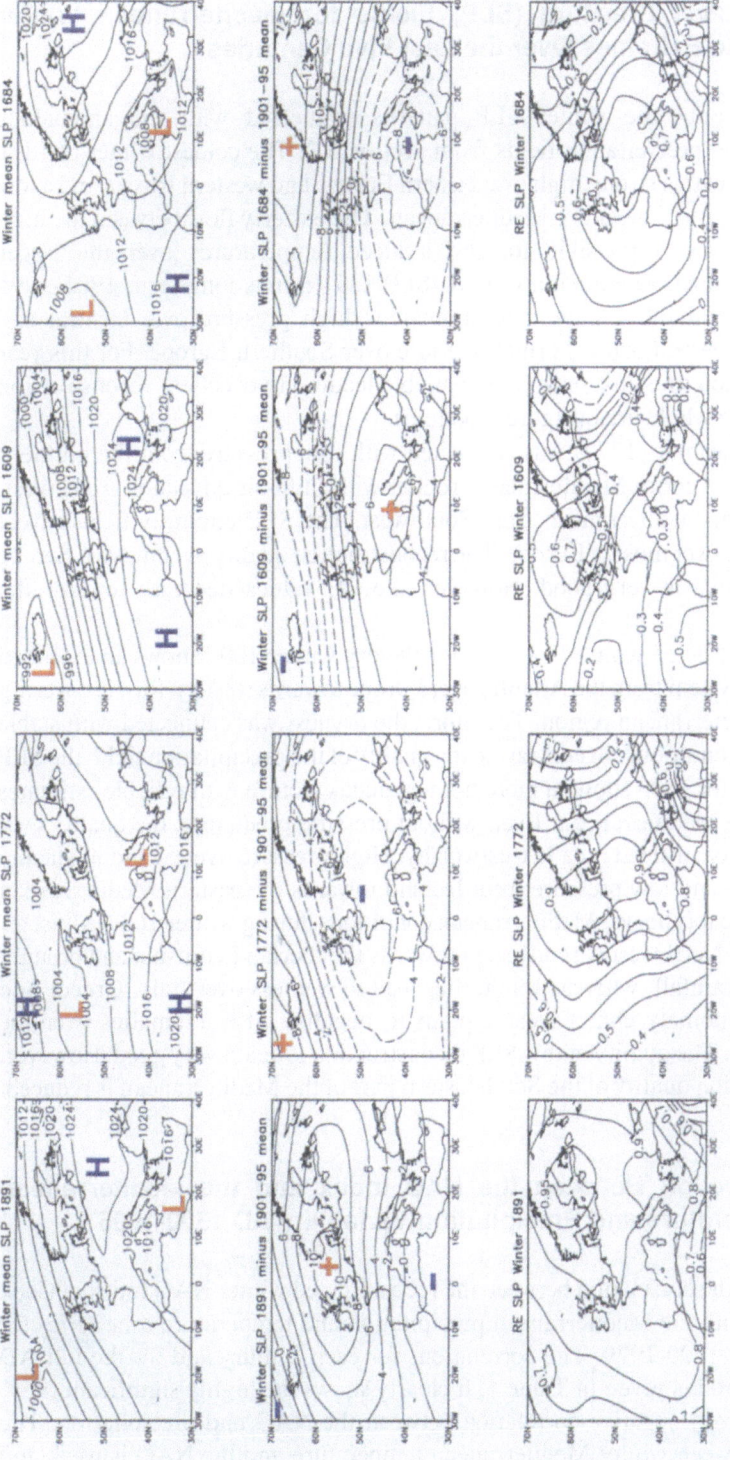

Fig. 5. Reconstructed Sea Level Pressure (SLP) fields. Top: Mean reconstructed SLP distribution for the coldest (1891), the warmest (1772), the driest (1609) and the wettest (1684) winter (DJF) over Europe (Luterbacher et al. 2001a) Middle: Seasonal SLP difference patterns of the four extreme Mediterranean winters (winter minus long-term seasonal SLP mean of 1901–1995). Continuous lines mark positive SLP deviations, and dashed lines negative SLP deviations. The contours are drawn at 2 hPa intervals. Bottom: Model performance (RE) for the four reconstructed winter SLP fields. For areas with RE > 0 meaningful reconstructions can be expected.

3.3 Sea Level Pressure (SLP) Fields for Mediterranean Winter Climate Extremes Over the Last Five Centuries

Figure 5 presents the winter SLP patterns connected with the anomalous temperature and precipitation fields from section 3.2. The coldest winter of 1891 was connected with a strong high over eastern Europe and western Russia and a low pressure system over the entire Mediterranean. The easterly flow between the high and the low was responsible for the reduced temperatures over the larger Mediterranean area. The anomaly chart (SLP 1891 minus long-term 1901-1995 mean; Figure 5 middle) shows the anomalous high pressure over Central and Northern Europe and below normal pressure over Southern Europe. For this year around 50 station pressure series are available leading to excellent reconstruction quality over the whole area (Figure 5 bottom).

The warm winter (1772) was connected with low pressure from the northern Atlantic over the entire Mediterranean region with mild air advection from West, over the eastern Mediterranean from Southwest. The SLP anomaly fields reveal negative values over most of Europe. The reconstruction quality over most of Europe and North Africa is very good. However, the RE values decrease towards the southeast.

The reconstructed winter SLP field for the dry winter 1609 shows an extended high pressure system from the Atlantic over Europe towards western Russia covering the whole Mediterranean region. Therefore, the dryness was connected with stable high pressure conditions. In contrast to the quality of the precipitation field, the SLP reconstruction for 1609 shows a more homogeneous picture with reliable estimates over the whole Mediterranean. Even without pressure predictors, the quality over most of the reconstructed area is trustworthy. High pressure over Scandinavia and western Russia and low pressure from Iceland towards the eastern Mediterranean was responsible for the wet Mediterranean conditions during winter 1684. Moist air from West and Northwest was advected towards the Iberian Peninsula and brought above normal rainfall, whereas southerlies lead to wetness over Italy, Greece and Turkey. The anomaly charts clearly point to negative SLP anomalies over the Mediterranean. The quality of this SLP reconstruction is again very good. However, the reconstruction quality of the Southeastern part of the Mediterranean is reduced.

3.4 Connection Between the NAO Index and the Mediterranean Temperature and Precipitation Variability AD 1500-1995

We calculated the correlation between the reconstructed winter NAO Index of Cook et al. (2001) with the Mediterranean precipitation and temperature time series for the period AD 1500-1979. The correlations for each century and for the full AD 1500-1979 period is given in Table 1. It clearly shows the highly significant (95% significance level) negative correlation between the NAO and precipitation. The connection between winter Mediterranean temperature and the NAO is weak and only slightly significant during the last centuries.

4 Discussion

The averaged winter Mediterranean temperature and precipitation time series (Figures 2 and 3) clearly showed periods with extended dry/wet and warm/cold periods. A physical and climatological interpretation of these time series requires information about their quality. The model performance of these reconstructions over the whole of Europe and North Africa is discussed in detail in Luterbacher et al. (2001a) and Luterbacher et al. (2001c). For the Mediterranean as a whole, the reconstruction quality increases from the 16th to the 20th century gradually. The increase of model performance is mainly a function of the number and the distribution of the predictors. For temperature generally more trustworthy estimates were obtained.

The overall model performance expressed by the mean RE (averaged over 2159 gridpoints) gives an indication how well, on average, the temperature and precipitation fields are reconstructed over the whole area. The time series indicate rather low RE values for the seasonal (AD 1500 to AD 1658) reconstructions, though still positive.

This mainly reflects the low number of predictors available for this early period. For the Mediterranean area there is only the precipitation predictor from Andalusia (Southern Spain, Rodrigo et al. 2001) from 1501 onwards available. Therefore, the averaged positive RE values for the early reconstructions are an indication that remote predictors (from Central and Eastern Europe) can significantly contribute to skillful temperature and precipitation reconstructions over the larger Mediterranean area.

From the beginning of the 18th century, with more temperature and precipitation predictors, the quality of the monthly reconstructions increases. Provisional results of time-dependent uncertainty ranges of the estimated temperature and precipitation series reveal similar performance than of e.g. Briffa et al. (2001a) which are based on tree-ring data. The current level of uncertainty in the first centuries can be reduced with the inclusion of further new early instrumental and proxy data from areas sensitive to the Mediterranean area.

The spatial model performance gives insight in which areas reconstructions are trustworthy. This has been demonstrated for the four winter extremes (Figures 4 and 5). Although the overall model performance for the wettest and driest winters implies low reconstruction quality (not shown), there are regions within the Mediterranean with skillful reconstructions. Therefore both, the overall and the spatial model performance have to be considered when reconstructed temperature and precipitation fields are climatologically interpreted.

A large fraction of precipitation and temperature variability can be explained by large-scale circulation anomalies when averaging in space and time over adequately large enough intervals (Trenberth 1990; Xu 1993; Qian et al. 2000; Xoplaki et al.

2001b). Thus, on a wider scale, changes in the temperature and precipitation distribution are influenced by changes in the atmospheric circulation, which in turn alter the storm tracks and the temperature and precipitation distribution.

Over the last five centuries the connection between the mean winter precipitation over the Mediterranean and the NAO reconstructed by Cook et al. (2001) turned out to be stable with highly negative correlations throughout the whole period (Table 1). This might be explained by a similar response of the Mediterranean precipitation behavior on the NAO phase. These findings are in accordance with van Loon and Rogers (1978), Lamb and Peppler (1987), Hurrell (1995, 1996), Cullen and de Menocal (2000) and Xoplaki et al. (2001b) who found negative correlations between the winter NAO and station precipitation over most of the Mediterranean area and its vicinity (except for the most southeastern part of the Mediterranean) for the 20th century. Our results suggest, that the close relationship between winter NAO and winter mean Mediterranean precipitation observed in the 20th century may have been robust at least back to AD 1500.

The low correlation between the winter NAO and the Mediterranean winter temperature time series over the last 496 years (Table 1) might be interpreted in terms of compensatory effects between the areas with negative correlations with the NAO (Eastern Mediterranean), areas with non significant correlations (Western and Central Mediterranean), and areas with positive correlations which has been found for parts of the 20th century (Cullen and deMenocal 2000; Xoplaki et al. 2001b). Xoplaki et al. (2001b) used around 250 station time series covering the 50 year period 1949-1999 and found, that the Eastern Mediterranean (Greece, Turkey, Libya, Egypt, Syria, Jordan, Cyprus, Israel) temperature is negatively correlated with the NAO, whereas stations mainly North of the Mediterranean Sea (North of 40°N and East of 5°E) show a positive connection to the NAO. Stations on the Iberian Peninsula, over Italy and Morocco, Algeria and Tunisia return non-significant correlations with respect to wintertime temperature. With a reduced data set back to 1900, the main features have been confirmed.

These results suggest to divide the Mediterranean region in various sub areas and conduct the analysis again. First results indicate, that the NAO plays an important role in winter variability and leaves a significant signature in temperature in the eastern Mediterranean whereas the connection with the western and central Mediterranean is small over the whole 496 years. It may be related also to not constant transfer functions over time and the extent to which climatic modes from the 20th century where the statistical model is fitted, represents the entire range of climate variability.

Mann (2001) formulated a yearly NAO index back to the middle of the 18th century in terms of large-scale surface temperature fields. He also found, that interannual temperature variability in the Near and Middle East region (30°-40°N; 20°-50°E) appears to be closely related to patterns of variation associated with this temperature based NAO.

However, other patterns seem to be of more importance on multidecadal and secular time scales. This is in agreement with Felis et al. (2000) who recently showed that coral records from the Northern Red Sea provides evidence that interactions between tropical and extratropical modes of the global climate system had an important control on Middle East climate variability on interannual and longer timescales since at least AD 1750. They found that colder periods are accompanied by more arid conditions in the northern Red Sea but increased precipitation in the southeastern Mediterranean region, whereas warm periods are accompanied by decreased rainfall in the latter and less arid conditions in the northern Red Sea.

Table 1. Correlation between the reconstructed winter NAO Index of Cook et al. (2001) with the averaged Mediterranean precipitation and temperature time series for the period AD 1500-1979. The asterix (*) denotes statistically significant on the 95% level

Period	NAOI - Precipitation	NAOI – Temperature
1500-1599	-0.26*	-0.05
1600-1699	-0.31*	-0.12
1700-1799	-0.48*	-0.22*
1800-1899	-0.50*	-0.36*
1900-1995 (1900-1979)	-0.65*	-0.32*
1500-1995 (1500-1979)	-0.44*	-0.22*

5 Conclusions and Outlook

We derived a winter mean Mediterranean precipitation and temperature time series from AD 1500-1995 through averaging the 2159 gridpoints over the larger Mediterranean land area (10°W-40°E; 30°N-47°N) (Figures 2 and 3). These time series are derived from principal component regression based reconstructions of temperature and precipitation fields over the land area (30° W - 40° E; 30° N - 70° N) based on long instrumental and proxy data (Luterbacher et al. 2001c). The two time series clearly showed several cold relapses and warm intervals as well as dry and wet periods on the decadal timescale, on which shorter-period quasi-oscillatory behaviour was superimposed.

Rather cool and wet conditions have been found at the end of the 16th century and during the Maunder Minimum (second half of the 17th century). It was rather dry and cool during the second part of the 19th century. The distinct trend towards Mediterranean winter dryness during the last few decades of the 20th century is unique in the whole 500 year precipitation time series.

The reconstruction quality is generally better for temperature than for precipitation. Seasonal reconstructions are less trustworthy than the monthly estimates, but they still show some skill over various Mediterranean areas allowing qualitatively and quantitatively statements about the temperature and precipitation course over the last few centuries.

The patterns of the coldest (1891), the warmest (1772), the driest (1609) and the wettest (1684) Mediterranean winters over the last few centuries have been shown together with the departures from the 20th long-term mean and the quality of the reconstructions. Clear temperature and precipitation anomalies have been found with inverse sign for Southern Europe/Northern Africa and Northern Europe. Although only a few predictors were available for the larger Mediterranean for the driest and wettest winters, there is potential for skillful reconstructions for some Mediterranean regions.

The reconstructed sea level pressure fields (Luterbacher et al. 2001a) related to these extremes clearly showed the expected anomalies with cold/warm, and moist/dry air advection towards the Mediterranean.

We found that the relationship between the reconstructed winter NAO and the mean winter Mediterranean precipitation remained stable over the last 5 centuries. In periods with stronger westerlies over the eastern North Atlantic, the Mediterranean area, as a whole, obtained less precipitation and vice versa.

The North Atlantic Oscillation it is not the most important signal of atmospheric variability concerning temperature over the averaged larger Mediterranean region. This mainly reflects the fact that the averaged mean temperature value includes both areas with a strong link to the NAO and others with no connection at all. Their competitive effects may be the reason for the low correlations over the last 500 years.

Preliminary results dividing the Mediterranean into sub-regions indicate close relation between the winter NAO and winter temperature over the Eastern Mediterranean whereas for the Western and Central parts this is not the case.

The presented continuous temperature and precipitation reconstructions over Europe, North Africa and the Near East are of great importance for all kind of climatic analyses dealing with natural variabilities on time scales from months to decades or centuries. They can be used to study the low and high frequency variability and the characteristics and extremes of climate and the relationship with the atmospheric circulation and can be compared with model-produced reconstructions of natural and forced (external and internal) variability for the last centuries.

Further investigations will include new predictor data in order to improve the reconstruction skill, to provide time-dependent uncertainty ranges of our estimates, to study the influence of forcing factors, and to study the low frequency climate variability for different climate sensitive regions along the Mediterranean for all the four seasons over the last centuries and the comparison with reconstructions from other parts of the Northern Hemisphere.

These reconstructed temperature and precipitation fields might be of use by researchers dealing with regional climate modelling, climate change impact assessment and ecosystem modelling.

Acknowledgements

The Global Climate Data (Version 1; gridded temperature and precipitation) has been supplied by the Climate Impacts LINK Project (UK Department of the Environment Contract EPG 1/1/16) on behalf of the Climatic Research Unit, University of East Anglia. The authors are grateful for using predictor data from various sources (GHCN2, NCAR, the IMPROVE, ADVICE, FLOODRISK and other projects) and many other persons providing their valuable instrumental or proxy data.

6 References

Alcoforado MJ, Nuñes MF, Garcia JC, Taborda JP (2000) Temperature and precipitation reconstructions in southern Portugal during the Late Maunder Minimum (1675 to 1715). The Holocene 10: 333-340

Barriendos M (1997) Climatic variations in the Iberian Peninsula during the Late Maunder Minimum (AD 1675-1715): an analysis of data from rogation ceremonies. The Holocene 7: 105-111

Briffa KR, Osborn TJ, Schweingruber FH, Jones PD, Shiyatov SG, Vaganov EA (2001a) Low-frequency temperature variations from a northern tree-ring density network. J Geophys Res 106:2929-2941

Briffa KR, Osborn TJ, Schweigengruber FH, Jones PD, Shiyatov SG, Vaganov EA (2001b) Tree ring width and density data around the Northern Hemisphere: Part 1, local and regional climate signals. The Holocene. In press

Camuffo D, Enzi S (1991) Locust Invasions and Climatic Factors from the Middle Ages to 1800. Theor Appl Climatol 43: 43-73

Camuffo D, Enzi S (1992) Reconstructing the Climate of Northern Italy from Archive Sources, pp. 143-154. In: Bradley RS, Jones PD (editors): "Climate since 1500 A.D.", Routledge, London

Camuffo D, Enzi S (1994) The Climate of Italy from 1675 to 1715, pp.243-254. In: Frenzel B (editor): "Climatic Trends and Anomalies in Europe 1675-1715", Paleoclimate Research, Special Issue 8, Fischer Verlag, Stuttgart

Camuffo D, Enzi S (1995) Climatic Features during the Spörer and Maunder Minima, pp. 105-125. In: Frenzel B (editor): "Solar Output and Climate during the Holocene", Paleoclimate Research, Special Issue 16, Fischer Verlag, Stuttgart

Camuffo D, Secco C, Brimblecombe P, Martin-Vide J (2000) Sea Storms in the Adriatic Sea and the Western Mediterranean During the Last Millennium. Clim Change 46: 209-223

Cook ER, D'Arrigo R, Mann M (2001) A well-verified, multi-proxy reconstruction of the winter North Atlantic Oscillation Index since AD 1400. Submitted

Cullen HM, deMenocal PB (2000) North Atlantic influence on Tigris-Euphrates streamflow. Int J Climatol 20: 853-863

D'Arrigo RD, Cullen HM (2001) A 350-year (AD 1628-1980) reconstruction of Turkish precipitation: Linkages to the NAO. Monograph volume of the Chapman conference on the North Atlantic Oscillation (NAO), AGU, American Geophysical Union

Felis T, Pätzold J, Loya Y, Fine M, Nawar AH (2000) A coral oxygen isotope record from the northern Red Sea documenting NAO, ENSO, and North Pacific teleconnections on Middle East climate variability since the year 1750. Paleoceanography 15: 679-694

Glaser R et al (1999) Seasonal temperature and precipitation fluctuations in selected parts of Europe during the sixteenth century. Clim Change 43: 169-200

Grove JM, Conterio A (1994) Climate in the Eastern and Central Mediterranean, pp. 275-285. In: Frenzel B (editor): "Climatic Trends and Anomalies in Europe 1675-1715", Paleoclimate Research, Special Issue 8, Fischer Verlag, Stuttgart

Grove JM, Conterio A (1995) The climate of Crete in the sixteenth and seventeenth centuries. Clim Change 30: 223-247

Grove AT (2001) The 'Little Ice Age' and its geomorphological consequences in Mediterranean Europe. Clim Change 48: 121-136

Guiot J (1991) The combination of historical documents and biological data in the reconstruction of climate variations in space and time. In Frenzel B, Pfister C, Gläser B (eds.) Paläoklimaforschung/Paleoclimate Research 7, Special Issue EFS Project "European Climate and Man" 2

Hurrell IW (1995) Decadal trends in the North Atlantic oscillation: Regional temperatures and precipitation. Science 269: 676-679

Hurrell JW (1996) Influence of variations in extratropical wintertime teleconnections on Northern Hemisphere temperature. Geophys Res Lett 23: 665-668

Jones PD, Briffa KR, Barnett TP, Tett SFB (1998) High-resolution palaeoclimatic records for the last millennium: interpretation, integration and comparison with General Circulation Model control-run temperatures. The Holocene 8:455-471

Kadioğlu M, Tulunay Y, Borhan Y (1999) Variability of Turkish precipitation compared to El Niño events. Geophys Res Lett 26: 1597-1600

Lamb PJ, Peppler RA (1987) North Atlantic Oscillation: Concept and application. Bull Amer Meteor Soc 68: 1217-1225

Lorenz EN (1956) Empirical Orthogonal Functions and Statistical Weather Prediction, M.I.T. Statistical Forecasting Project Report No. 1, Contract AF 19, (604)-1566

Luterbacher J, Xoplaki E, Dietrich D, Rickli R, Jacobeit J, Beck C, Gyalistras D, Schmutz C, Wanner H (2001a) Reconstruction of Sea Level Pressure fields over the Eastern North Atlantic and Europe back to 1500. Clim Dynamics, in press

Luterbacher J, Xoplaki E, Dietrich D, Jones PD, Davies TD, Portis D, Gonzales-Rouco JF, von Storch H, Gyalistras D, Casty C, Wanner H (2001b) Extending Highly Resolved NAO Reconstructions Back to 1500. Atmos Sci Lett, doi:10.1006/asle.2001.0044

Luterbacher J et al. (2001c) European temperature variability over the last 500 years. Submitted

Mann, ME (2001) Large-scale climate variability and connections with the Middle East in past centuries. Clim Change, in press.

Martin-Vide J, Barriendos M (1995) The use of rogation ceremony records in climatic reconstruction: a case study from Catalonia (Spain). Clim Change 30: 201-221

New M, Hulme M, Jones PD (1999) Representing Twentieth-Century space-time climate variability. Part I: Development of a 1961-1990 mean monthly terrestrial climatology. J Climate 12: 829-856

New M, Hulme M, Jones PD (2000) Representing Twentieth-Century space-time climate variability. Part II: Development of 1901-1996 monthly grids of terrestrial surface climate. J Climate 13: 2217-2238

Pfister C, Kington J, Kleinlogel G, Schüle H, Siffert E (1994) The creation of high resolution spatio-temporal reconstructions of past climate from direct meteorological observations and proxy data. Methodological considerations and results, pp. 329-376. In Frenzel B, Pfister C, Glaeser B (eds) *Climatic trends and anomalies in Europe 1675-1715*, Paleoclimate Research, Special Issue 8, Fischer Verlag, Stuttgart.

Pfister C, Luterbacher J, Schwarz-Zanetti G, Wegmann M (1998) Winter air temperature variations in western Europe during the Early and High Middle Ages (AD 750-1300). The Holocene 8: 535-552

Price C, Stone L, Huppert A, Rajagopalan B, Alpert P, (1998) A possible link between El Niño and precipitation in Israel. Geophys Res Lett 25: 3963-3966Qian B., Corte-Real J, Xu H (2000) Is the North Atlantic Oscillation the most important pattern for precipitation in Europe? J Geophys Res 105: 11901-11910

Qian B, Corte-Real J, Xu H (2000) Is the North Atlantic Oscillation the most important pattern for precipitation in Europe? J Geophys Res 105:11901-11910

Rodó X, Baert E, Comin FA (1997) Variations in seasonal rainfall in Southern Europe during the present century: relationships with the North Atlantic Oscillation and the El Niño-Southern Oscillation. Clim Dynamics 13: 275-284

Rodrigo FS, et al. (2001) A reconstruction of the winter North Atlantic Oscillation Index back to AD 1501 using documentary data in Southern Spain. J Geophys Res 106:14805-14818

Serre-Bacher F, Guiot (1987) Summer temperature changes from tree-rings in the Mediterranean area during the last 800 years. In Berger W, Labeyrie L (eds.), Reidel, Dordrecht, pp 89-98

Serre-Bacher F, Guiot J, Tessier L (1992) Dendroclimatic evidence from southwestern Europe and northwestern Africa. In Bradley R, Jones PD (eds.) Climate since A.D. 1500. Routledge, London

Touchan R, Meko D, Hughes MK (1999) A 396-year reconstruction of precipitation in southern Jordan. J Americ Wat Res Ass 35: 49-59

Trenberth KE (1990) Recent observed interdecadal climate changes in the Northern Hemisphere. Bull Amer Meteorol Soc 71: 989-993

van Loon H, Rogers JC (1978) The seesaw in winter temperatures between Greenland and northern Europe. Part I: General descriptions. Mon Wea Rev 106: 296-310

Xoplaki E, Luterbacher J, Burkard R, Patrikas I, Maheras P (2000) Connection between the large-scale 500 hPa geopotential height fields and precipitation over Greece during wintertime. Clim Res 14: 129-146

Xoplaki E, Maheras P, Luterbacher J (2001a) Variability of Climate in Meridional Balkans during the periods 1675-1715 and 1780-1830 and its impact on human life. Clim Change 48: 581-615

Xoplaki E et al. (2001b) Winter precipitation and summer temperature variability over the whole Mediterranean basin during the second half of the 20th century. In preparation

Xu JS (1993) The joint modes of the coupled atmosphere-ocean system observed from 1967 to 1987. J Climate 6: 816-838

Long Term Variability of Sea Level and Atmospheric Pressure in the Mediterranean Region

F. Raicich

1 Introduction

The Mediterranean is a bordering region that feels the dynamics of the circulation of both the mid latitudes and the Tropics. The seasonal cycle regulates the transition from the winter regime, which is dominated by the mid-latitude westerly flow, although strongly modified by local orography, to the summer regime, when the westerlies weaken and a meridional circulation develops over the eastern basin. This paper summarizes some features of atmospheric pressure and sea level variability in the Mediterranean region at different time scales over the last 130 years.

2 Data and Methods

Time series of atmospheric pressure (AP) and sea level (SL) were formed for locations in the Mediterranean region. The locations were selected on the basis of the availability of SL time series, which were taken from the Permanent Service for Mean Sea Level data bank (Spencer and Woodworth, 1993). Time series were formed on a regional basis using the available stations, preferably if associated to a Revised Local Reference (RLR); data from nearby stations were intercompared and normalized to each other to obtain as complete and long series as possible. As a result eight SL time series were obtained, constructed from the time series of Tarifa (representing Alboran Sea), Alicante, Marseille and Genoa (Northwest Mediterranean), Trieste and Venice (North Adriatic), Split (East Adriatic), Varna and Burgas (West Black Sea), Sevastopol (itself) and Alexandria (South Levantine). In the case of South Levantine, based on Alexandria data, no RLR is available, however the series was retained since it appears to be homogeneous and is the only one available in that important region of the Mediterranean Sea. Finally, East Adriatic and West Black Sea were excluded from further analyses since the time series are very similar to and shorter than North Adriatic and Sevastopol, respectively. For the selected location (Alboran, NW Med, N Adriatic, Black Sea and S Levantine) time series of AP were extracted by interpolation from the UK MetOffice $5° \times 5°$ gridded data set (Jackson, 1986). In this paper variability is examined for interannual and interdecadal time scales. All the time series were filtered to separate those time scales by means of Ng and Young's (1990) technique; the cutoff period was set at about 10 years. However, since the SL is affected by a "secular" trend, time scales longer than about 100 years have been preliminarily removed from the time series. Due to the absence of information, the effects of possible vertical land motion on the SL time series could not be checked and corrected, therefore, in the discussion, it is assumed that SL changes are really related to SL only, but this assumption has to be taken cautiously. For all the parameters standardized monthly and seasonal anomalies were computed relative to the period 1958-87, for which enough SL data is available for all locations to establish a common climatological period.

3 Discussion

The undetrended SL time series show the well known increasing trend of about 1-1.5 cm/decade, but also reveal that during the last 30-40 years the tendency has changed, with a SL decrease rate of 0. 5 -1.5 cm/decade. By comparing SL and AP trends, at least part of the trend inversion can be ascribed to the recent AP increase that affected the Mediterranean region mainly in winter. In fact, overall, an inverse relationship is observed between AP and SL at all locations. It is related to the static inverse barometer effect (IB) and is more pronounced in winter and in the NW Med and Adriatic, where in this season the water column is generally barotropic. In other seasons and/or locations SL changes are affected to a large extent also by baroclinic processes, as the presence of stratification, particularly in summer, and, in the case of Alboran, the water exchange through the Gibraltar Strait. In winter the IB is effective both at interannual and, although weaker, interdecadal time scales at all location except Alboran, where at interdecadal time scales the IB is stronger than at interannual time scales. In summer the IB is negligible at interdecadal time scales, indicating that baroclinic processes play the most important role, and in other seasons it is observed only in the Adriatic.

AP and SL in the Mediterranean region are teleconnected with large scale patterns of the atmospheric circulation characteristic of surrounding regions. In winter a major climatic pattern is the North Atlantic Oscillation (NAO, Hurrell, 1995). The AP-NAO teleconnection is highest at the western end of the Mediterranean (Alboran) and decreases eastwards, although remaining significant all over the basin. This occurs both at interannual and interdecadal time scales. Concerning the SL-NAO connection, it is mainly an indirect effect of the AP-SL correlation. In summer the Mediterranean participates in the circulation of the Tropics (Ward, 1996; 1998; Raicich et al., 2001). It is in fact anticorrelated at interannual time scales with the interannual anomalies of the precipitation regimes of India and the Sahel, defined by July-September rainfall, with different characteristics for the western and eastern Mediterranean basins. In the eastern basin AP is anticorrelated with both precipitation regimes in July-September, while in the western basin it is highest positively correlated in September-November, that is later than the main rainfall period. SL is highest anticorrelated with both precipitation regimes in September-November all over the basin. This means that, while in the western basin the teleconnection can be considered indirect, being mediated by the IB, in the eastern basin SL may exhibit a delayed response to the summer circulation via baroclinic processes, such as heat storage.

Teleconnections with India and Sahel precipitation regimes are also found with the Mediterranean Pressure Index (MPI, Raicich et al., 2001), that is an East-West pressure difference related to the geostrophic component of the Etesian wind regime, typically affecting the Mediterranean in summer.

References

Hurrell JW (1995) Decadal trends in the North Atlantic Oscillation: Regional temperatures and precipitations. Science, 269:676-679.

Jackson M (1986) Operational Superfiles in Met. 0. 13. Met. 0. 13 Technical Note 25.

Ng CN, Young PC(1990) Recursive estimation and forecasting of non-stationary timeseries. J. Forecasting, 9:173-204.

Raicich F, Pinardi N, Stevanato A, Navarra A (2001) Teleconnections between Indian Monsoon and Sahel rainfall and the Mediterranean. International Journal of Climatology, submitted.

Spencer NE, Woodworth PL (1993) Data holdings of the Permanent Service for Mean Sea Level (November 1993). Bidston, Birkenhead, Permanent Service for Mean Sea Level, 81 pp.

Ward MN (1996) Local and remote climate variability associated with East Mediterranean sea surface temperature anomalies. Proceeding, Conf on Mediterranean Forecasting, La Londe Les Maures, France, European Science Foundation, pp. 16-25.

Ward MN (1998) Diagnosis and short-lead time prediction of summer rainfall in Tropical North Africa at interannual and multidecadal timescales. J. Climate, 11, 3167-3191.

Annual and Seasonal Century Scale Trends of the Precipitation in the Mediterranean Area During the Twentieth Century

A. Douguédroit and C. Norrant

1 Introduction

A global climate warming between 0.4 and 0.8 °C has been assessed during the last century (IPCC WGI Third Assessment Report, 2001), but much uncertainty remains about the evolution of the precipitation during the same period; the mediterranean case is widely unknown. Here is intended to determine the precipitation century-scale trends in the Mediterranean area and to search for the representativity of regional trends versus station ones and of last century trends versus longer ones.

2 Data and Method

The study concerns the Mediterranean area from the Atlantic coast to the Near East using annual and seasonal data (except summer which is dry in most of the area), from 1915 to 1988. 41 stations are unequally spread over the area but shorter series have been added to fill spatial gaps. Mann-Kendall test has been used to check the homogeneity of the stations.

Rotated Principal Component Analyses (RPCA: Richman 1985), R type, have been applied with the stations as variables and the precipitation totals of the seasons or the year as observations. Classes obtained after rotation have been mapped according to their loadings. They represent the regionalisation of the precipitation evolution. Their scores have been used as the precipitation history of each region. Linear trends have been calculated as linear regressions on scores and corresponding 7 years moving averages. The signification of the trends at the 5 % level have been checked with Spearman –u(r)- and Kendall –u(t)- tests. Trends have also been calculated and tested for isolated stations and station series longer than a century.

3 Regionalisation of the Precipitation

According to the seasons, the Mediterranean lands from the Atlantic coast to the

Near East are divided into 6 to 8 regions which represent between 62 and 67% of explained variance. The general features of the regional distribution of the precipitation are common to the year and the three rainy seasons, from autumn to spring. Two regions extending in latitude over the area concern its western and eastern limits, the Atlantic region and the East Mediterranean. In-between, the Mediterranean is divided between north and south coasts, each subdivided into several regions (Figure 1).

4 Regional Trends

All the trends of the regional precipitation (the scores of the RPCA factors) are non significant according to Spearman and Kendall tests except for the annual evolution in Greece. But some trends of the moving averages are significant for the year or one or two seasons in some regions (Table 1 and Figure 2).

5 Comparison Between Regional and Local Trends

Annual regional and local (station) trends have been compared to determine the interest of regional trends calculated from scores obtained by RPCA. In regions where score trends are non significant the following different possibilities can occur:

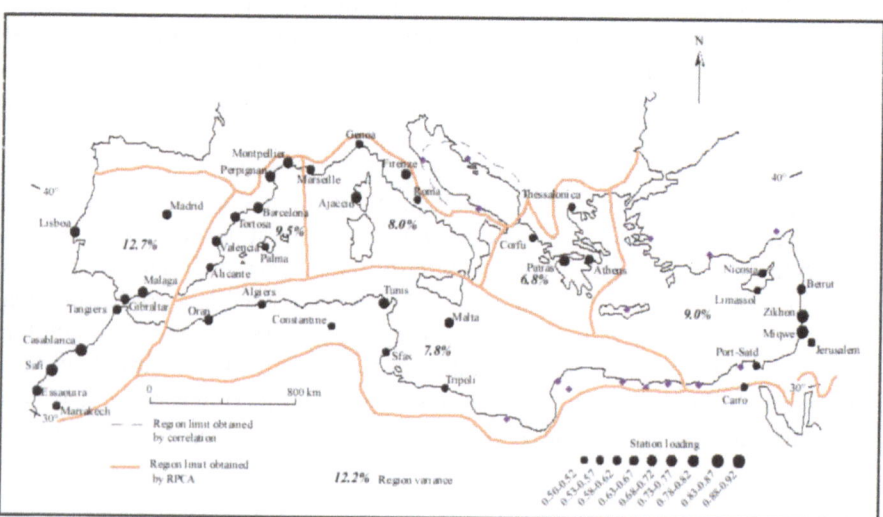

Fig. 1. Regionalisation of the annual precipitation (diamond-shaped points: stations with shorter series)

Table 1: Regional trends of annual and seasonal precipitation. +: increasing trend, –: decreasing trend. On the right: Spearman test, on the left: Kendall test. In bold: significant test.

Regions	Year scores		Year moving averages		Autumn scores		Autumn moving averages		Winter scores		Winter moving averages		Spring scores		Spring moving averages	
Atlantic region	+	+	**+**	**+**	–	–	**–**	**–**	+	+	**+**	**+**	–	–	–	–
North Africa	+	+	**+**	**+**	+	+	+	+	–	– (Libya –)	–	– (Libya –)	–	–	–	–
Near East	–	–	–	–	+	+ (Egypt –, Cyprus +)	**+**	**+** (Egypt +, Cyprus +)	**–**	**–**	**–**	**–**	+	+ (Cyprus and Cairo –)	**+**	**+** (Cyprus and Cairo –)
Gulfs of Lions and Valencia	–	–	–	–	–	–	–	–	–	–	–	–	+	+	**+**	**+**
Greece	**–**	**–**	**–**	**–**	+	+	+	+	–	–	**–**	**–**	+	+	+	–
Gulf of Genoa	–	–	–	–	**–**	– (Italy –)	**–**	**–** (Italy –)	+	+	**+**	**+**	–	–	**–**	**–**

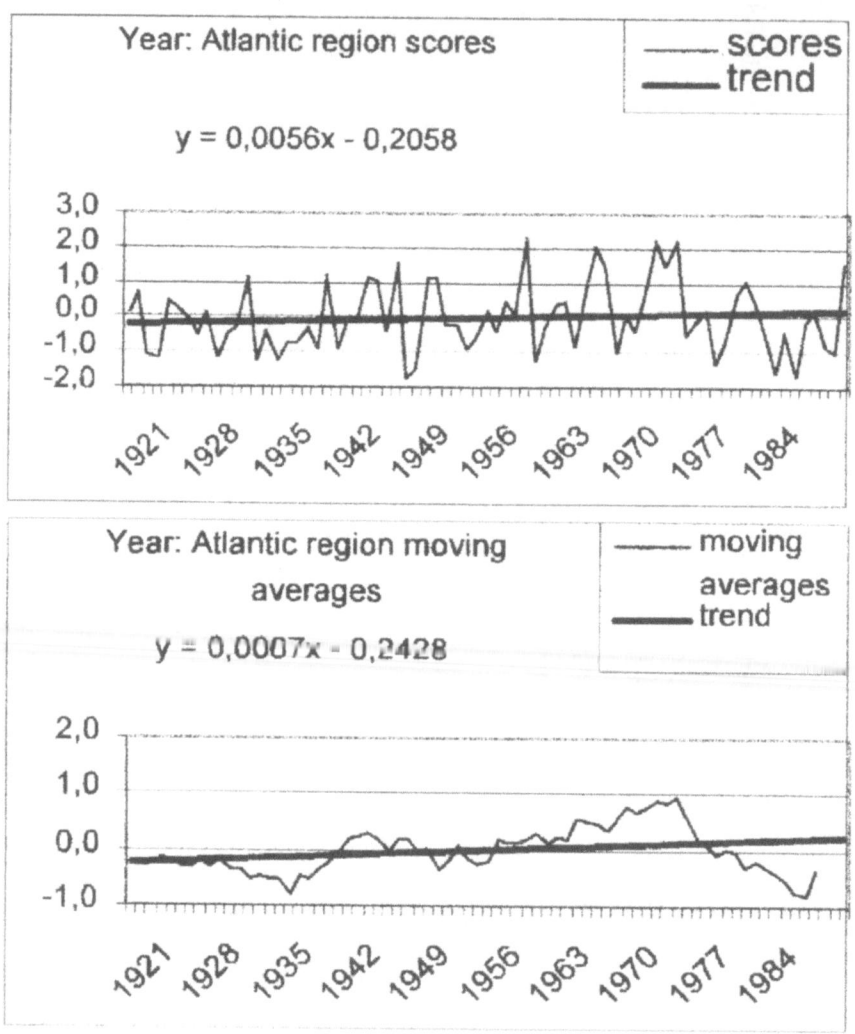

Fig. 2. Precipitation trends of the Atlantic region

▸ All the stations have no significant trend (North Africa),
▸ one station has a significant trend which is decreasing as in the region (Nicosia in the Near East, Palma in the Gulfs of Valencia and Lions), or
▸ one or two stations have a significant trend which is opposite to the regional one, but these stations are not as many as the stations having a similar trend as the whole region (Gibraltar and Tangiers in the Atlantic region).

In Greece, three among the four retained stations have non significant decreasing trends while the region has a significant decreasing trend.

When using the moving averages of the scores, the number of stations with significant trends much increases as for the regions. But relations between regional and local trends are the same as previously.

6 Comparison Between Centuries Old and Longer Trends

When station series longer than the twentieth century are available, they have been compared with the previous ones. Different possibilities occur:
* No change, significant and non significant trends remain the same (Miqwe, Algiers, Gibraltar, etc.),
* trends remain non significant but their signs change (Safi),
* several trends which were not significant during the last century become significant when the series are longer (Essaouira, Tunis, etc.).

7 Conclusion

During the twentieth century, most of the Mediterranean area had a non significant evolution of its precipitation and only Greece had a significant linear rainfall decrease at the century scale (even if each of the Greek stations had only a non significant decreasing evolution). A few isolated stations belonging to the other regions with non significant century scale trends present significant ones.

So, we can conclude that when rainfall trends are calculated from scores of a factor obtained by RPCA, they perform as averages of the area stations.

When scores are transformed by moving averages, their linear trends can be significant when score trends are not.

Some century scale trends of isolated stations become significant only if they are calculated with series longer than a century.

References

IPCC (2001) Contribution of Working Group I to the Third Assessment report of the Intergovernmental Panel on Climate Change. In: Houghton JT, Ding Y, Griggs DJ, Noguer M, Van der Linden PJ, Xiasu D (eds.). Third Assessment Report: Climate Change 2001 (The Scientific Basis). Cambridge University Press, UK, 944 p

Richman MB (1985) Rotation of principal components. Journal of Climatology 6, pp 293-335

12 Years Mediterranean Satellite Data Set and Analysis

Dirk Koslowsky

Abstract

Research conducted in the framework of the EU projects EFEDA, RESRAPS, RESMEDES, and RESYSMED stresses the need to (i) correlate results obtained at specific sites with a regional or large scale overview at the conditions in the Mediterranean Basin and (ii) to expand short term ground based experiments with operational observation of at least a few key quantities to periods of climatological relevance. Remote sensing techniques can definitely contribute to this task. NOAA AVHRR data are still the only source with a satisfying compromise between spatial and time resolution covering a period longer than a decade of years. To serve these purposes in all regions of the basin the *"Mediterranean Extended Daily One Km AVHRR Data Set* (MEDOKADS)" was developed. It provides as basic quantities reflectances and radiances in different spectral regions from which products like albedo, vegetation index, land and sea surface temperatures, cloud cover and cloud classification are derived. Twelve years of data covering the western part and six years of basin wide data allow to identify regional changes and trends related to variability of the climate. Efforts are ongoing to make the full data set directly accessible and to update it in nearly real-time which is a prerequisite for a monitoring system for the Mediterranean environment.

1 Mediterranean Extended Daily One-km AVHRR Data Set (MEDOKADS)

1.1 Design of the Data Set

The data received by the HRPT receiving station of the Free University of Berlin are used to process a daily precisely navigated data set covering the Mediterranean and the surroundings (Koslowsky, 1998, Bolle, 1998). The geographic area of interest, even if not completely filled by data, is defined by the corner coordinates:

55^0N/10^0W 55^0N/42^0E 27^0N/10^0W 27^0N/42^0E.

in latitude/longitude representation (geographic projection) with $1/100^0$ resolution. The total set of data contains 2800 lines and 5200 columns. It is subdivided for faster access by regional users into four overlapping subframes of identical 1536 pixels line length defined by the coordinates of the upper left corners at 10^0W, 5^0E, 15^0E,and 27^0E. This covers respectively the Iberian Peninsula, Italy, the Balkans, and the eastern Mediterranean.

With the calibrated AVHRR data of its channels 1 to 5 the following supplemental information is delivered: observation geometry (local zenith distance and azimuth of satellite and sun, scattering angle), an origin indicator (link to date and pass number), bitmaps (land-water, cloud mask), local time, NDVI, broad band reflectance, and split-window land/sea surface temperatures as well as cloud classification (currently in preparation). With theses data a congruent 16 channel data set is build up. Daily as well as ten days composite data sets according to the maximum local NDVI's are generated. For the years 1989 to 1994 only the western and the central part so far are processed while the whole Mediterranean is covered for the period 1995 until now amounting to about 2 Tbyte of data.

Data handling and access was found to be one of the most important prerequisite for a successful evaluation of such large data sets. As long as the data were stored only on about 500 magnetic tape cartridges it was a very time consuming job to run evaluations throughout the full archives. Thus efforts were made to built up a large array of magnetic disks in order to provide the data set in fast random access. The actual capacity is now at 2.5 Tbyte capable to store the actual state of the daily 12 years MEDOKADS data set and will be expanded to about 4 Tbyte till the end of the year 2001 to include the missing data for the eastern part of the years 1989 to 1994.

1.2 Inter-Calibration of NOAA 11 and NOAA 14

A critical issue is the necessary absolute calibration of the measured signals. No calibration information of the two short wave channels of the AVHRR are supported by the instrument. Preflight calibrations were found to be not applicable and thus targets at the earth' surface have to be used for this task as well as to control the degradation of the sensor response.

Reflectances, albedo and the NDVI are very sensitive to calibration errors and changes of the environment can be detected if artificial trends like sensor degradation are removed with great reliability. Even more than 20 years after the first mission of an AVHRR instrument these problems are not settled finally. A detailed study was performed (Koslowsky, 1996, 1997) to quantify the influences of sensor degradation, atmospheric attenuation by changing local observation time and influence of the bidirectional reflectance distribution function of the desert

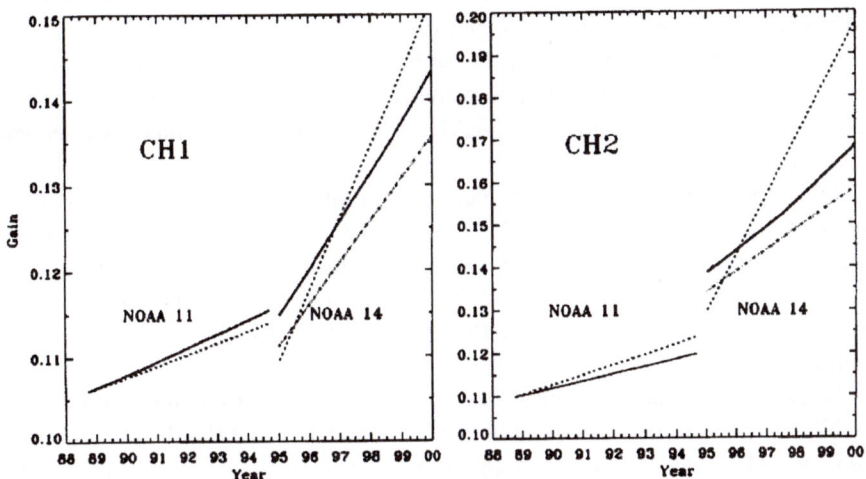

Fig. 1. Calibration gain for NOAA 11 and NOAA 14 AVHRR channel 1 (CH1) and 2 (CH2). Solid: Berlin, dashed: NOAA, dash-dotted: NOAA after correction

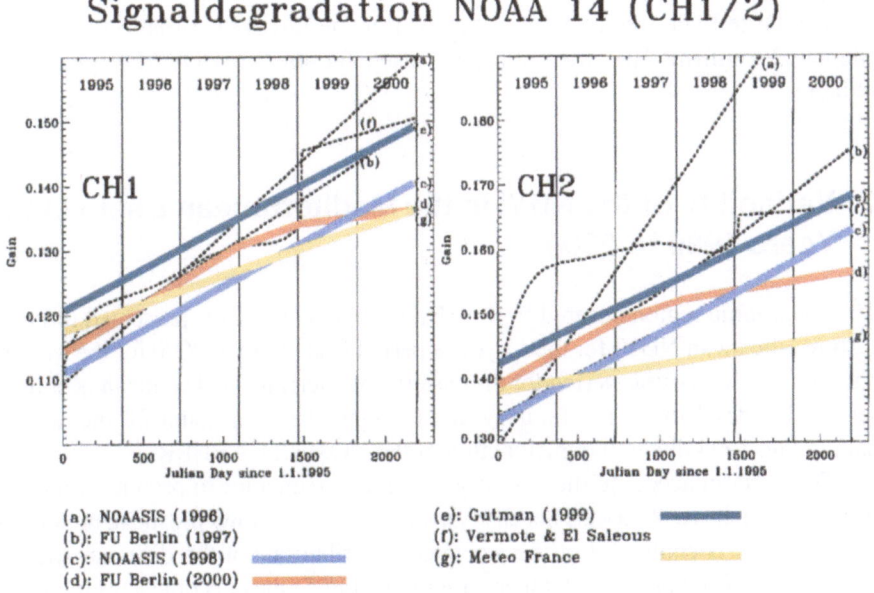

Fig. 2. Gain for channel 1 and 2 of the AVHRR of the NOAA 14 satellite using degradation formulas of different authors

calibration target. The combined effect of these factors, which can be observed directly, is called *effective signal degradation.*

The local observation times of the NOAA 14 satellite, launched in December 1994 and NOAA 11, launched in September 1988, show a close agreement. This post-launch calibration coefficients could be derived and are presented together with those for NOAA 11 in Fig. 1.

In comparison to NOAA 11 the degradation rates are considerably higher and rather comparable to those of NOAA 7 and NOAA 9. A large difference is found between the pre-launch calibration coefficients and the observed in-flight values, too. Thus the apparent sensor degradation in the first weeks of operation was very strong possibly caused by contamination of the radiometric optics by outgassing of the rocket engine.

Not so many studies are published concerning short wave channel calibration of the NOAA 14 AVHRR. But already in 1996 it was felt that the degradation rates used by NOAA for the calibration of the Level 1b data seem to be to high (Koslowsky,1997). Revised calibration coefficients are used for Level 1b product since December 1998. Evaluations by the author in 1999 and 2000 showed that a linear extrapolation of the degradation rates are not longer applicable and lower values were selected. An inter-comparison with results of other authors is shown in Fig. 2.

The first in-flight calibration and initial degradation coefficients for the NOAA 16 AVHRR instrument, the successor of NOAA 14, requires a data series of at least half year and will be derived in May 2001.

The large scale trend analysis studies over the last twelve years described in chapter 2 confirms that the calibration used for the MEDOKADS product is reliable.

2 Variability of the NDVI in the Mediterranean Basin in the Years 1989 to 2000

The geographic region covered by the MEDOKADS data set is presented in Fig 3. It shows the mean NDVI for the 12 year's period from 1989 to 2000 for the western part and for the six year period for the eastern Mediterranean. The mean is built up for the decades 1 to 26, i.e. January to mid September to account for the missing data in late 1994 due to the misfunction of the NOAA 11 AVHRR.

The random access to the data discussed above enables to produce a number products which are otherwise not easily available. Fig. 4 shows the mean broadband albedo which is composed of the weighted contributions of the different spectral reflectances. The maximum temperature maps (Fig.5) may become a useful tool to assess temperature trends. The vegetation as visualized by the normalized difference vegetation index (NDVI) is an indicator of climate variability. Fig. 6 presents the data of the occurrence of the NDVI maximum in two years, 1995 and 1996 which can be interpreted as the phase of the vegetation wave in different years. The considerable differences in the phase of the vegetation wave of successive years indicate such evaluations as useful tools for the analysis of land surface processes. The deviations of the ten year's mean values from the average of these years is presented in Fig. 7 for the Iberian Peninsula. It provides evidence that in the beginning of the nineties there was a trend towards a reduction of the biomass production while the vegetation recovered in the last years of the century. This is an example which makes it evident that long data series of the identical kind of data are necessary to identify real trends and to separate them from interannual variations. In all examples given here the data retain the full AVHRR resolution.

Even if the twelve year's period is rather short for trend analysis, it is worthwhile to study the development of NDVI and reflectances. The main interesting information is given by the overall differences and the local gradients rather than by the absolute values. Interesting features can be found if the development is represented by the slopes of regression lines for each individual pixel position at full resolution for the period under consideration. Fig. 8 shows the NDVI development for twelve years in the western part of the Mediterranean and central Europe. The slope of the regression line is given in percent per year and presented in colours. It is interesting to note that these slopes only in a few positions (such as the south-east coast of Spain and Ebro valley in opposit directions) depart significantly from zero which is presented in yellow color. In the upper part of Fig. 9 the development for the period from 1989 to 1995 is shown for the same area and in the lower part of Fig. 9 for the period 1995 to 2000 in the whole Mediterranean Basin and central Europe.

Most interesting are the striking inverse developments not only for the Ibearian Peninsula and northwest Africa but also for western and central Europe. This example gives an impression of the larger scale synergies of climate variability.

The moderate spatial resolution of about 1 km furthermore allows to zoom into local events like forest fires as presented in Fig. 10.

3 The MEDOKADS Data Set As A Substantial Part of A Remote Sensing Data Network for A Mediterranean Research and Applications Network

The MEDOKADS data was developed at the Free University of Berlin where the HRPT AVHRR data receiving station has been working since nearly 20 years. To ensure the further access to the archived data in case of a failure of the system and to extend the data set in a near real-time manner, efforts are under way to built up a network of national institutions working with AVHRR data. On the one hand there have been made arrangements between the receiving stations at the German Remote Sensing Data Center (DFD) of the DLR, GKSS and Berlin to supply each other with

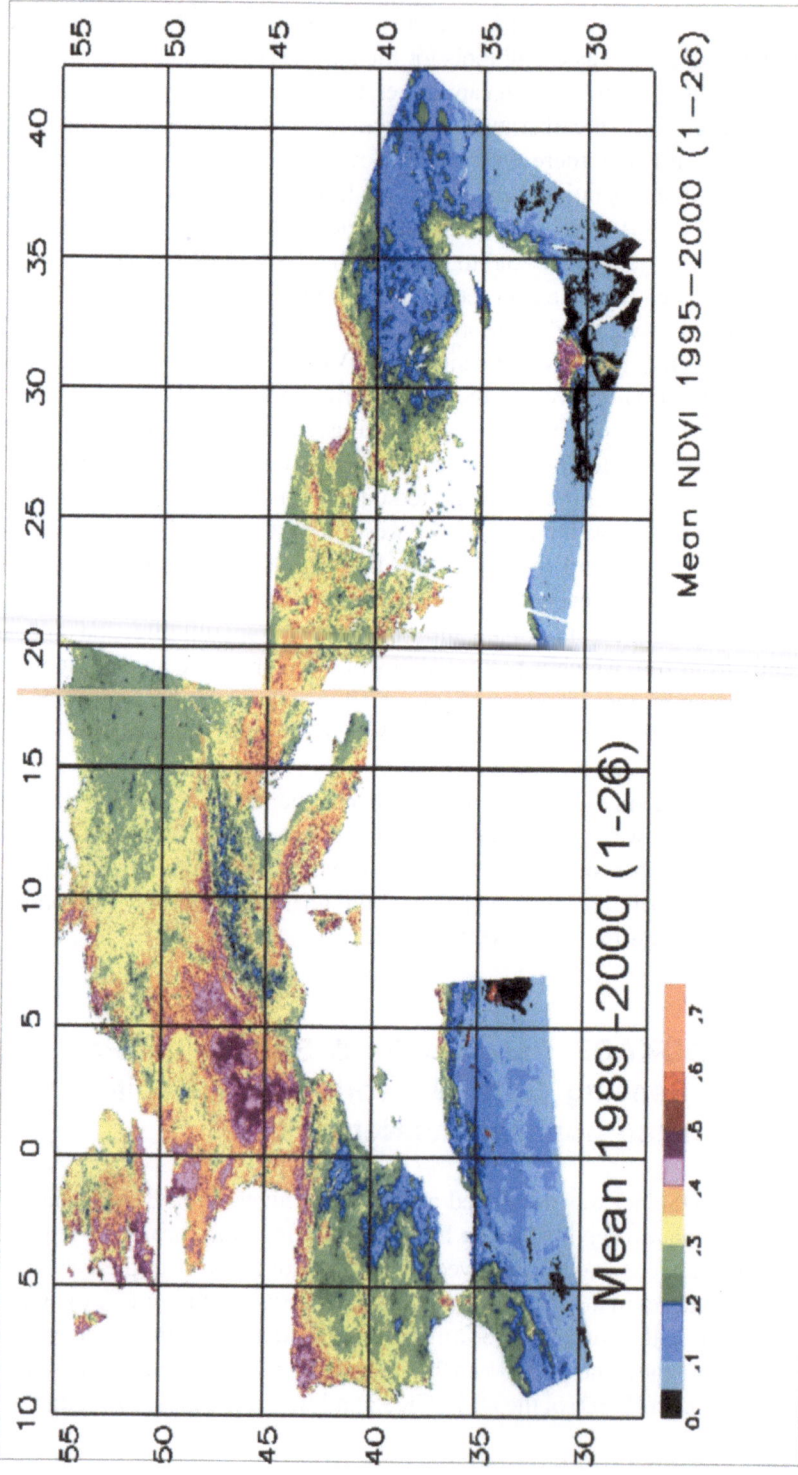

Fig. 3. Mean NDVI January to September for the period 1989 to 2000 for the western and 1995 to 2000 for the eastern Mediterranean area covered by the MEDOKADS data set

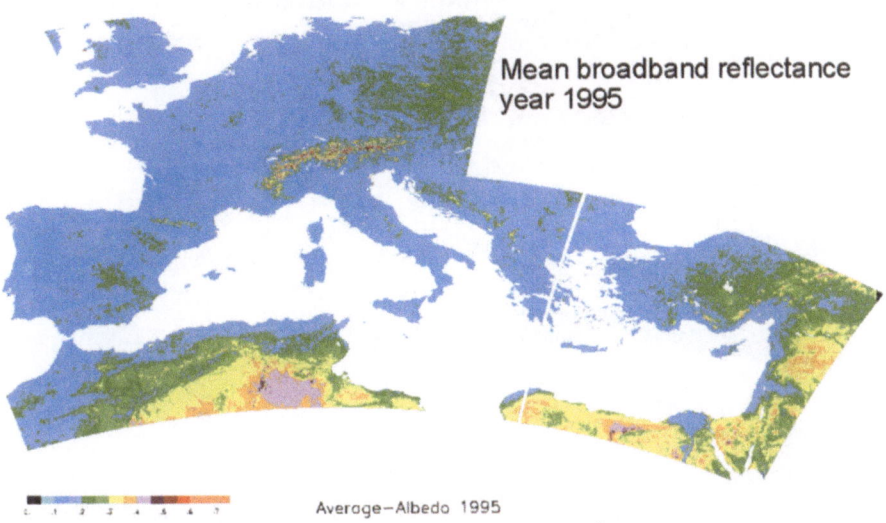

Fig. 4. Mean broadband reflectance for the year 1995

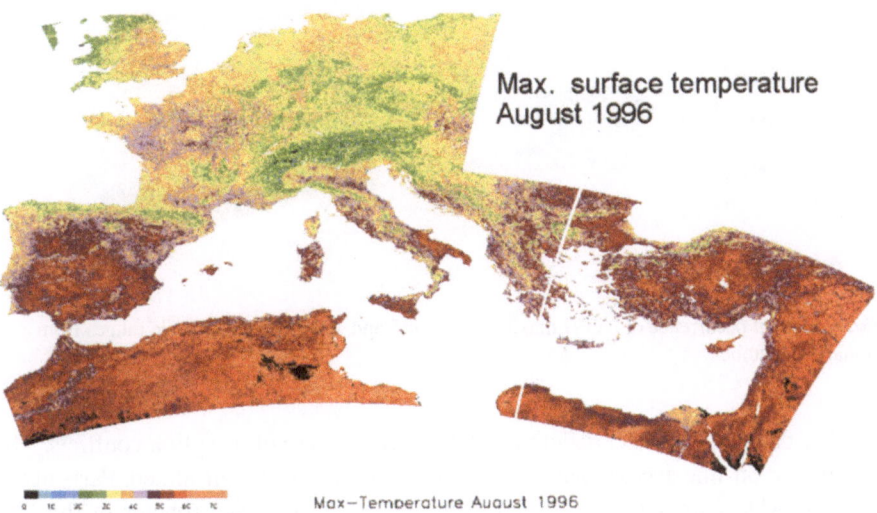

Fig.5. Maximum surface temperature in August 1996

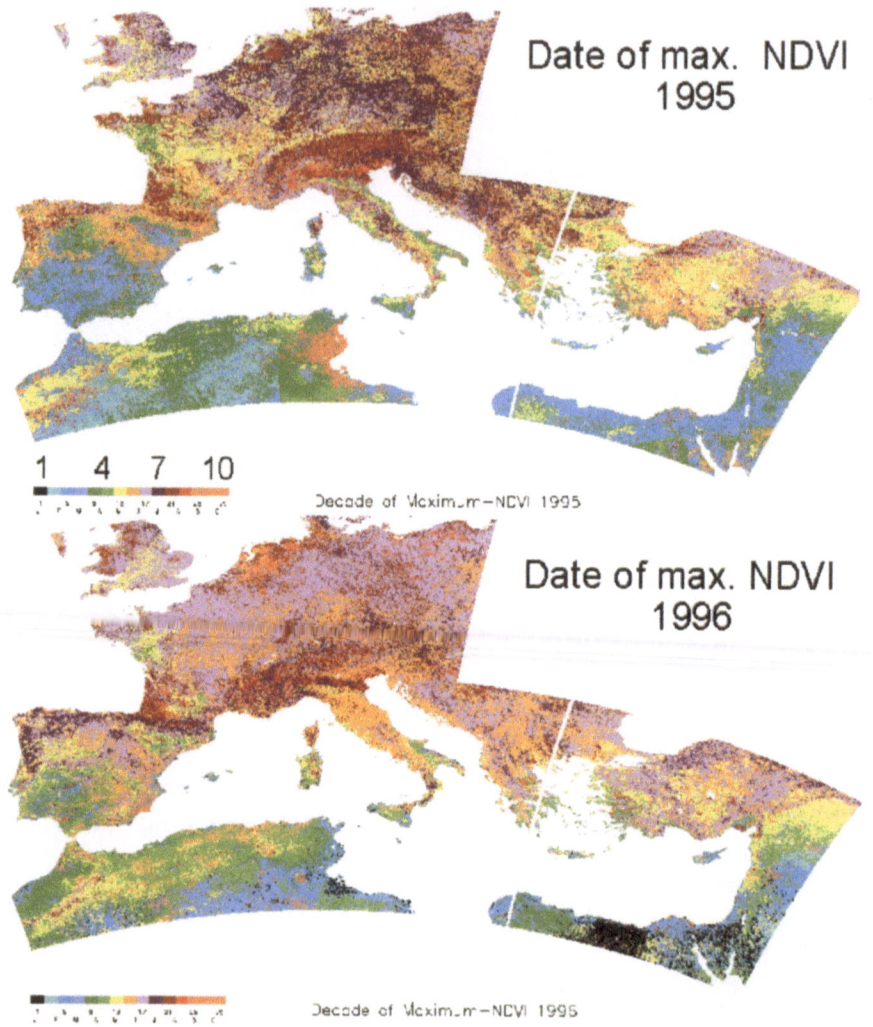

Fig. 6. Date of occurrence of NDVI maximum in 1995 and 1996. The color scale ranges from January to October

missing data and to select different satellites in the case of reception conflicts. The support for on-line access to archived data in addition is decentralized. Parts of the MEDOKADS data set have already been stored on the tape roboter of the PIK (Potsdam Institute for Climate Impact) and work is under way to copy not only the full data set to an on-line accessible tape roboter at the IMK-FZK (Institute of Meteorology at the Forschungszentrum Karlsruhe) but also the whole raw data archives of Berlin which amounts to about 20 Tbyte of data (Fig.10). Sharing the

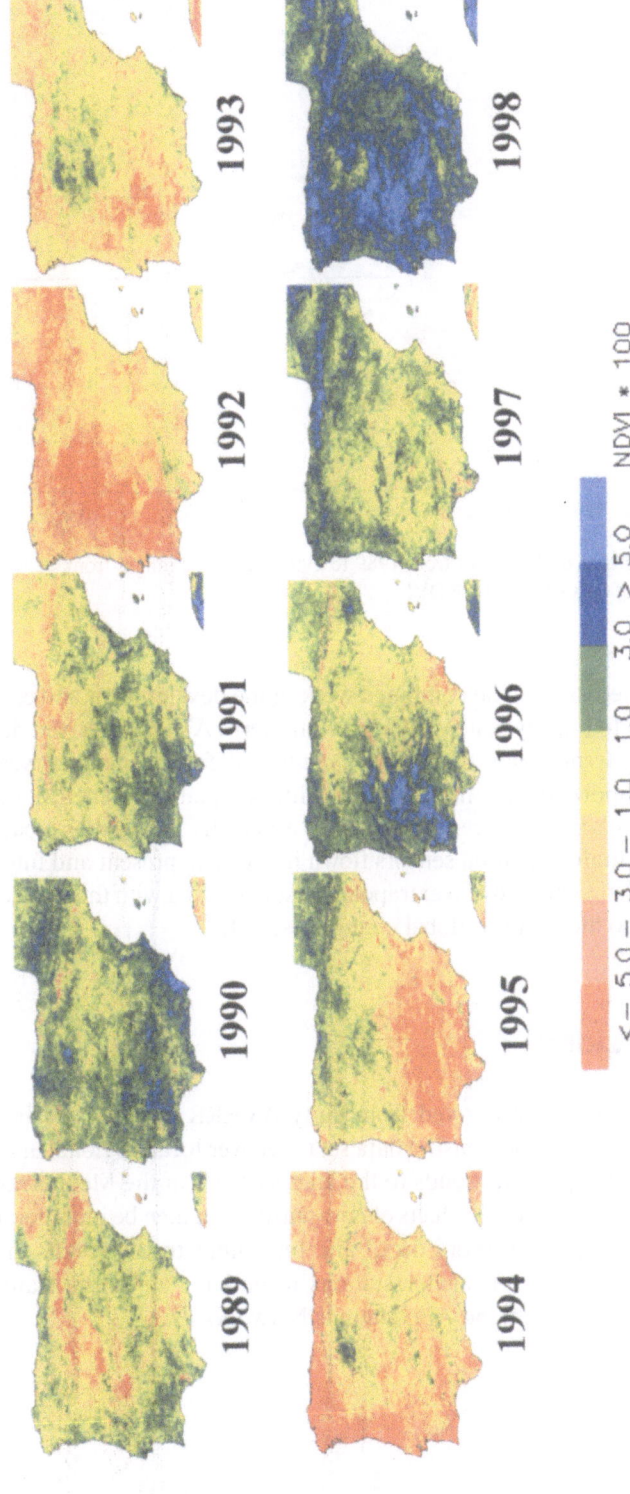

Fig. 7. Deviations of the annual mean NDVI of the Iberian Peninsula over the years 1989 to 1998 from the mean of the ten years

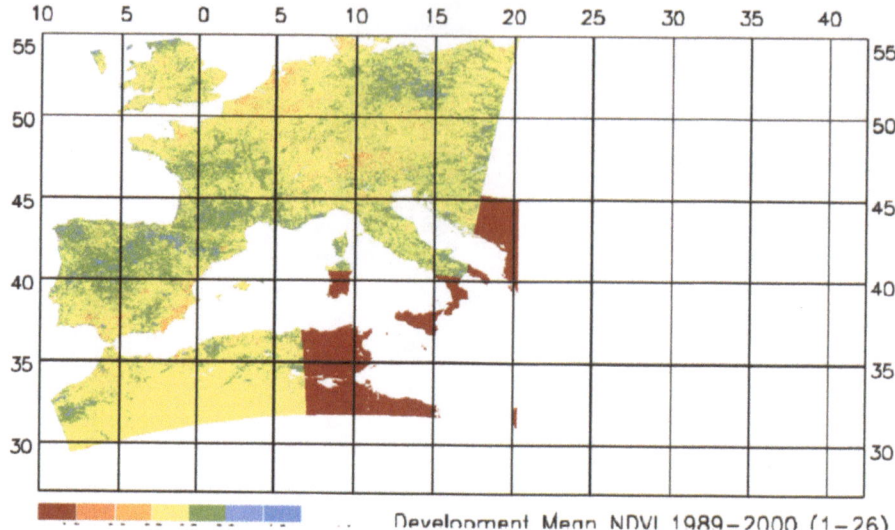

Fig. 8. Development of the NDVI from 1989 to 1995. The scale ranges from <1 % /year to >1 % /year at a scale of 0 to 100 for the NDVI

same data is necessary to do combined work in the development and easy exchange of algorithms to improve and extend the long-term AVHRR derived data series.

This long-term series that will be continued for the next 10 years will get increasing importance as the only data with acceptable geometric and temporal resolution for earth's surface survey. It will especially function as a backbone with links to other high resolution sensors flown in the past, present and future, and it is to be hoped that it will allow to extrapolate results gained with the new sophisticated sensors back to the past for global change research.

4 Conclusions

The MEDOKADS data derived from daily AVHRR data will be best suited to function as a basic remote sensing data set to recover local, regional and large scale singularities, changes and trends to the land surfaces of the Mediterranean. High resolution sensor data and products of new sensors can then be incorporated to study the conspicuous areas in more detail and link them to the results of the in situ observation by the net of anchor stations to be built up in the framework of a Mediterranean Research and Applications Network.

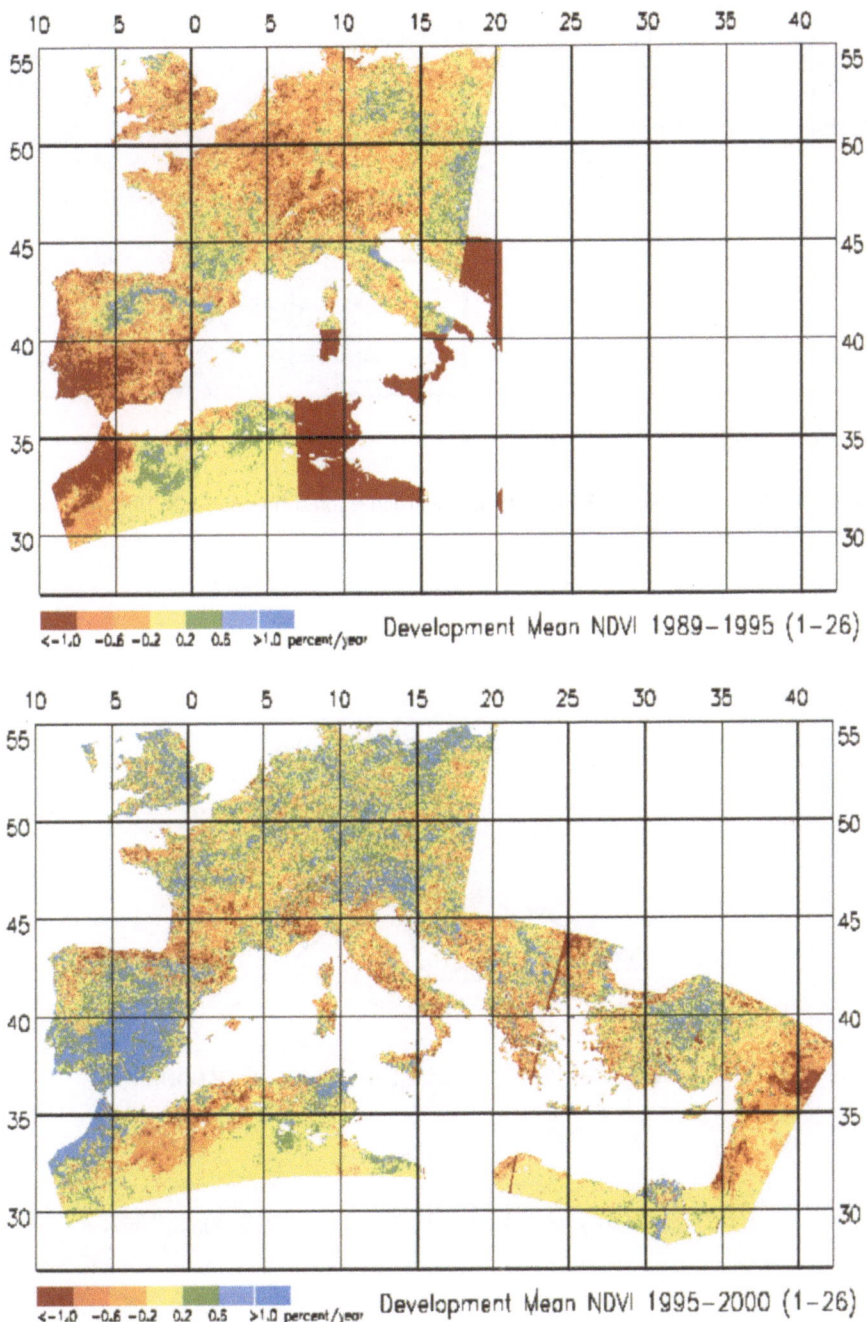

Fig. 9. NDVI development 1995 to 2000

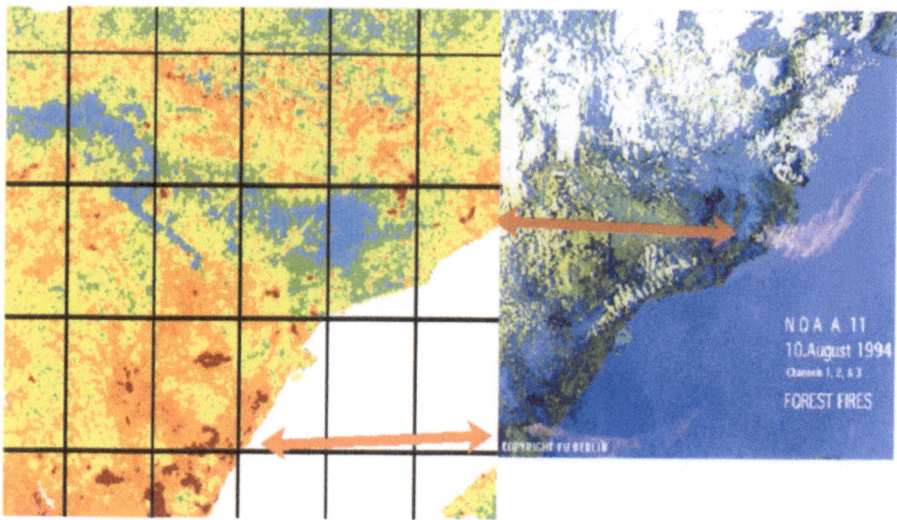

Fig.10. NDVI development 1989 to 1996 (left) and active forest fires on 11. 8. 1994 with smoke trails (left) with quenched ones (black).

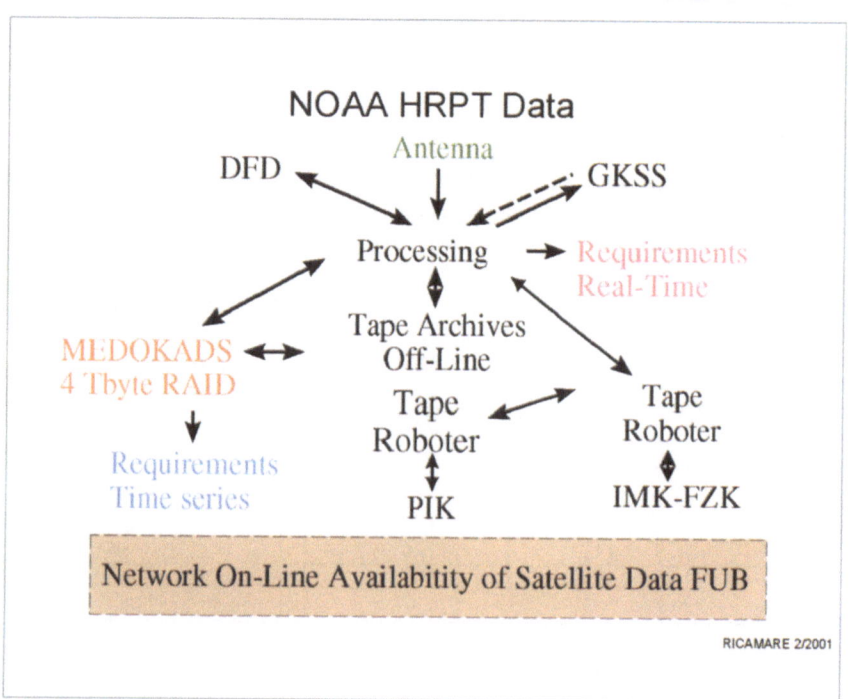

Fig. 11. Network for HRPT reception and on-line data access

References

Bolle H-J, André J-C, Arrue JL et al. (1993) EFEDA: European Field Experiment in a Desertification-threatened Area, Ann. Geophys. II:173-189

Bolle H-J (ed.) (1998) Remote sensing of Mediterranean desertification and environmental changes (Resmedes), Final Report. European Commission, EUR 18352, ISBN 92-827-4040-40

Koslowsky D (1996) Mehrjährige validierte und homogenisierte Reihen des Reflexionsgrades und des Vegtationsindexes von Landoberflächen aus täglichen AVHRR-Daten hoher Auflösung. Dissertation. Freie Univeristät Berlin, Institut für Meteorologie, Met.Abh, Neue Folge, 9(1)

Koslowsky D (1997) Signal degradation of the AVHRR shortwave channels of NOAA 11 and NOAA 14 by daily monitoring of desert targets. Adv. Space Res. 19(1):1355-1358

Koslowsky D (1998) Improved long-term AVHRR reflectance and NDVI time series using empirically derived nadir correction factors. Preceedings 9th Conf. Sat. Meteor. and Oceanogr. UNESCO, Paris; France, 25-29 May, AMS, Boston, MA, 313-316

Rao CRN, Chen J (1994) Post-launch calibration of the visible and near infrared channels of the Advanced Very High Resolution Radiometer on NOAA-7,-9, and -11 spacecraft. NOAA Technical Report NESDIS 78, NOAA/NESDIS, Washington, D.C.

Rao CRN, Chen J (1996) Post-launch calibration of the visible and near-infrared channels of the Advanced Very High Resolution Radiometer on the NOAA-14 spacecraft. Int. J. Remote Sensing 17(14):2743-2747

Rao CRN, Chen J (1999) Revised post-launch calibration of the visible and near-infrared channels of the Advanced Very High Resolution Radiometer (AVHRR) on the NOAA-14 spacecraft. Int. J. Remote Sensing 20:3485-3491

Chapter 5
Regional Aspects

Spatial and Temporal Variations in Precipitation and Aridity Index Series of Turkey

M. Türkeş

Abstract

Annual and seasonal precipitation series and annual aridity index series of Turkey were investigated with respect to spatial and temporal variations for the period 1930-1993. Analysis of normalized precipitation anomalies was also performed for the 1994-2000 period. Semi-arid and dry sub-humid climatic conditions are dominant over the continental interiors and continental Mediterranean region of Turkey. Normalized annual and winter precipitation series have tended to decrease over a considerable part of Turkey since the early 1970s. For the normalized annual and winter precipitation anomaly series, wet conditions generally occurred during the 1940s, 1960s, late 1970s, early 1980s and mid-late 1990s, whereas dry conditions generally dominated over the early-mid 1930s, early-mid 1970s, mid-late 1980s, early 1990s, and 1999/2000 in most of Turkey. Spring precipitation series generally indicated an upward trend from the mid 1940s to the late 1960s at many stations and to 1980s at some stations. This period was generally followed by a downward trend at many stations. Significant decreasing trends showed up in the annual precipitation series of 15 stations and in the winter precipitation series of 14 stations, mostly over the Mediterranean rainfall region. Summer rainfall series have tended to increase significantly at 7 stations. There has also been a general tendency from humid conditions of around the 1960s towards dry sub-humid climatic conditions in the aridity index values of many stations of Turkey. At some stations over the Aegean part of the Mediterranean region, there has been a significant change from humid conditions to dry sub-humid or semi-arid climatic conditions.

Key Words: Turkey, Precipitation, Aridity Index, Seasonality, Variability, Trend and Persistent.

1 Introduction

Increased frequencies and intensities of the drier conditions in about last twenty-five years may have been related to dominance of the anticyclonic circulation during the same time period, due to mainly increased atmospheric geopotential heights and decreased cyclone activity over the Turkey region (Türkeş, 1998a). Furthermore, various climate models predict a decrease in precipitation for future changes over many parts of subtropics including eastern Mediterranean Basin and Turkey, particularly in the winter or generally during cool part of the year (ECSN, 1995; UKMO, 1995; Jacobeit, 1996; UKMO/DETR, 1999). In addition to the long-lasting summer dryness of the Mediterranean type climate, considerable persistent dry conditions during about last three decades and secular decreasing trends in the annual and particularly in the winter precipitation totals make the natural and socio-economic system of Turkey more vulnerable to the projected changes in the mean climate. Projected changes in the climatic variability in addition to the mean climate and changes in land use, which should be considered in the future especially with respect to the land use management, regional planning, forestry and agricultural activities, would also make the existing conditions progressively worse.

This presentation has been prepared mainly based on the author's previous studies (Türkeş, 1996a, 1996b, 1998a, 1998b and 1999) and on some new and additional analyses, for the RICAMERA Workshop on the 'Development of a concerted data assessment, assimilation and validation for "Global Change" research' held at Casablanca - Morocco, February 21-24, 2001. The scope of the presentation is:

1. to give detail information on spatial distribution of mean precipitation totals, precipitation variability and seasonality of precipitation, and to examine the arid lands of Turkey by using an aridity index;
2. to evaluate the long-term variations including trend and persistence in normalized precipitation and aridity index series, and periodicity in aridity index series of Turkey, for the 1930-1993 period; and,
3. to analyse the spatial and temporal variations of normalized precipitation anomalies, with respect to the dry (drier than normal) and wet (wetter than normal) conditions for the 1994-2000 period over Turkey.

2 Data

The Turkish State Meteorological Service (TSMS) is responsible for the synoptic (weather and upper air), climatological and marine meteorological observations, and the weather and sea forecasts and warnings in Turkey. Aims and activities of the TSMS can be summarized as follows: to open and operate all kinds of meteorological stations or units over Turkey; to make various observations and weather forecasts for all users and sectors; to support particularly to the sectors of agriculture, forest, tourism, transport, energy, health and armed forces; to broadcast weather and sea forecasts and warnings; and to make researches to describe and understand Turkey's weather and climate. Observations and works related to hydrology and water resources in Turkey are carried out by other governmental organizations.

For the time-series analysis of precipitation and aridity index series, this study was carried out mainly by using monthly precipitation totals and monthly mean temperatures, which were recorded at the 91 and 55 stations of the TSMS, respectively, during the period 1929-1993. The data of 1929 was only used in computing 1930 winter season. The lengths of records vary from 54 to 64 years for precipitation data and from 42 to 64 years for temperature data. Division of the year was made according to the basic seasons, as winter (December to February), spring (March to May), summer (June to August) and autumn (September to November). The precipitation data set had been broadly explained already by Türkeş (1995, 1996a). Missing values in monthly station records were replaced with the estimates from nearby stations, only if the gaps in a station's record did not consist in more than five per cent of the total number of monthly values in that station's record. This was performed by using the normal ratio method, as described in Singh (1992). In this procedure, three nearby reference stations were chosen by taking into account the highest correlation coefficients (Pearson's r), between the base station to be filled and the possible reference stations.

Statistical evaluations of homogeneity for the annual and seasonal precipitation series, and annual aridity index and temperature series were made by using the non-parametric Kruskal-Wallis test for the homogeneity of means of the sub-periods (Sneyers, 1990). For this study, homogeneity means that there is no jump in the climatic series of observations. Jumps consist of non-climatological abrupt falls and/or abrupt rises in the series. In order to make a more powerful assessment, statistical results from the homogeneity test were also controlled by means of the plotted graphs combined with the information available from the station history file. Monthly precipitation and monthly mean temperature series were also subjected to

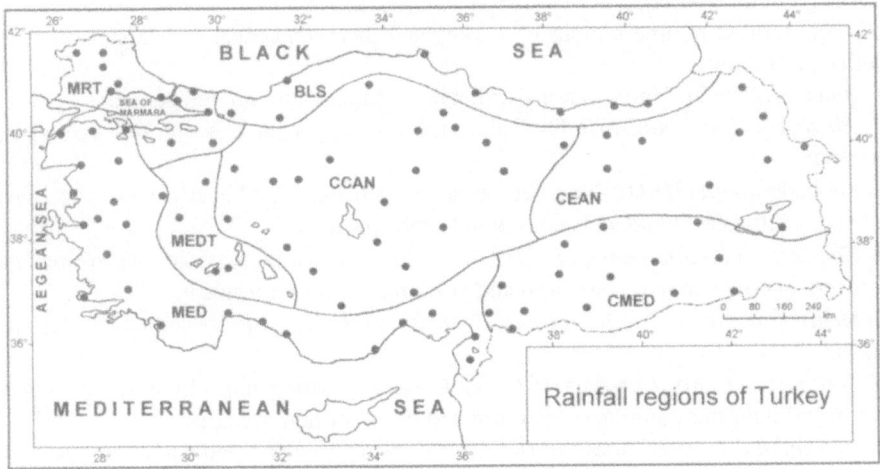

Fig. 1. Geographical distribution of Turkey's rainfall regime regions and 91 stations used in time-series analysis (see the text for the names of the regions) (Türkeş, 1998a)

the homogeneity test, and checked by the station files. These station history files were prepared especially for the studies of precipitation and temperature, and include necessary information, such as changes of meteorological instruments and of height and location of the stations for about 80 stations, in order to detect inhomogeneity in the series. This type of information is useful to decide whether a jump arose from a relocation of the station, or it happened due to a low-frequency fluctuation (or a strong persistence) in the series. For instance, by using additional information from the station history files, we found that significant inhomogeneities in the annual and seasonal air temperature series of Antalya, Anamur and Fethiye stations had occurred owing to relocation of the stations.

However, it was seen that monthly series could not be a good indicator for some inhomogeneity types, such as in the series that were exposed to a location change without considerable change in station height. This situation also happens when a station relocation occurs from a relatively inland site to a new site on coast or on near-coast. Effect of the local climatic conditions of a station could be different in course of the year, when a relocation of the station occurs. Thus, to make a best decision on whether a climatic series is appropriate for the time-series analysis or not, information from both objective and subjective analyses for the seasonal series should be examined.

According to the results of the objective and subjective homogeneity analyses, annual and seasonal precipitation series of 91 stations, and mean annual temperature and aridity index series of 55 stations were found to be homogeneous, with respect to abrupt changes in the series (Türkeş, 1995, 1996a and 1999). Regarding the time-series analyses, this study uses the precipitation series of these 91 stations. Geographical distribution of the 91 stations over the rainfall regime regions of Turkey is shown in Fig. 1. The rainfall regime regions of Turkey were originally designated by taking into account the seasonality of Turkish precipitation and the physical geographical conditions of Turkey. The delimitation of these regions and the description of their rainfall characteristics were widely discussed and presented by Türkeş (1996a, 1998a).

The rainfall regime regions of Turkey and their fundamental characteristics are defined as follows:

1. *Black Sea (BLS):* Uniform rainy with a maximum in autumn; temperate.
2. *Marmara Transition (MRT):* Quite uniform rainy with a warm and light rainy summer.
3. *Mediterranean (MED):* Markedly seasonal with a cool and heavy rainy winter and a hot dry summer; humid and semi-humid subtropical.
4. *Continental Mediterranean (CMED):* Seasonal with a rainy winter and spring and a severe hot dry summer; semi-arid and dry semi-humid subtropical.
5. *Mediterranean to Central Anatolia Transition (MEDT)*: Moderate rainy winter and spring.
6. *Continental Central Anatolia (CCAN):* Cool rainy spring and cold rainy winter, and warm and light rainy summer; semi-arid and dry semi-humid steppe.
7. *Continental Eastern Anatolia (CEAN):* Cool rainy spring and early summer with a very cold and snowy winter; dry semi-humid and semi-humid steppe and highland.

Additional eight and thirty-five stations having relatively shorter series of observations were included to this main 91 and 55 stations, respectively, for

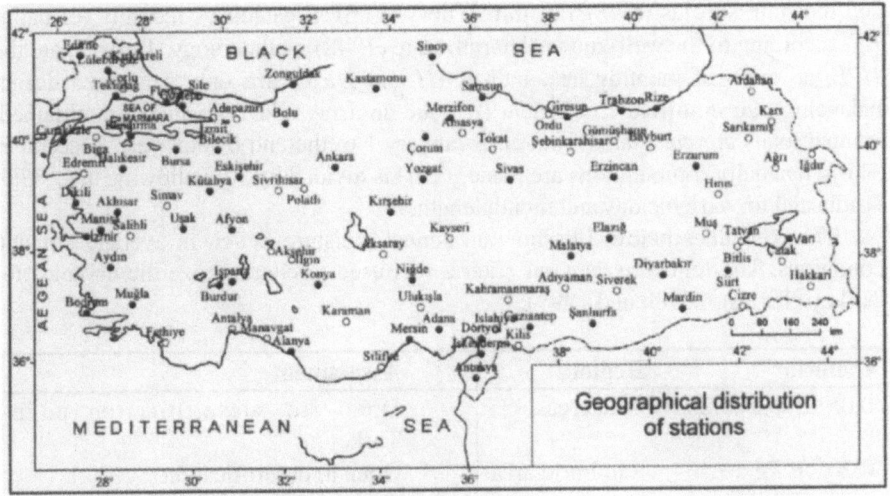

Fig. 2. Locations and names of all stations used in the study

examination of mean precipitation and aridity conditions of Turkey. Precipitation data set of 1994-2000 period for these 99 stations was also used for evaluating the spatial and temporal characteristics of dry and wet conditions over this period in Turkey.

The locations and names of all stations used in this study are also shown in Fig. 2, in which solid circles denote the 55 stations for which variations and trends of the aridity index values were investigated.

3   Method of Analysis

For the purposes of the United Nations Convention to Combat Desertification (UNCCD), arid, semi-arid and dry sub-humid areas were defined as "areas, other than polar and sub-polar regions, in which the ratio of annual precipitation to potential evapotranspiration falls within the range from 0.05 to 0.65" (UNCCD, 1995). In the present study, the same Aridity Index (AI) was adopted as the base method for determining dry land types and thereby delineating boundaries, and showing changes in aridity conditions in Turkey. Following the UNEP (1993), AI could be written as

$$AI = P / PE \qquad (1)$$

where P is the annual precipitation total (mm) and PE the potential evapotranspiration (mm). PE values were calculated with the WATBUG program, which was developed by Willmott (1977) for calculation of the climatic water budgets. With this program, water budgets can be computed on a monthly (or daily) basis; and "look-up" tables are not needed as all relationships are explicitly specified. The required input is minimal; for example, air temperature, precipitation and a few initial parameters. Some of outputs that the WATBUG produces are as follows: unadjusted potential evapotranspiration (UPE) in mm; adjusted PE (APE) in mm; soil moisture storage (ST) in mm; actual evapotranspiration (AE) in mm; soil moisture deficit (DEP) and soil moisture surplus (*SURP*) in mm. The WATBUG calculates monthly (or daily) *PE*, according to the well-known Thornthwaite (1948) methodology. To calculate the *UPE*, an array of monthly heat indices (*H* and *HEAT* are only calculated during balancing), an empirical coefficient (*A*), and an array of *UPE* values are obtained. Annual totals are calculated from each January 1 to the end of that year (December 31). When daily computations are made, *PE(I)* is divided by 30. Following this, *PE(I)* is adjusted for variable day and month lengths.

The *AI* values below 1.0 show an annual moisture deficit in average climatic conditions. The following general criteria was used to characterize the drylands for Turkey (Türkeş, 1998b and 1999):

Criteria	Region	Assessment
$0.05 \leq P/PE < 0.20$	Arid areas	Open to desertification (no in Turkey)
$0.20 \leq P/PE < 0.50$	Semi-arid areas	Open to desertification
$0.50 \leq P/PE < 0.65$	Dry sub-humid areas	Open to desertification

Normalized precipitation anomaly series were used in the study. A normalised precipitation anomaly (A_{sy}) for a long series of a given station is computed with

$$A_{sy} = (P_{sy} - \overline{P_s}) / \sigma_s \qquad (2)$$

where P_{sy} is total precipitation amount (mm) for a station s during a year y (or a season); $\overline{P_s}$ and σ_s are long-term mean and standard deviation of annual (or seasonal) precipitation for that station, respectively.

The Mann-Kendall rank correlation test (Sneyers, 1990) was chosen to detect any possible trend in the series of normalized annual and seasonal precipitation anomalies and of annual *AI* values, and to test whether or not they are statistically significant. The Mann-Kendall test statistic $u(t)$ is a value that indicates direction (or sign) and statistical magnitude of a secular trend in a series. When the value of $u(t)$ is significant at the 5 per cent significance level, it can be decided whether it is an increasing or a decreasing trend depending whether on $u(t) > 0$ or $u(t) < 0$. A one per cent level of significance was also taken into consideration. In addition to this, partial and short-period trends, and a change point or beginning point of a trend in the series are also investigated by using the time-series plot of the $u(t_i)$ and $u'(t_i)$ values. In order to have such a time-series plot, sequential values of the statistics $u(t)$ and $u'(t)$ are computed from the progressive analysis of the Mann-Kendall test. In Sneyers methodology, the values of $u(t_i)$ are automatically computed for all values of i, because computing the main statistic $u(t)$ requires these values of $u(t_i)$. Following Sneyers (1990), this procedure can be formulated as follows: first the original observations are replaced by their corresponding ranks y_i, which are arranged in ascending order. Then for each term y_i, the number n_k of terms y_j preceding it ($i > j$) is calculated with ($y_i > y_j$), and the test statistic t_i is written as

$$t_i = \sum_{k=1}^{i} n_k \qquad (3)$$

Distribution function of the test statistic t_i has a mean and a variance derived by

$$E(t_i) = i(i-1)/4 \quad and \quad \text{var}(t_i) = [i(i-1)(2i+5)]/72 \qquad (4)$$

Values of the statistic $u(t_i)$ are then computed as

$$u(t_i) = \frac{[t_i - E(t_i)]}{\sqrt{\text{var}(t_i)}} \qquad (5)$$

Finally, the values of $u'(t_i)$ are similarly computed backward, starting from the end of the series. With a trend, inter-section of these curves enables the beginning of a trend in the series to be located, approximately. Without any trend, the time-series plot of the values $u(t_i)$ and $u'(t_i)$ indicates curves that overlap several times (Sneyers, 1990; Türkeş 1996 and 1999). A nine-point Gaussian filter is also used as a low-pass filter to visually investigate characteristics of the long-period fluctuations in the series (WMO, 1966).

Another form of non-randomness in the climatic series is persistence, which is defined as "a tendency for successive values of the series to 'remember' their antecedent values, and to be influenced by them" (WMO, 1966). Lag-one serial correlation coefficient (L-1SC) was used to examine the nature and magnitude of the possible persistence in the long time-series. By using one-sided test of the normal distribution, the null hypothesis of the randomness against the serial correlation is rejected for large values of $(r_l)_t$ with

$$(r_1)_t = \frac{-1 \pm t_g \sqrt{N-1}}{N-1} \qquad (6)$$

where t_g = 1.645 for the 5 per cent level of significance and 2.330 for the 1 per cent level. The serial correlation coefficients for the second and third lags were also checked to determine whether the persistence in the series is a simple Markov type persistence. If the L-1SC coefficient significantly differs from zero, and if lag-two and lag-three serial correlations approximate the square and cube of the L-1SC, respectively, the series is assumed to contain a Markov type persistence (WMO, 1966).

The spectral (power spectrum) analysis, as described in WMO (1966), was applied to the normalized aridity index values to detect dominant and hidden cycles within the observed fluctuations. A detailed description of this approach can be found in various text books: Blackman and Tukey (1958); WMO (1966); Jenkins and Watts (1968). Maximum lag was chosen as about one-third of the record length (e.g., 21 years for a 64 year long aridity series). Final spectral estimates were calculated by smoothing raw spectral estimates with three-term weighted average of the Hanning method. Then a procedure of the tests of statistical significance, which was proposed by WMO (1966), was used for an objective assessment of the spectral estimates from the power spectrum analysis.

The method of Kriging was used in order to produce contours of the spatial distribution maps. Detailed explanations of the method can be found in Delfiner and Delhomme (1975), Journel and Huijbregts (1978), Cressie (1991), and Hevesi et al.

(1992a, b). This method was applied to the stations' data by means of a mapping package.

4 Results

4.1 Mean Precipitation and Aridity Conditions

4.1.1 General

The climate of Turkey, which is mainly characterized by the Mediterranean macro climate, results from the seasonal alternation of the mid-latitude frontal depressions, with the polar air masses, and subtropical high pressures, with the subsiding maritime tropical and continental tropical air masses. Continental tropical airstreams from the Northern Africa and Arabian deserts dominate particularly throughout the summer, by causing the long-lasting warm and dry conditions over Turkey, except in the Black Sea region and the north-eastern Anatolia.

4.1.2 Mean Precipitation

Mean annual precipitation totals range from below 500 mm over the continental interiors and eastern margin of the Eastern Anatolia region to above 1000 mm along the western Mediterranean, and the western and eastern Black Sea coastal areas (Fig. 3). Annual precipitation below 400 mm extends over a large area of the Continental

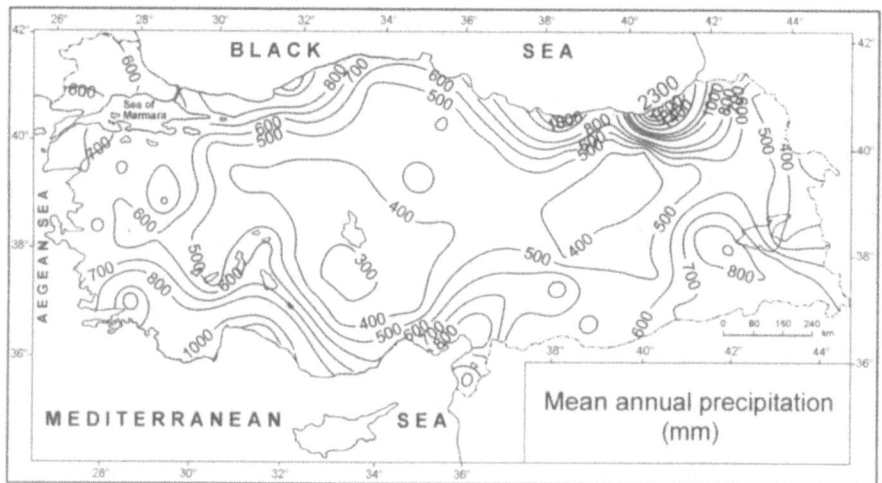

Fig. 3. Geographical distribution of mean annual precipitation totals over Turkey (re-plotted from Türkeş, 1999)

Central Anatolia region, especially over the Konya sub-region. Zonality of the long-term mean precipitation totals is most pronounced for winter (Türkeş, 1998a).

Mean winter precipitation decreases from coastal belts to interiors by varying from above 650 mm along the eastern Black Sea and the western Mediterranean to below 100 mm over continental Central Anatolia and Eastern Anatolia regions (Fig. 4a). The western Mediterranean sub-region is the most rainy area of Turkey in winter, with about 754 mm mean precipitation total at the Manavgat station. High precipitation amounts along the Black Sea and the Mediterranean coastal zones are associated mainly with the North-eastern Atlantic originated mid-latitude depressions and the Mediterranean depressions, respectively (Türkeş, 1998a). Orographic rains on the windward slopes of the Taurus Mountains and the Northern Anatolia Mountains also contribute to increase of the precipitation amounts.

Mean spring precipitation amounts range from below 150 mm over the Aegean coast with the northern Marmara and the Continental Central Anatolia connected with the middle Mediterranean coast to above 300 mm on the eastern Black Sea coast and the south-east corner of the country (Fig. 4b). Most of the country gets a precipitation ranging from below 150 mm to about 200 mm in spring. In addition to the frontal rains, higher precipitation amounts over the wetter areas with a 200-350 mm mean are attributed to high topography and their exposures to the dominant air streams, and the local convective activities, respectively (Türkeş, 1998a). Summer rainfall amounts increase from below 5 mm over the Syrian border to above 450 mm on the eastern Black Sea sub-region of Turkey (Fig. 4c). Approximately half of the country receives a mean rainfall less than 50 mm in summer. Higher summer rainfall over the eastern Black Sea and the high north-eastern Anatolia is related closely with the post-frontal and orographically induced rainfall in addition to the frontal rains. Mean autumn precipitation totals generally decrease from coastal areas to interiors, as in winter (Fig. 4d). Eastern Black Sea is the wettest area, with about 795 mm seasonal mean precipitation at the Rize station. Precipitation amounts of about 100 mm cover much of the country, particularly continental interiors. With respect to amount and distribution pattern of precipitation, the mean autumn precipitation is quite similar to mean winter precipitation over the Black Sea region, with a strong gradient along the Northern Anatolia Mountains. Mean autumn precipitation over the Mediterranean rainfall region differ from winter precipitation, because the Mediterranean type frontal cyclones are not so active in autumn yet as much as in winter.

4.1.3   Seasonality of Precipitation

Maps for the geographical distribution of percentage contribution of mean seasonal precipitation amounts to mean annual total are shown in Fig. 5. Precipitation is more seasonal over the western and southern regions of Turkey, with a winter maximum above 40 per cent, whereas it is generally uniform over the Black Sea region (Fig. 5a). Spring precipitation contributes more than 30 per cent of the annual total over most of the continental interiors (Fig. 5b). In summer, maximum precipitation is concentrated over the north-eastern Anatolia, while the minimum occurs over the southern parts of the Continental Mediterranean region, with less than 5 per cent of the annual total (Fig. 5c). Contribution of the autumn precipitation is above 30 per

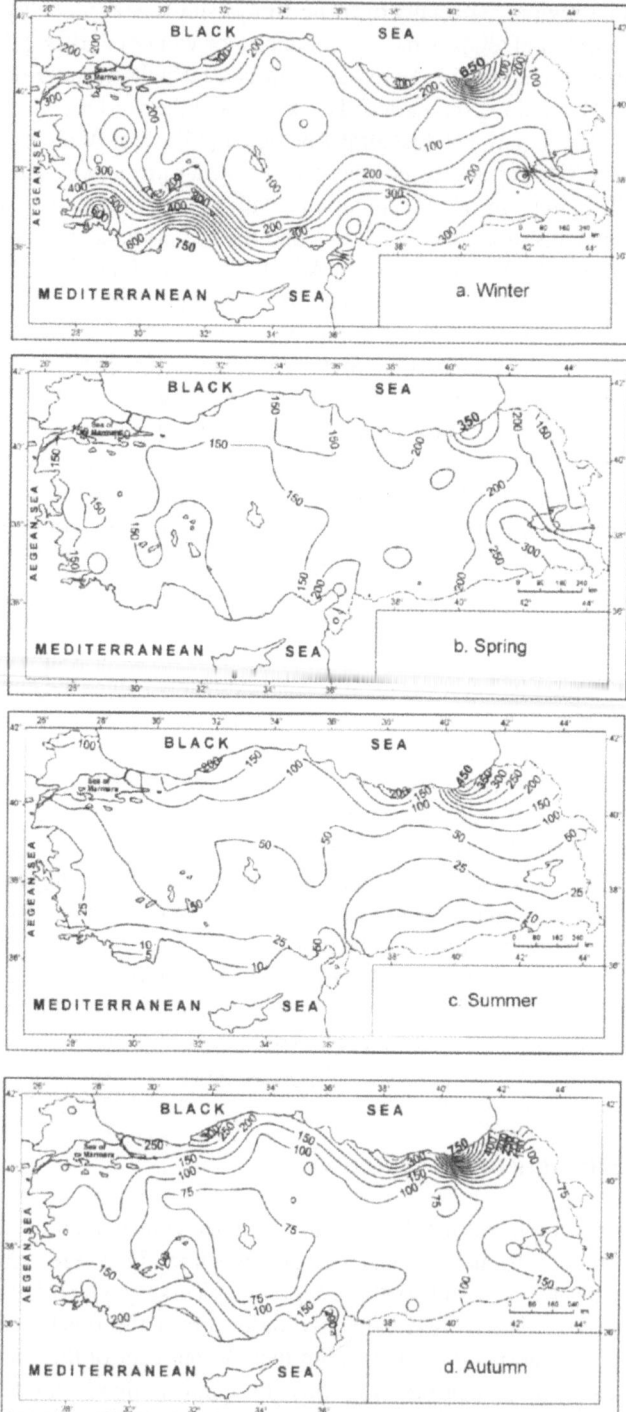

Fig. 4. Geographical distributions of mean seasonal precipitation totals
(mm) over Turkey.

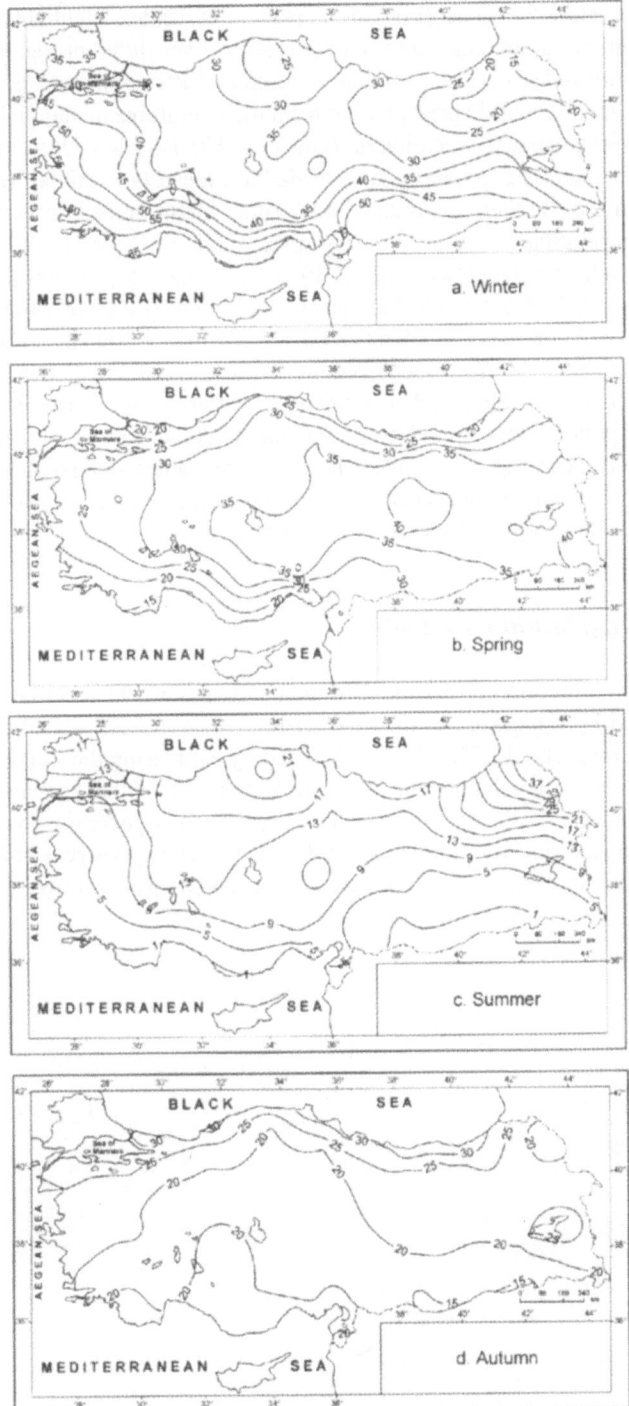

Fig. 5. Geographical distributions of percentage contribution of mean seasonal precipitation amounts to mean annual total in Turkey.

cent along the Black Sea coast (Fig. 5d). A seasonality index highlights this contrast of the precipitation amounts between the seasons over different regions of Turkey. The seasonality index (*SI*) at a station is computed by summation of the absolute deviations of mean monthly precipitation from the overall mean and dividing it to the long-term average annual precipitation (Glantz 1987; Türkeş 1998a). Areas with *SI* values greater than 0.55 generally have a Mediterranean type rainfall regime, which coincide with the Mediterranean and Continental Mediterranean regions of Turkey (Fig. 6). Contribution of the mean spring precipitation amount to the mean annual precipitation total ranges between about 30 per cent and 35 per cent in the majority of the stations of the Continental Mediterranean region. If the contribution of spring precipitation is taken into consideration, it is seen that seasonality of most stations over the Continental Mediterranean region, however, differs somewhat from those over the Mediterranean region. This percentage pattern of the spring precipitation is similar to the pattern that is found over the continental interior regions rather than that of the Mediterranean region. On the other hand, *SI* values smaller than about 0.35 generally correspond to a uniform and quite uniform rainfall regime over the Black Sea region and the northern Marmara, respectively.

4.1.4   Precipitation Variability

The spatial distribution of variability in annual and seasonal precipitation amounts was examined by a measure termed the coefficient of variation (*CV*) and shown in Fig. 7 and 8, respectively. The *CV* is calculated by expressing the standard deviation as a percentage of the long-term average. Variability of the annual precipitation decreases from the southern part of the country, which is generally characterized by the Mediterranean rainfall regime, to the Black Sea coast, where a uniform rainfall regime is dominant (Fig. 7). The *CVs* are greater than 25 per cent over a great part of

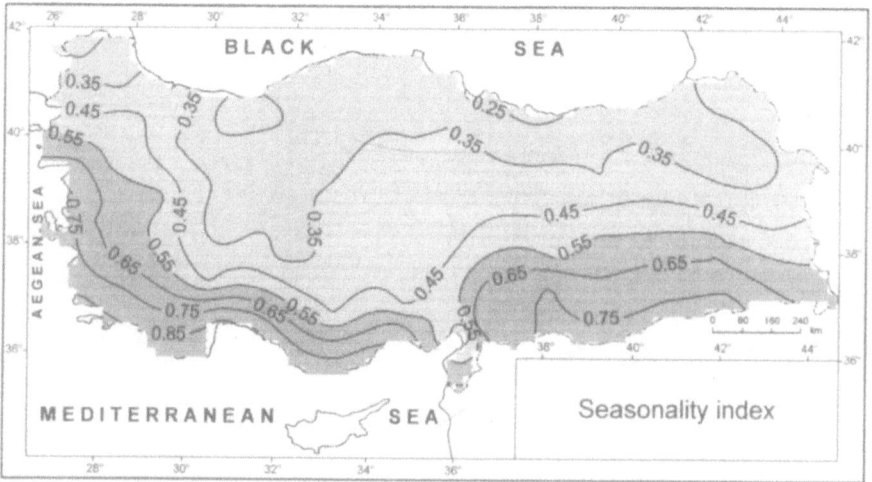

Fig. 6. Geographical distribution of the Sesonality Index (*SI*) of precipitation over Turkey (re-plotted from Türkeş 1999)

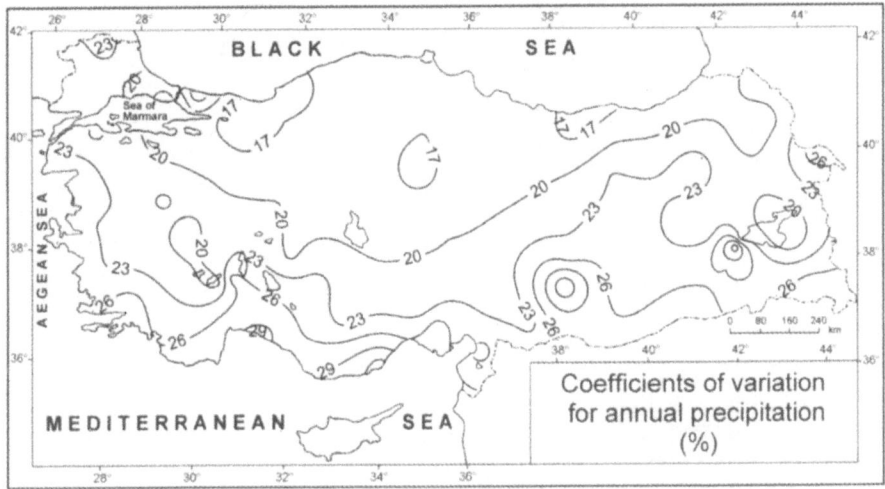

Fig. 7. Geographical distribution of coefficients of variation for annual precipitation totals in Turkey (re-plotted from Türkeş, 1999)

the Mediterranean region and almost all over the Continental Mediterranean region.

The *CV*s of winter precipitation are well above 35 per cent over most of the Marmara, Mediterranean and Continental Eastern Anatolia regions (Fig. 8a). Variability of summer rainfall is above 80 per cent over the Mediterranean and Continental Mediterranean regions, and below 35 per cent on the eastern Black Sea Coast (Fig. 8c).

4.1.5   Aridity Index

The spatial distribution of the aridity index (*AI*) is shown in Fig. 9. Areas having values of $0.65 \leq AI < 0.80$ where an annual moisture deficit exists, are shown to be concentrated in the surroundings of the semi-arid and dry sub-humid areas, although humid lands of Turkey are excluded from the study. Dry sub-humid climatic conditions extend over most of the Continental Central Anatolia and Continental Mediterranean regions, some part of the eastern Mediterranean, and the eastern and western parts of the Continental Eastern Anatolia region. Semi-arid climatic conditions are dominant only over the Konya Plain of the Central Anatolia and at the Iğdır district of Eastern Anatolia.

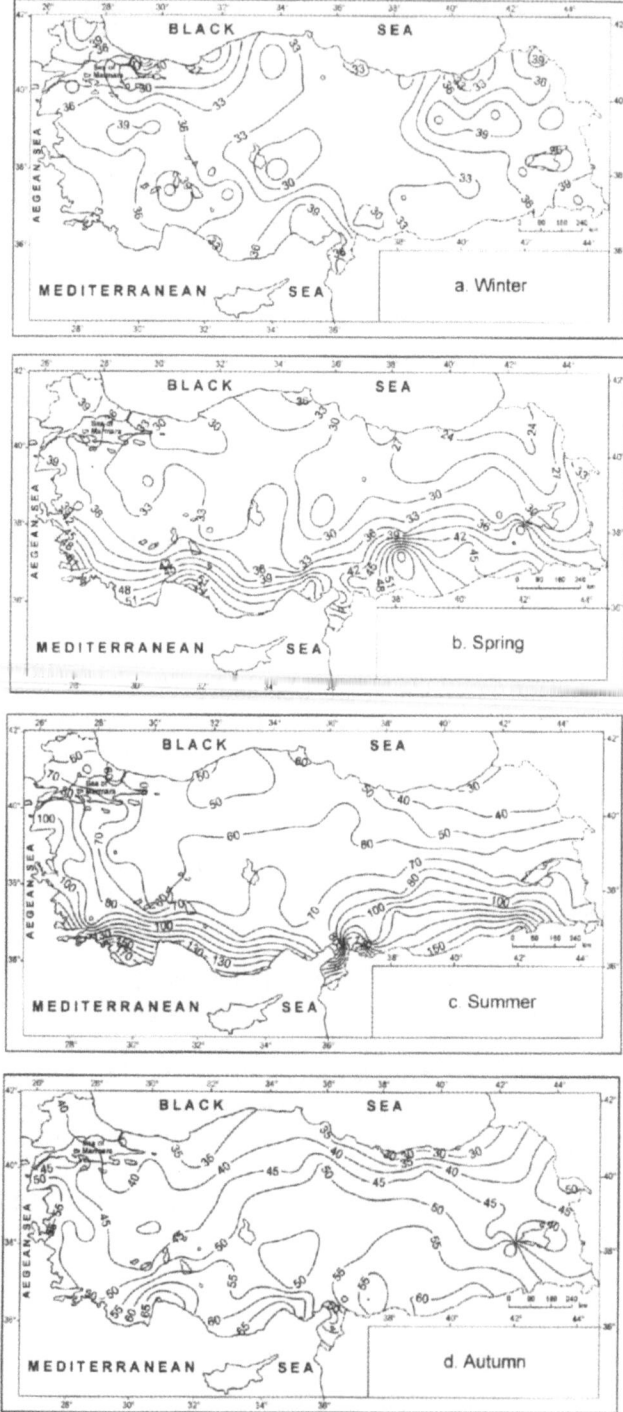

Fig. 8. Geographical distributions of coefficients of variation (CV in percent) for seasonal precipitation totals in Turkey

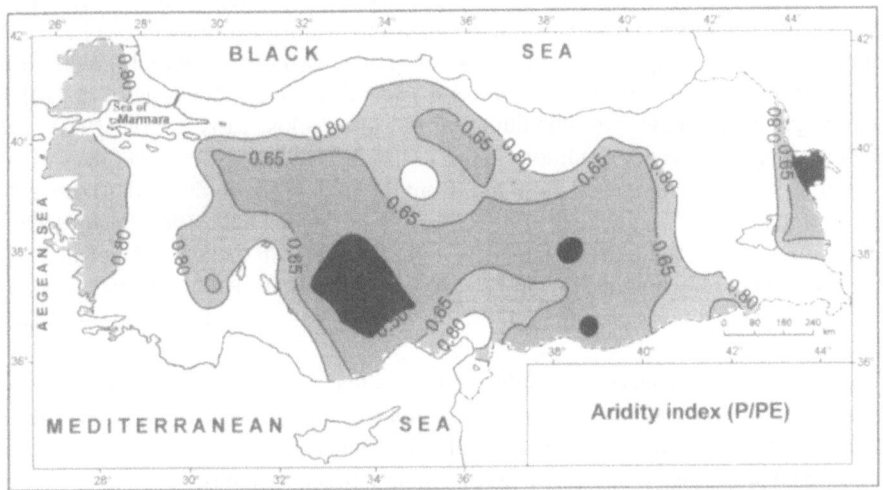

Fig. 9. Geographical distribution of aridity index values over Turkey (re-plotted from Türkeş, 1999)

4.2 Variations and Trends in Precipitation and Aridity Index Series

4.2.1   Annual and Seasonal Precipitation Series

Long-period fluctuations. Long-term fluctuations in annual and seasonal normalized precipitation anomaly series of 91 stations are examined by means of the 9-point Gaussian filter to eliminate the variations shorter than about ten-year periods. Time-series plots for variations in winter and spring precipitation anomaly series are shown in Fig. 10.1-2 for the selected 15 stations that are representative for all rainfall regime regions of Turkey. Results from visual examinations on the time-series plots are summarized as follows.

In winter, a general upward trend with a low-frequency fluctuation dominated by wetter anomalies are followed by a general downward trend, which is dominated by longer runs of drier anomalies from the late 1960s at many stations and from the late 1970s at some stations (Fig. 10.1). Stations of the BLS region generally experience the wet conditions mostly in the late 1930s, 1940s, 1960s, late 1970s and early 1980s, along with the dry conditions in the 1970s, mid-late 1980s and 1990s. At the stations of the MED region, wet conditions generally occur in the 1940s, 1960s and early-mid 1980s, whereas dry conditions appear in the early-mid 1930s, around 1970s, late 1980s and 1990s. At the stations of the CMED and the MEDT regions, fluctuation patterns are quite similar to the patterns of the MED region, with the wet conditions in the 1940s, mid 1950s, mid-late 1960s, late 1970s and early 1980s and with the dry conditions in the early-mid 1930s, early-mid 1970s, mid-late 1980s and 1990s. At the stations of the CCAN region, wet conditions are evident in the early-mid 1940s, 1960s, mid-late 1970s and early 1980s, and dry conditions are evident in the early-mid 1930s, 1950s, early 1970s, mid-late 1980s and 1990s.

Spring series generally indicate a high-frequency variation with an unchanged mean at some stations and about slightly increasing mean at many stations (Fig. 10.2). Stations of the BLS region generally experience increased precipitation from the early 1940s to mid-1960s. Thereafter there is a general decrease. Stations of the MED region generally depict wet conditions in the late 1940s, early 1950s, 1970s and 1980s, along with the dry conditions in the 1940s and 1990s. Stations over the CMED region show a general increasing trend, from the mid-1940s to the mid-1970s, with the wet conditions in the late 1940s, early-mid 1950s, late 1960s and 1970s. Thereafter drier conditions dominate. At the stations of the CCAN region, wet conditions generally dominate over the early 1950s, mid-late 1960s and 1970s, along with the dry conditions over the 1930s, early-mid 1940s, mid-late 1950s and early 1960s.

Summer rainfall series generally tend to increase at some stations and to show unchanged situation at many stations. The longer runs of negative anomalies are a dominant characteristic, even though individual highest positive anomalies are also observed in this season (not shown here). Autumn precipitation series generally show a random run of anomalies at different time periods at many stations, mostly with an unchanged mean, along with a some degree of low-frequency fluctuation at some stations. Autumn precipitation series also indicate an increase during the last 10 years at many stations, except those over the MED and MEDT regions (not shown here).

As an expected result of the general decrease with an apparent low-frequency fluctuation in winter precipitation, annual precipitation series of many stations also indicate a low-frequency fluctuation with a decreasing mean over the study period (not shown here). Annual precipitation is generally decreasing at many stations on the Black Sea coast and over the MED region. Long-period fluctuations are evident particularly in the stations over the MED, MEDT, CCAN and CEAN regions. Most of the years from the early 1970s are generally characterized by a longer run of the dry anomalies.

Secular trends. Numbers of the stations having a significant secular trend in normalized annual and seasonal precipitation anomaly series are given in Table 1. The time-series plots of sequential values of the statistics $u(t)$ and $u'(t)$ from the Mann-Kendall rank correlation test are also shown in Fig. 11.1-2 for winter and spring precipitation anomaly series of the selected 15 stations of Turkey. Results from both the Mann-Kendall test statistics and visual interpretations of the time-series plots of sequential values of the statistics $u(t)$ and $u'(t)$ for 91 stations are summarized as follows:

- In spring, a general upward trend with high year to year variations is dominant in many stations except the stations over the BLS Region, where a general downward trend appears. The upward trend was evident during the period from the mid-1940s to the late 1960s at many stations and to the 1980s at some stations (Fig. 11.2). This increased precipitation period has been generally followed by a downward trend at many stations. Only 4 stations experience any significant trend in spring precipitation series.
- In summer, 7 stations, almost all of which take place in the CMED and CCAN regions, indicate a significant upward trend.

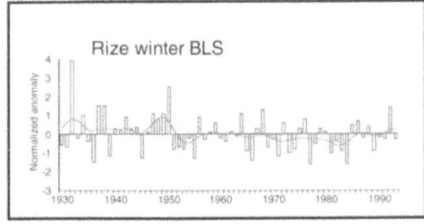

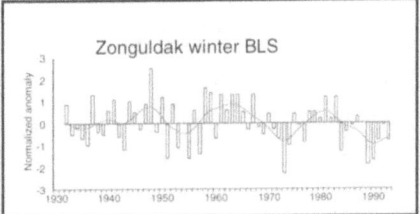

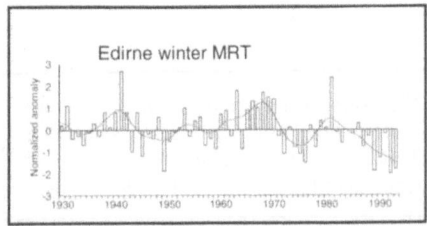

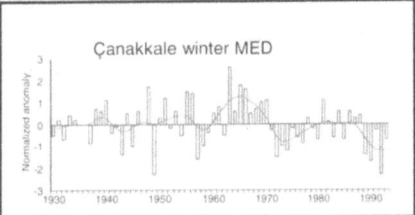

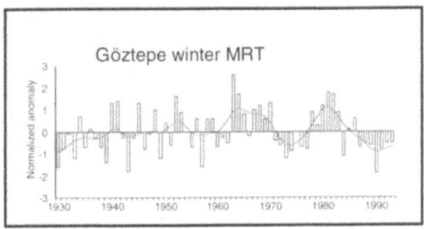

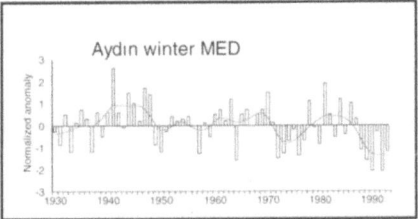

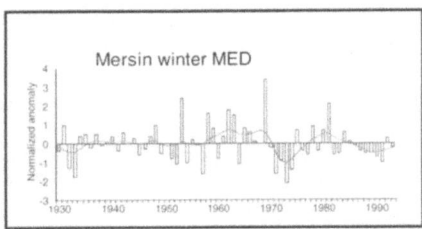

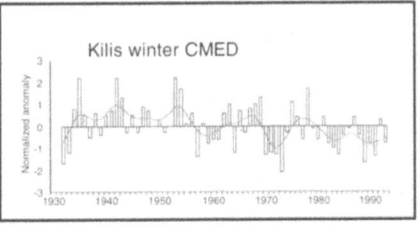

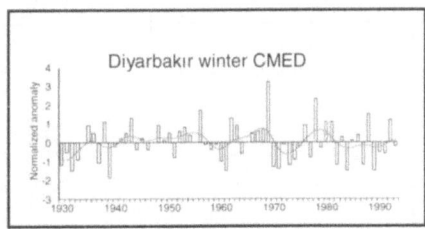

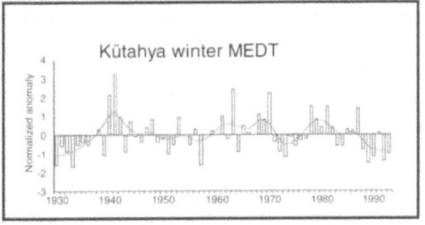

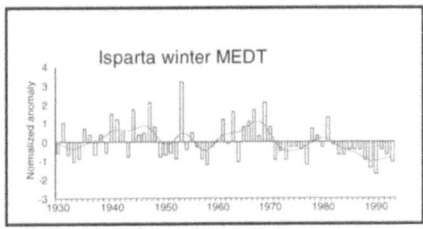

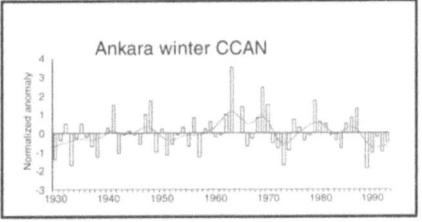

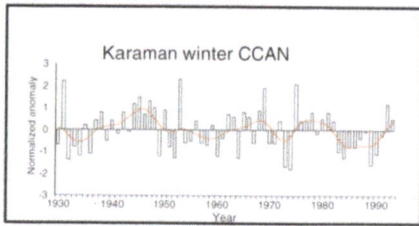

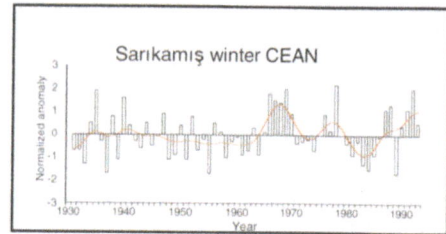

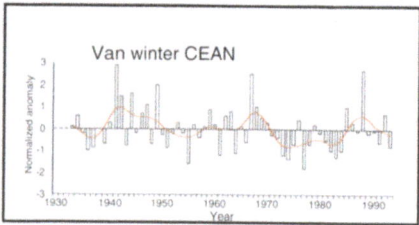

Fig. 10.1. Variation of normalized winter precipitation series of selected stations in Turkey with smoothed line by the nine-point Gaussian filter (——) with padded ends.

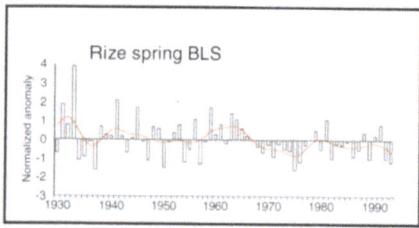

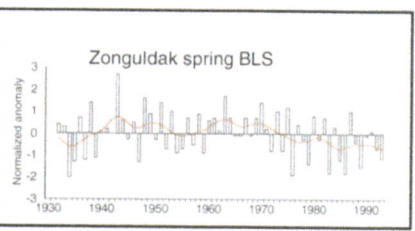

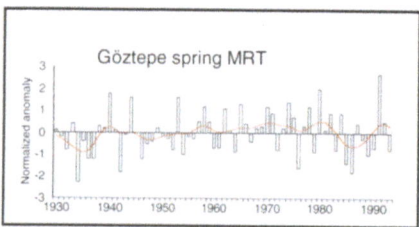

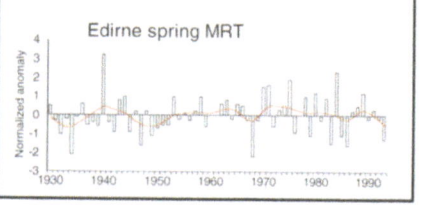

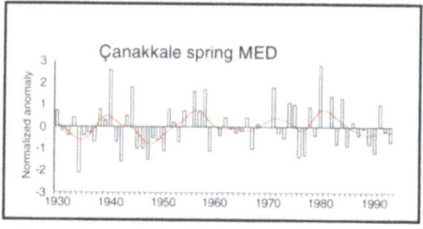

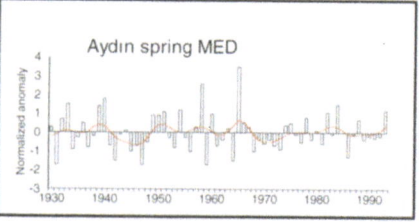

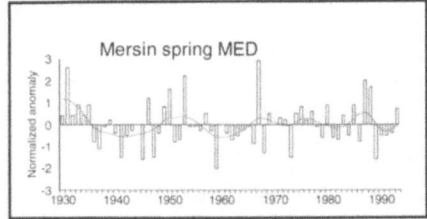

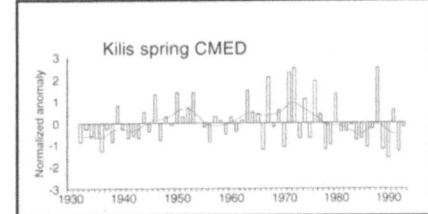

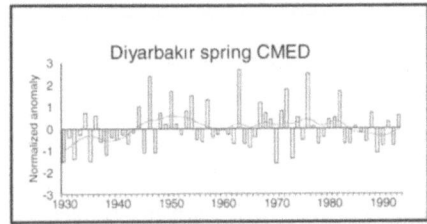

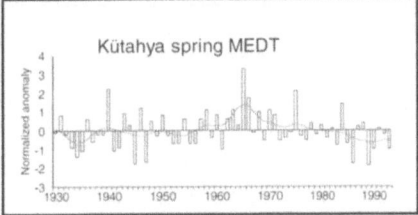

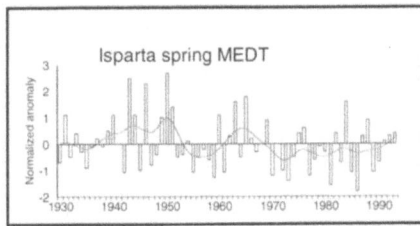

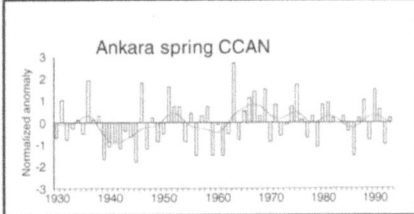

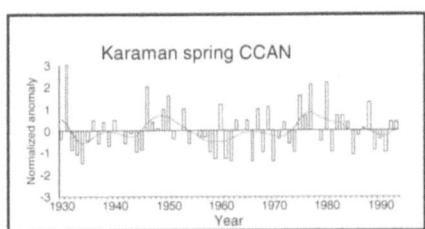

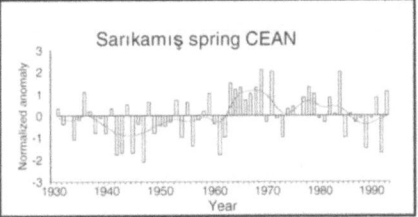

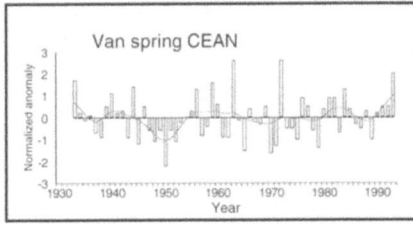

Fig. 10.2. Variations of normalized spring precipitation series of selected stations in Turkey, with smoothed line by the nine-point Gaussian filter (————) with padded ends.

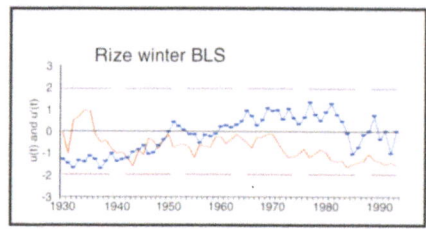

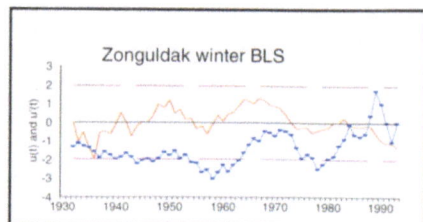

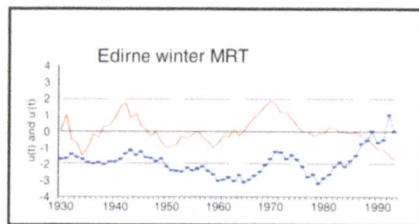

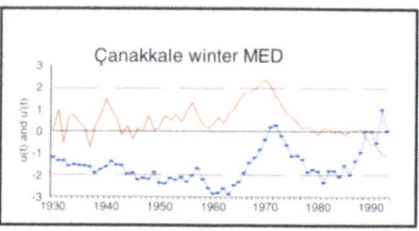

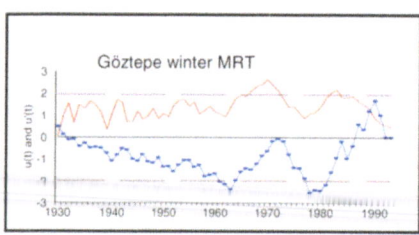

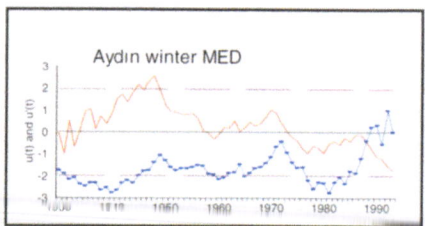

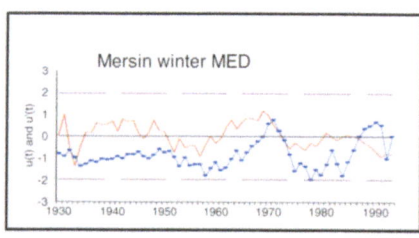

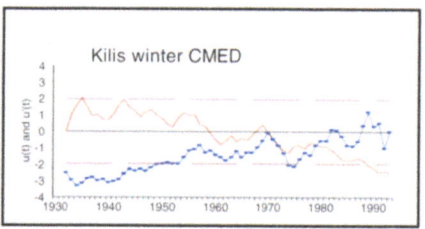

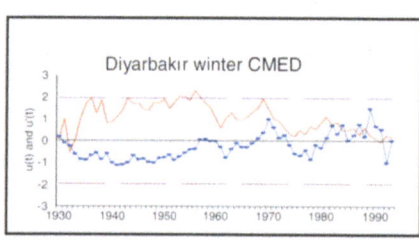

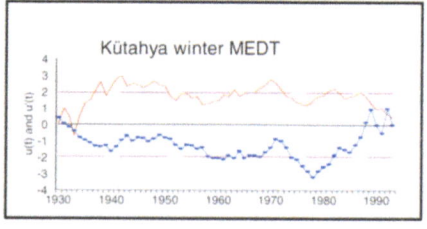

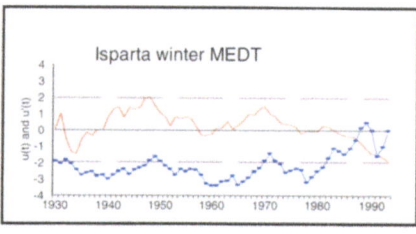

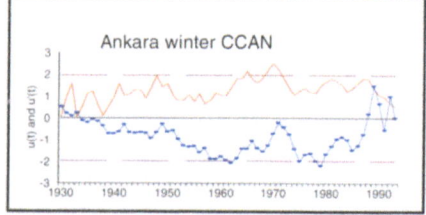

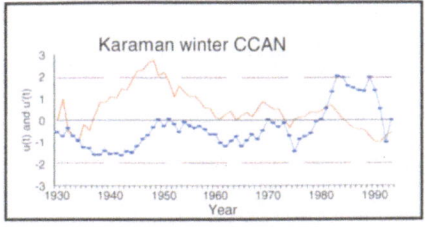

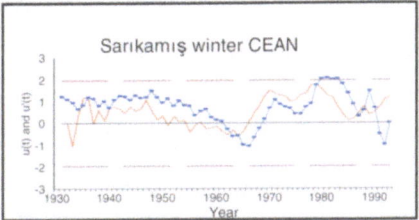

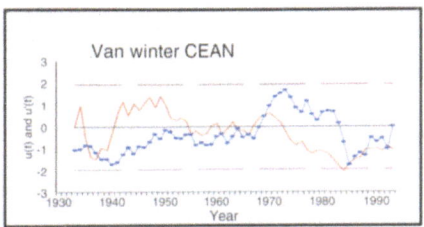

Fig. 11.1. Trends of normalized winter precipitation series of selected stations in Turkey from sequential values of the statistics u(t) (———) and u'(t) (—•—) of the Mann-Kendall test, with the critical value

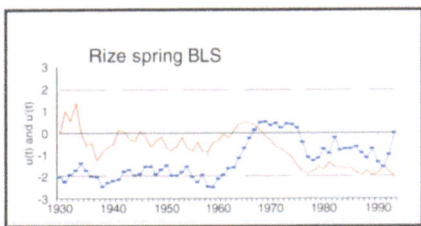

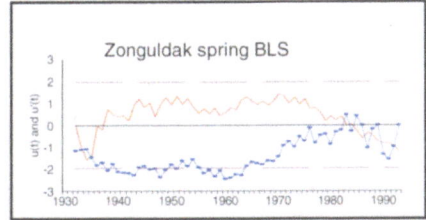

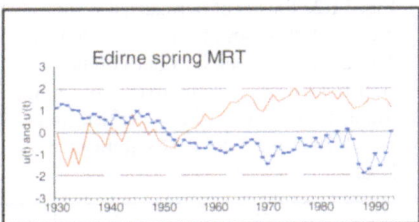

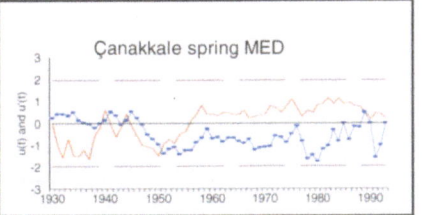

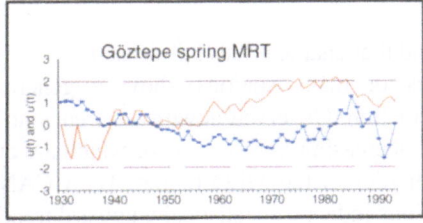

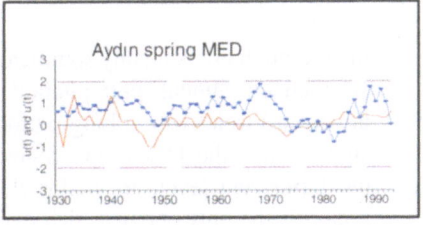

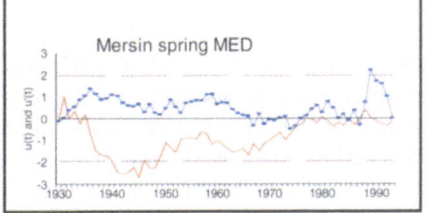

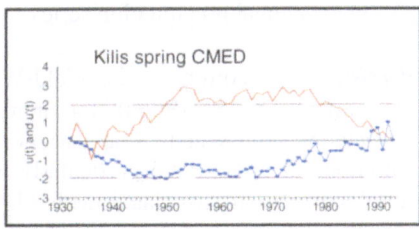

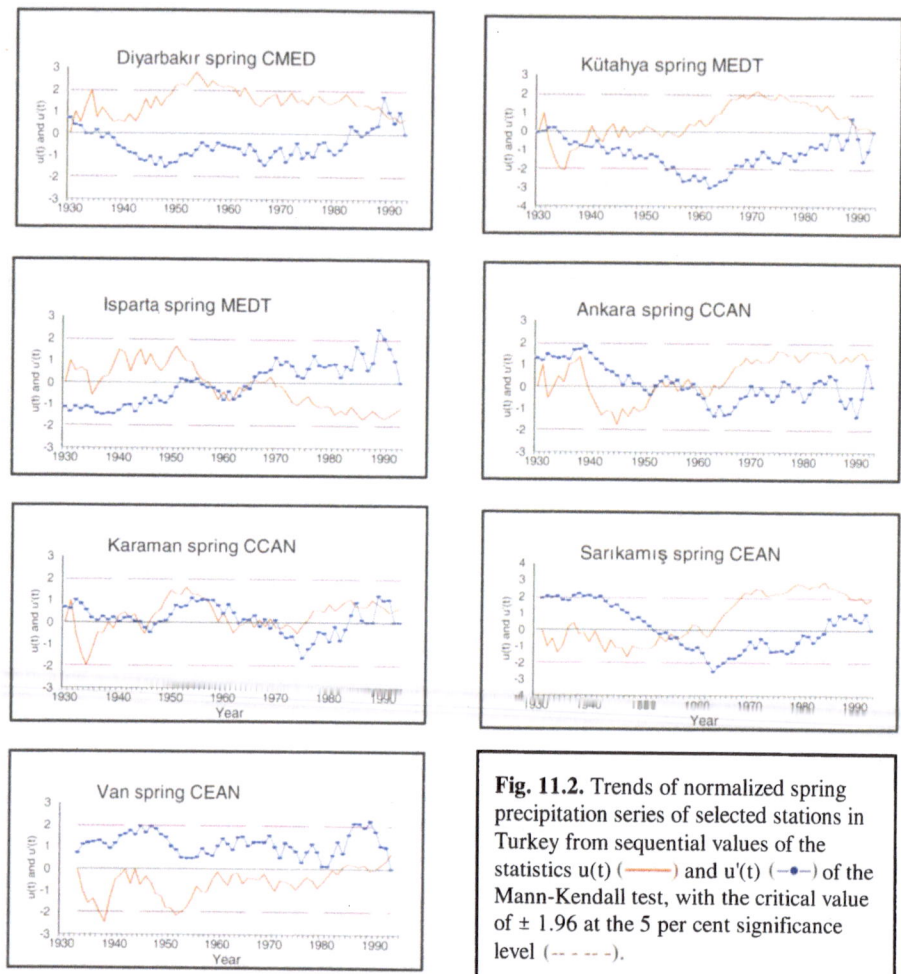

Fig. 11.2. Trends of normalized spring precipitation series of selected stations in Turkey from sequential values of the statistics u(t) (———) and u'(t) (—•—) of the Mann-Kendall test, with the critical value of ± 1.96 at the 5 per cent significance level (-- -- --).

- In autumn, there is no any significant trend that characterizes this season.
- Normalized annual precipitation series of many stations show a general downward trend with a long-period fluctuation. Observed downward trends are significant at the 15 stations. Significant decreasing trends, most of which are at the 1 per cent level, are major characteristics of the MED region. The CCAN region is the only one in which two stations indicate a statistically upward trend in the annual precipitation series.

Persistence. According to the Wald-Wolfowitz serial correlation analysis performed by Türkeş (1995, 1996a) for normalized precipitation anomalies, significant persistence from year to year variations was one of the dominant characteristic of variations in annual and particularly in winter series, whereas significant negative serial correlation associated with high year to year oscillations was evident in variations of spring series.

In order to assess the nature and magnitude of the observed persistence in variations of precipitation anomaly series, Türkeş (1998a) computed the lag-one serial correlation (L-$1SC$) coefficient, r_1, for 91 stations and rainfall regions of Turkey. However, only station-based results are given here. The L-$1SC$ is positive at most of the stations in winter, except at the stations of the BLS region. There is a significant persistence in winter precipitation variations from year to year at 31 stations, 8 of

Table 1. Number of the stations indicating a significant trend and/or serial correlation in the normalized precipitation series at the 5 or 1 per cent level, according to the Mann-Kendall (M-K) and lag-one serial correlation (L-$1SC$) tests (Türkeş, 1998a). (+): upward trend from the M-K test and positive correlation coefficient from the L-$1SC$ test; (−): downward trend from the M-K test and negative correlation coefficient from the L-$1SC$ test.

Rainfall Region	Winter M-K +	Winter M-K −	Winter L-1SC +	Spring M-K +	Spring M-K −	Spring L-1SC −	Summer M-K +	Summer L-1SC +	Summer L-1SC −	Autumn L-1SC +	Autumn L-1SC −	Annual M-K +	Annual M-K −	Annual L-1SC +
BLS	1			1		2		1				1		
MRT			6					1				1		2
MED	8		8	1		4	1	1	2	1	1	6		4
CMED	3		5	1		4	4		1	1		1		2
MEDT	1		1					1				1		2
CCAN	1	1	6			6	2	1				2	4	6
CEAN	1		5	1		2			1		1	1		1
Total	16		31	4		18	7	9		4		17		17

which are in the stations with winter-maximum Mediterranean type rainfall regime (Table 1). Annual precipitation series of 17 stations also depict a significant positive correlation. However, the L-$1SC$s are negative at almost all stations in spring, 18 of which are statistically significant. This statistic indicates that spring precipitation series contains a high-frequency oscillation. Both negative and positive L-$1SC$s are found for summer rainfall series, in which only 9 of them differ significantly from zero. Majority of autumn precipitation series seems to be statistically random against the serial correlation.

A simple Markov type persistence is found approximately in winter precipitation anomaly series of 11 stations, when the exponential relationships are that of $r_2 \cong r_1^2$ and $r_3 \cong r_1^3$ are considered. The Markov persistence is the dominant characteristic particularly in those winter precipitation series of the stations, which exist over the colder and more continental north-eastern part of the CEAN region. Desired relationships for the Markov persistence are also seen to be satisfied approximately in summer series of four stations, in one autumn series, and in the annual series of four stations. On the other hand, the significant negative L-$1SCs$ in spring series are associated closely with a marked short-period oscillation rather than a persistence, which comes out in the spectral analysis.

4.2.2 Aridity Index Series

Fluctuations and trends. As an expected result of the decreasing trends in annual precipitation of most stations, annual AI series have generally tended to decrease at many stations of Turkey. However, significant downward trends are found at some stations of the MED region (Table 2). At the stations of the Aegean part of the MED region with a significant decreasing trend, there is a marked change from the humid or near humid conditions of the 1960s to the dry sub-humid or semi-arid climatic conditions of the mid and late 1980s and early 1990s (Fig. 12). In contrast, annual AI values tend to increase significantly towards humid or semi-humid climatic conditions at 5 stations, which are located over the northern part of the CCAN region of Turkey (Table 2).

In addition to the secular trends from the Mann-Kendall $u(t)$ statistics, the beginning of secular or partial trends in AI series, the duration of significance and the change points were investigated by means of time-series plots of sequential values of the statistics $u(t)$ and $u'(t)$ from the Mann-Kendall test. Plots of the selected stations are shown in Fig. 13. When the curve of the $u(t_i)$ values exceed the absolute value of the critical value, which is 1.96 at the 5 percent significant level, a significant period of trend is indicated. On the plots of Balikesir, Akhisar, Dikili and Bodrum, few values of $u(t_i)$ passed the 5 per cent significant level, although the $u(t)$ statistics of these stations indicated a significant decreasing tendency in their AI series. Thus, as pointed out by Sneyers (1990), it may be assumed that, if the trend is real, its effect is very recent, and it is recommended to wait for later observations to confirm the trend. For an abrupt change, or for the beginning point of a secular trend, $u(t_i)$ and $u'(t_i)$ curves would exhibit the same behaviour at the change point over the series. The two lines should also intersect at that point. This clearly shows up on the plots of Afyon, Ankara and Merzifon stations. Upward trends in AI series of Ankara and Merzifon began in 1957 and 1970, and become significant during the periods of 1971-1993 and 1979-1993, respectively.

Persistence and periodicity. Türkeş (1999) applied the L-$1SC$ test to the annual AI values, and found followings. The AI series of eight stations indicate a significant positive L-$1SC$ (Table 2), although most AI series are dominated by a positive serial correlation. By considering the exponential relationships $r_2 \cong r_1^2$ and $r_3 \cong r_1^3$, it was found that only the AI series of Konya station in the CCAN region is characterized

Table 2. Number of the stations indicating a significant trend and serial correlation in the annual aridity index series of 55 stations at the 5 or 1 per cent level, according to the Mann-Kendall trend and the L-$1SC$ tests (Türkeş, 1999). (+): upward trend from the Mann-Kendall test and positive correlation coefficient from the L-$1SC$ test; (-): downward trend from the Mann-Kendall test.

Rainfall Region	Aridity Index			
	Mann-Kendall Test		Serial Correlation Test	
	+	−	+	−
BLS				
MRT			1	
MED		5	3	
CMED			1	
MEDT		1	1	
CCAN	5		2	
CEAN		1		
Total	5		8	

by a simple Markov type persistence. Hence, the appropriate 'null' hypothesis continuum to the spectrum was assumed to be the Markov 'red' noise only for Konya. The rest of the series of 55 stations, having non-significant positive L-$1SC$ or significant positive L-$1SC$ without Markov-type persistence and significant but negative L-$1SC$, was formulated by a 'white' noise continuum. The 'white' noise continuum and its 0.90 and 0.95 confidence levels that are plotted superposed on the spectrum are shown in Fig. 14 for annual AI series of selected stations. The significant spectral peaks are also specified by their period values in years.

Results of the estimated spectral peaks that exceed the 0.90 and 0.95 confidence limits of the 'white' noise continuum of the spectrum are not shown in a table, but can be summarized as follows:

Stations of the Black Sea rainfall region are mostly characterized by spectral peaks with cycles shorter than 3 years. Spectral peaks having 2.5 year quasi-biennial oscillation (QBO) deviate from the 'white' noise continuum at the 0.95 confidence level at the Rize, Samsun and Bolu stations, first two of which are located on the coastal belt of the Black Sea. In case of significant spectral peaks, relatively frequent peaks occur at shorter periods of 2.2-2.3 years (QBO) at Adana and 2.5-2.6 years at Dörtyol, and at longer periods of 15-20 years at rest of the Mediterranean stations. At the stations of the MRT region, shorter cycles of 2.5-3.5 years in which 3.0-3.3 year periods are significant at the 0.95 confidence level, and longer cycles of 18-19 years generally dominate within the observed variations of the aridity index series. Stations in the CCAN, CEAN and CMED regions have different periodicity in their aridity index series. QBOs with the cycles of 2.5-2.6 years at Eskişehir and 2.1-2.2 years at Diyarbakır exceed the 0.95 confidence limits of the 'white' noise continuum. Diyarbakır also has significant biennial oscillation at the 0.90 confidence level.

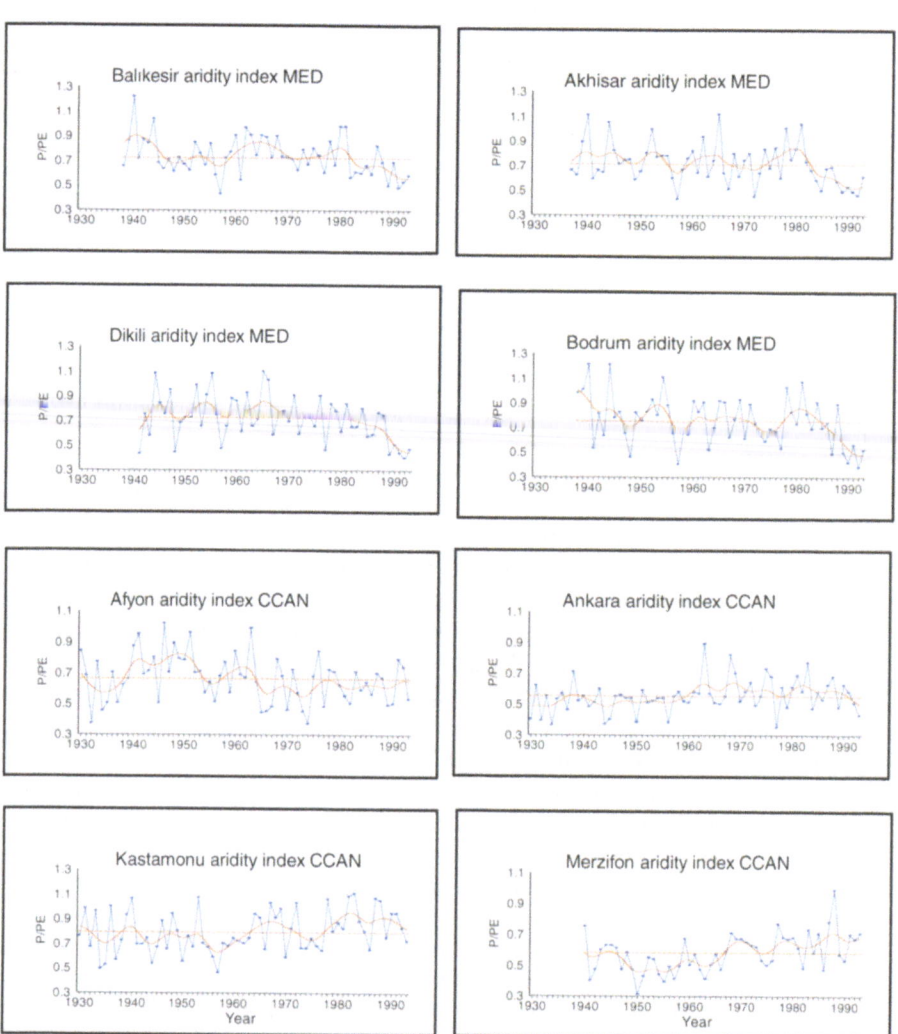

Fig.12. Variations of annual *AI* series of selected stations in Turkey, with smoothed line by the nine-point Gaussian filter (——) with padded ends and long-term average (- - - -) (re-plotted from Türkeş, 1999)

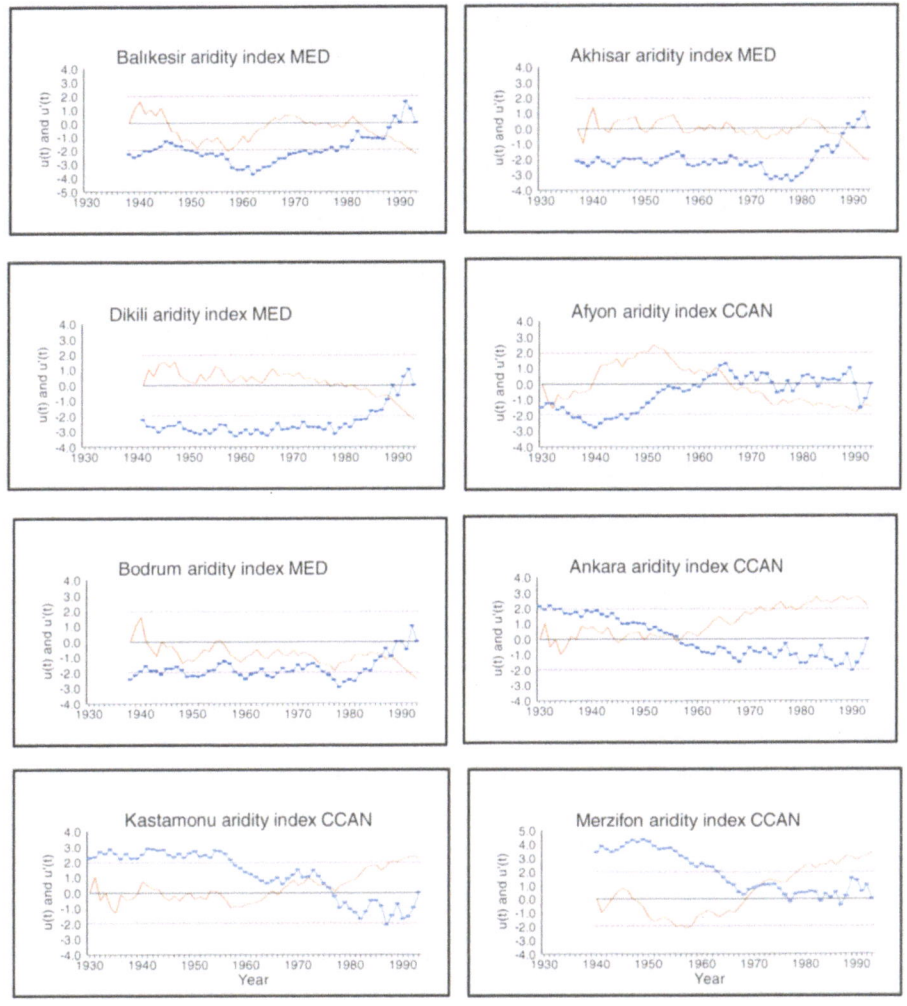

Fig. 13. Trends of annual *AI* series of selected stations in Turkey from sequential values of the statistics u(t) (————) and u'(t) (   ) of the Mann-Kendall test, with the critical value of ± 1.96 at the 5 per cent significance level (- - - - -) (re-plotted from Türkeş, 1999)

4.2.3   An Assessment of Precipitation Anomalies During the Period 1994-2000

The main section of this study was performed based on the precipitation and aridity index series for the 1930-1993 period. It was already seen that dry conditions dominated mostly during the period from mid-late 1980s to the early 1990s over the considerable parts of Turkey. A new analysis has also been made for the precipitation data of 1994-2000, in order to evaluate the spatial and temporal variations of dry and

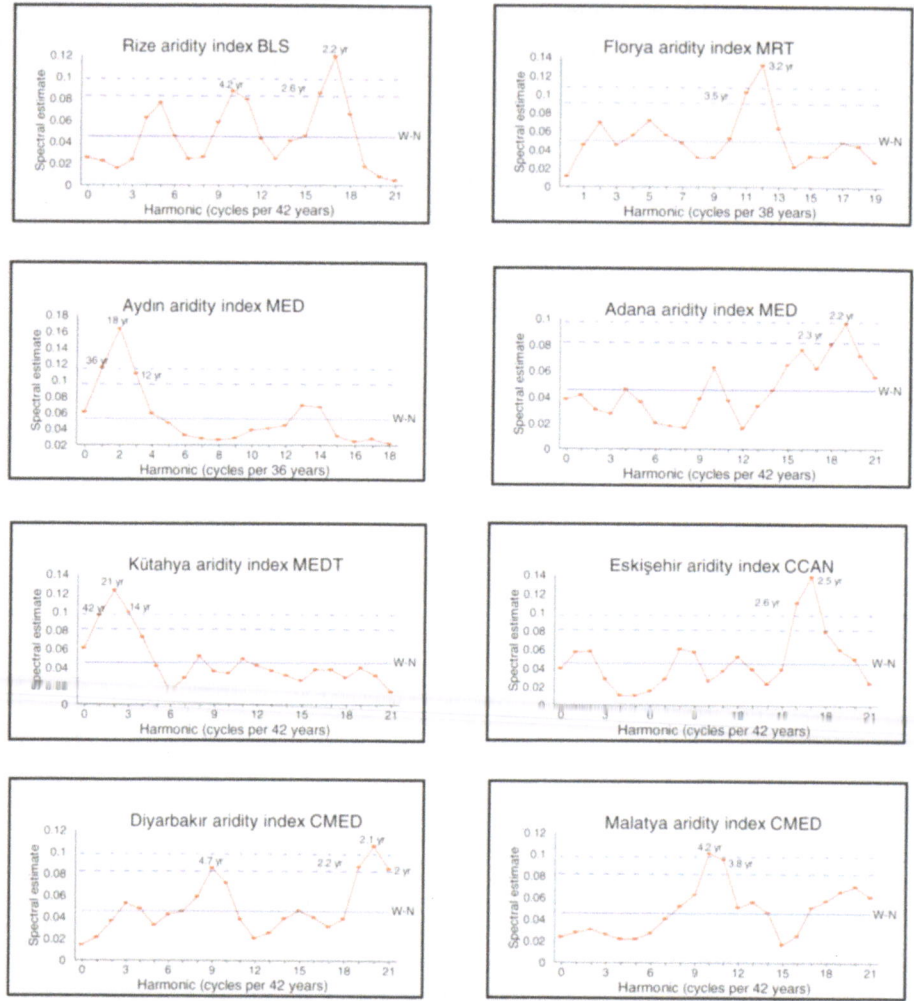

Fig. 14. Power spectrum plots of annual aridity index series of selected stations in Turkey. (—•—), spectral estimates; (———), 'white' noise (W-N) continuum with the 0.90 (– – – –) and 0.95 (– – -) confidence limits of the spectrum (re-plotted from Türkeş, 1999).

wet conditions over this period in Turkey. To make such an assessment, a normalized precipitation anomaly index, which was previously used for Turkey, has been taken into account. Description and classification of this index can be found in Türkeş (1996b).

During the period between the years of 1994 and 1998, precipitation of Turkey has experienced a recovering situation towards near normal and/or above normal conditions over most of the country, except the Continental Eastern Anatolia region.

For instance, in 1997, annual precipitation was generally near and/or above normal over most of Turkey (Fig. 15a). Much below normal precipitation anomalies dominated over the Continental Eastern Anatolia and eastern part of the Continental

Mediterranean, whereas wetter than normal conditions occurred over the western and middle parts of the Black Sea region, and Marmara and Continental Central Anatolia regions. Extremely above normal precipitation conditions was observed over the western Black Sea and Marmara regions. In 1998, above normal precipitation conditions showed a spatial coherence over most of Turkey (Fig. 15b). Precipitation anomalies that are very much above normal were apparent throughout most of the continental interiors of the country. Below normal precipitation conditions dominated over eastern and southern parts of the Continental Eastern Anatolia and Continental Mediterranean regions. Very much and extremely below normal precipitation anomalies influenced the south-eastern corner of Turkey.

Immediately following this recovery period, most areas of Turkey, except the western and middle parts of the Black Sea region and partly Continental Central Anatolia, have experienced another dry period lasting up to now, including January of 2001.

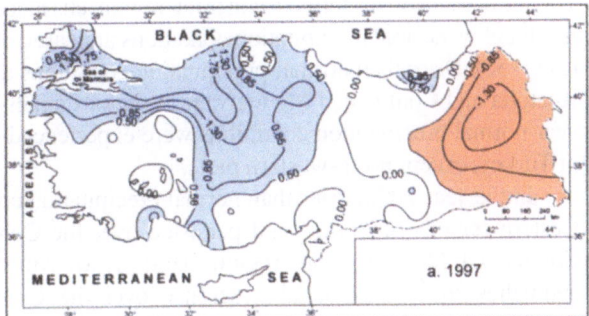

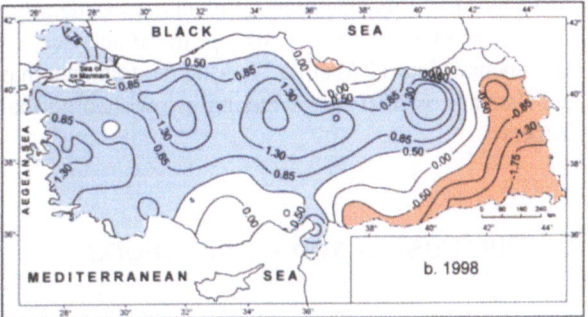

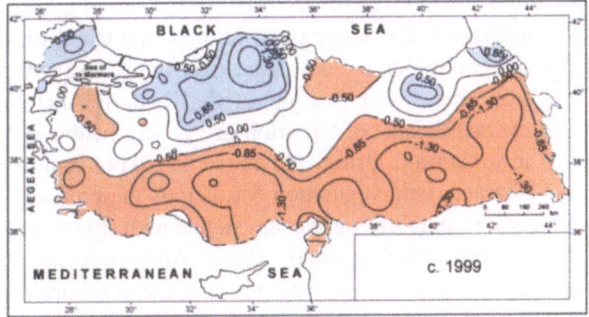

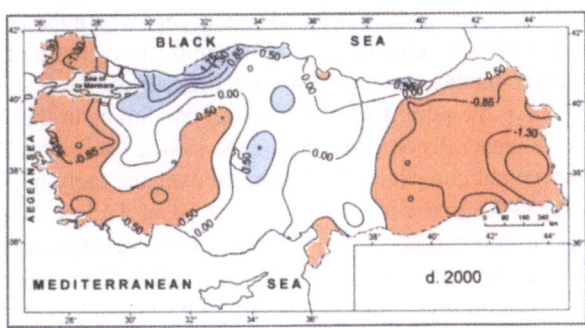

Fig. 15. Geographical distributions of normalized precipitation anomalies over Turkey, for the years of (a) 1997, (b) 1998, (c) 1999, and (d) 2000

In 1999, a well-defined coherent region characterized with below normal and much below normal precipitation conditions appeared over the southern half of Turkey (Fig. 15c). Drier than normal precipitation conditions affected the Continental Eastern Anatolia and Continental Mediterranean regions again. In this year, drier than and/or near normal precipitation conditions were experienced almost in all months over most of Turkey, except north-western parts.

In the year 2000, drier than normal precipitation conditions were evident over the western regions of Turkey and persisted over the Continental Eastern Anatolia and Continental Mediterranean regions (Fig. 15d). Much below normal precipitation anomalies showed a spatial coherence over the Continental Eastern Anatolia and Continental Mediterranean regions. Monthly assessment of the precipitation anomalies indicated that drier than normal conditions generally dominated throughout the year 2000 over most of Turkey, except the Black Sea, eastern part of the Marmara and Continental Central Anatolia. During the four-month period between October of 2000 and January of 2001, areas having below than normal precipitation covered most of Turkey.

5 Discussion and Conclusions

Dry sub-humid climatic conditions extend over most of the Continental Central Anatolia and Continental Mediterranean regions, some parts of the eastern Mediterranean, and some parts of the Continental Eastern Anatolia region. Semi-arid climatic conditions are dominant over the Konya sub-region of the Central Anatolia, and over the eastern margin of the Eastern Anatolia region. In the arid lands of Turkey, coefficients of variation are generally above 20 per cent annually and 30 per cent in winter and spring, and above 45-50 per cent in autumn and summer.

Characteristic vegetation formations of semi-arid and dry sub-humid regions of Turkey are generally steppe and steppe with sparse trees, and steppe with trees and dry forests, respectively (Atalay, 1994). Anthropogenic steppes with dry forests are the dominant vegetation formations over the semi-arid and dry sub-humid parts of the Continental Eastern Anatolia region. These sparse vegetation covers protect and stabilize the land surface and soil, except where the land is completely degraded, and when climatic changes result in significant increase of aridity conditions, or decrease of precipitation amounts, and water resources.

For the period 1930-1993, wet conditions in the annual and winter precipitation series generally occurred during the 1940s, 1960s, late 1970s and early 1980s, whereas dry conditions generally dominated over the early-mid 1930s, early-mid 1970s, mid-late 1980s and early 1990s over most of Turkey. When the 1970-1993 period was taken into account, it was found that severe and widespread dry conditions occurred especially in the years of 1973, 1977, 1984, 1989 and 1990 (Türkeş, 1996b).

Spring precipitation series of many stations generally indicated an upward trend from the early 1940s to the late 1960s or to the 1980s and this period was generally followed by a downward trend. The longer runs of dry anomalies were dominant characteristics of the summer. Autumn precipitation series generally show a random run of anomalies mostly with a stable mean at many stations, along with a some degree of low-frequency fluctuation at some stations. Significant decreasing trends showed up in the annual precipitation series of 15 stations and in the winter precipitation series of 14 stations, mostly over the Mediterranean region. Summer rainfall series tended to increase significantly in 7 stations. The annual and winter precipitation anomalies of 17 and 31 stations, respectively, had a significant positive L-$1SC$. Spring precipitation anomalies of 18 stations depicted a significant negative L-$1SC$. The Markov type persistence showed up in the winter precipitation series of 11 stations.

When the additional analysis on the spatial and temporal variations of normalized precipitation anomalies for the 1994-2000 period was taken into account, it has been found that a marked recovery period characterized by near normal and/or above normal precipitation conditions occurred during the mid-late 1990s (1994-1998) over most of the country, except the Continental Eastern Anatolia region. Immediately following this recovery period, much of Turkey, except the western and middle parts of the Black Sea region and partly Continental Central Anatolia, have experienced another dry period lasting up to now, including January of 2001. As a result of last period of drier than normal precipitation conditions, water stress and shortage reached a critical point again in Turkey, for not only agricultural and energy production but also water resources including irrigation, drinking water and other hydrologic systems.

For the period 1930-1993, there has been a general tendency of a shift from humid conditions during the 1960s to the dry sub-humid climatic conditions of the early 1990s, in the aridity index series of many stations in Turkey. Significant downward trends in the AI series show up at the stations of the Mediterranean region. At some stations of the Aegean part of the Mediterranean region, there is a significant change from humid conditions to dry sub-humid or semi-arid conditions. Aridity index series of the Black Sea rainfall region are mostly characterized by spectral peaks with cycles shorter than 3 years. Relatively frequent peaks occur at shorter periods of 2.2-2.3 years (QBO) particularly at the eastern Mediterranean stations, and at longer periods of 15-20 years at rest of the Mediterranean stations. At the stations of the MRT region, shorter cycles of 2.5-3.5 years in which 3.0-3.3 year periods are significant, and longer cycles of 18-19 years generally dominate within the observed variations of the aridity index series. Stations in the CCAN, CEAN and CMED regions have different periodicity in their aridity index series. QBOs with the cycles of 2.5-2.6 years at Eskişehir and 2.1-2.2 years at Diyarbakır exceed the 0.95 confidence limits of the 'white' noise continuum.

The Continental Mediterranean region and continental interiors of Turkey could be aridlands that are affected by desertification processes, owing to the climatic factors that may lead to the desertification. Climatic factors would include existing semi-arid and dry sub-humid climatic conditions, long-lasting summer dryness of the air and soil, particularly in the Continental Mediterranean region with high temperature, high precipitation variability, and low and erratic rainfall amount. Significant trends towards drier than normal conditions in annual and winter precipitation, and towards dry sub-humid or semi-arid climatic conditions may have been increasing desertification processes in the Mediterranean region of Turkey. When other natural (especially geomorphologic and pedologic) and anthropogenic factors, such as forest fires, recent miss-use of agricultural lands and, are also taken into account, this region could be considered as areas that may be more vulnerable to desertification processes in the future.

The quantity and quality of water and land supplies, especially agricultural lands in Turkey have already been affected by rapid population growth and industrialization, as well as by changes in demands, technology, and socio-economic and legislative conditions, as in the other developing countries. Thus, options for dealing with the possible impacts of the climate change on water resources, drought and desertification should include efficient management of existing water and land supplies, and forecasting systems for droughts and monitoring of the desertification processes including soil erosion and changes in vegetation formations and/or covers.

Acknowledgements: The author would like to thank to Utku M. Sümer of the TSMS for his support in programming of some computations, and to Yurdanur Türkeş of the TSMS for her valuable comments.

References

Atalay İ (1994) Türkiye Vejetasyon Coğrafyası (Vegetation Geography of Turkey). Ege Üniversitesi Basımevi, İzmir

Blackman RB, Tukey JW (1958) The Measurement of Power Spectra. Dover Publications, New York.

Cressie NAC (1991) Statistics for Spatial Data. John Wiley, New York

Delfiner P, Delhomme JP (1975) Optimum interpolation by kriging. In: Davis JC, McCullagh MJ (eds) Display and Analysis of Spatial Data. John Wiley, New York pp. 96-119.

ECSN (1995) Climate of Europe: Recent Variation, Present State and Future Prospects. European Climate Support Network (ECSN), Nijkerk (the Netherlands)

Glantz MH (1987) Drought, famine and the seasons in Sub-Saharan Africa. In: Climate and Human Health. Proceedings of the Symposium in Leningrad, Volume I, WCAP-No., 1, World Meteorological Organization, Geneva, pp. 217-232

Hevesi JA, Flint AL, Istok JD (1992a) Precipitation estimation in mountainous terrain using multivariate geostatistics, I, Structural Analysi. J. Appl. Meteorol., 31:661-676

Hevesi JA, Flint AL, Istok JD (1992b) Precipitation estimation in mountainous terrain using multivariate geostatistics, II, Structural Analysis, J. Appl. Meteorol., 31:677-688

Jacobeit J (1996) Atmospheric circulation changes due to increased greenhouse warming and its impact on seasonal rainfall in the Mediterranean area. In Climate Variability and Climate Change Vulnerability and Adaptation, Proceedings of the Regional Workshop in Praha, September 11-15, 1995, pp 71-80

Jenkins GM, Watts DG (1968) Spectral Analysis and its Applications. Holden-Day, San Francisco

Journel AG, Huijbregts CJ (1978) Mining Geostatistics. Academic Press Inc., London

Singh VP (1992) Elementary Hydrology, Prentice-Hall Inc., New Jersey

Sneyers R (1990) On the Statistical Analysis of Series of Observations. WMO Technical Note 43, World Meteorological Organization, Geneva

Thornthwaite CW (1948) An Approach toward a rational classification of climate. Geog, Rev., 38:55-94

Türkeş M (1995) Türkiye'de yıllık ve mevsimlik yağış verilerindeki eğilimler ve dalgalanmalar', Türkiye Ulusal Jeodezi-Jeofizik Birliği ve Türkiye Ulusal Fotogrametri ve Uzaktan Algılama Birliği Kongreleri Bildiri Kitabı, 694-706, Harita Genel Komutanlığı, Ankara

Türkeş M (1996a) Spatial and temporal analysis of annual rainfall variations in Turkey. Int. J. Climatol, **16**, 1057-1076

Türkeş M (1996b) Meteorological Drought in Turkey: A Historical Perspective, 1930-1993, Drought Network News, University of Nebraska, **8**, 17-21

Türkeş M (1998a) Influence of geopotential heights, cyclone frequency and Southern Oscillation on rainfall variations in Turkey. Int. J. Climatol, **18**, 649-680

Türkeş M (1998b) İklimsel değişebilirlik açısından Türkiye'de çölleşmeye eğilimli alanlar', DMİ/İTÜ II. Hidrometeoroloji Sempozyumu Bildiri Kitabı, 45-57, Devlet Meteoroloji İşleri Genel Müdürlüğü, Ankara.

Türkeş M (1999) Vulnerability of Turkey to desertification with respect to precipitation and aridity conditions. Tr. J. of Engineering and Environmental Science, 23:363-380

UKMO (1995) Modelling Climate Change 1860-2050. Report published coincide with the COP-I to the UN/FCCC, Berlin, March 27 to April 7 1995, UK Meteorological Office, the Hadley Centre for Climate Prediction and Research

UKMO/DETR (1999) Climate Change and Its Impacts, Stabilisation of CO_2 in the Atmosphere. United Kingdom Meteorological Office and Department of the Environment, Transport and the Regions (UKMO/DETR), the Hadley Centre for Climate Prediction and Research, Bracknell.

UNCCD (1995) The United Nations Convention to Combat Desertification in those Countries Experiencing Serious Drought and/or Desertification, Particularly in Africa Text with Annexes, UNEP, Geneva

UNEP (1993) World Atlas of Desertification, the United Nations Environment Programme (UNEP), London.

Willmott CJ (1977) WATBUG: A FORTRAN IV Algorithm for Calculating the Climatic Water Budget. Water Resources Centre, University of Delaware, Newark, Delaware

WMO (1966) Climatic Change. WMO Technical Note, 79, World Meteorological Organization, Geneva.

Circulation Types and Their Influence on the Interannual Variability and Precipitation Changes in Greece

P. Maheras, C. Anagnostopoulou

Abstract

The data used in this study are daily precipitation amounts of 20 stations equally distributed over Greece and daily (12 TU) 500hPa geopotential heights, (20° to 65°N and 20° to 50°E) for the period 1958-1997. A decrease of annual, winter and autumn precipitation over the whole country was found, significant over several parts of Greece. An automatic classification scheme of the atmospheric circulation affecting Greece, between 1958-1997, is presented using spatial methods of topology and geometry. A short description of the synoptic characteristics of the 14 classified circulation types is given as well as the probability and the amount of precipitation associated with each type. It is found that the six anticyclonic types presented, during the rainy period (October to March), a cumulative frequency equal to 39.2% and give a rather small (8.2%) contribution to the "rainy" period precipitation amounts. On the other hand the four more "rainy" circulation types, namely the C, C_{SW}, C_{NW} and C_W together represent 37.8% of all the rainy period days and account more than 72% of the observed daily precipitation. The results obtained accentuate the existence of the strong links between the interannual variability of seasonal precipitation and interannual variability of seasonal frequency of the circulation types. The anomalous low or high precipitation amount during the hydrological year show a strong relationship with the anomalous low or high frequency of "rainy" circulation types. Multiple regression models developed for the simulation of the annual cycle of rain days and the corresponding precipitation, as well as selected extreme precipitation events for Greece stations. It is concluded that there is an agreement between the simulated annual cycle as well as the extreme rainfall events with the observations, both in terms of rainy days and precipitation amounts. It suggests that the precipitation regime in Greece including interannual variability, trend and extremes can be explained in terms of variability of a small number of circulation types patterns.

1 Introduction

Greece is located in the Eastern Mediterranean basin in the south Balkans. Its climate is highly diverse because of the complex terrain, together with the strong continental and maritime influences (Xoplaki et al., 2000) contributing to difficulties in weather forecasting, especially precipitation. The precipitation pattern over the Greek area shows a sharp gradient between the western part, where precipitation is 2 to 3 times higher than other regions. The higher precipitation totals in the west are related to the atmospheric circulation in association to the Mediterranean Sea surface temperature distribution and the complex topography of the region. More specifically, as air masses move towards the east passing over the sea, they pick up moisture in their lower layers and become unstable. Condensation of the water vapour occurs as a result of topographic forcing by the land and the mountains, such as the Pindus in NW and centre Greece, Olympus and the mountains of Crete, leads to heavy rainfall on the windward side.

Although the global circulation and regional climate factors can explain the spatial distribution of rainfall, as well as seasonal variability, they could not account for its interannual variability since it has different origin. Different techniques have been used to relate Greek monthly precipitation and atmospheric circulation indices, namely correlation coefficients analysis (Kozuchowski et al., 1992), zonal and meridional indices analysis (Kutiel et al., 1996a; 1996b) and Canonical Correlation analysis (Xoplaki et al., 2000). Despite the strong association identified between the monthly or seasonal circulation indices and monthly or seasonal precipitation in Greece, it would be, however unreasonable to expect the daily precipitation regime in Greece to be totally controlled by one or two modes of the atmospheric circulation variability. In our opinion, only an exhaustive study focusing on a large number of years and at the same time includes an individual study of types circulation will allow to make progress in the understanding of the mechanisms that govern the rainfall regime over Greece.

The main aim of this paper is to present an automatic classification scheme of circulation types over Greece (Maheras et al., 2000a; 2000b) and to study the links

Fig. 1. The location of 20 stations distributed over Greece

between the atmospheric circulation and the rainfall in Greece, namely the interannual variability, as well as any long-term trends and extremes.

2 The Spatial Distribution of Precipitation in Greece

In order to study the spatial distribution of precipitation in Greece, daily and monthly precipitation data from 20 stations distributed evenly over Greece (Fig. 1) was used that covers a 40-years period from January 1958 to December 1997.

Kriging interpolation methods were employed to describe the spatial patterns of precipitation. It should be noted that the interpolated values based on the Kriging methods, over the mountainous regions as well as over the Eastern Aegean Sea and western Turkey are less certain because there are no stations in these areas and thus, should be considered with caution. This should be kept in mind for all the following maps in this study.

Meridional Aegean Sea (Cyclades) has long been recognised as the driest region in Greece (Fig. 2). Of more practical importance is the fact that in many parts of the region the annual average rainfall does not exceed the 400 mm. In contrast, the Ionian Island and western Greece reveals a remarkable maximum of more than 900 mm per year. Precipitation is unevenly distributed throughout seasons with the winter being the rainiest season. During winter the spatial distribution of rainfall amount (Fig.3)

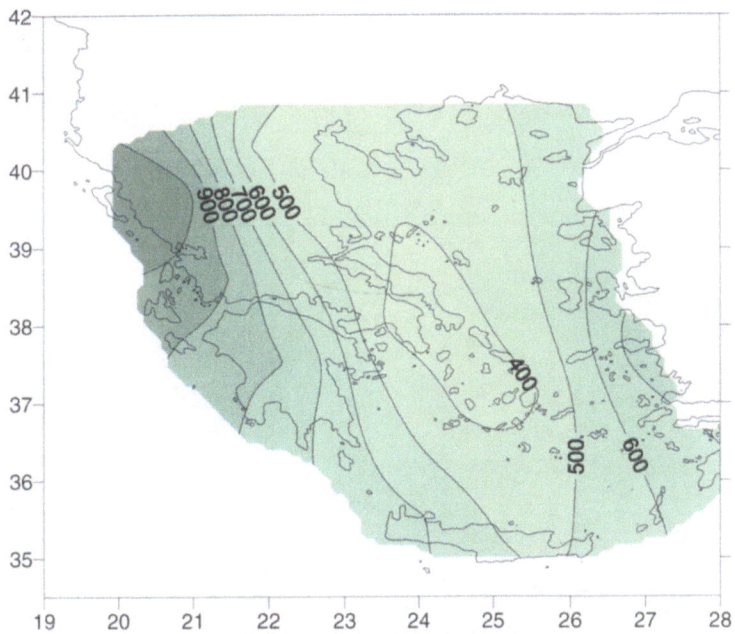

Fig. 2. Geographical distribution of annual precipitation amounts in mm/year

shows the maximum in the western part of Greece, whereas the minimum is located in central and northern Greece.

A second maximum with values up to 300mm can be found over eastern Aegean Greek Islands. In spring the maximum is located in north-west of the country (more than 220mm) whereas the minimum is found in meridional Aegean Sea (less than 90mm). During summer there is an almost zonal distribution of precipitation, with the maximum being located in the north part of Greece (~100mm). The autumn distribution of precipitation resembles the annual one, where the maximum (more than 350mm) is located in west and the minimum (~110mm) in Cyclades.

The geographical distribution of the number of rainy days per year reveals a maximum over Ionian Sea that is much more than 100 days per year (Fig. 4). The lowest values are observed along the Aegean Sea where average is 70 wet days per year. The seasonal spatial distribution of wet-days is summarised in Fig. 5, where the winter maxima and summer minima are clearly depicted.

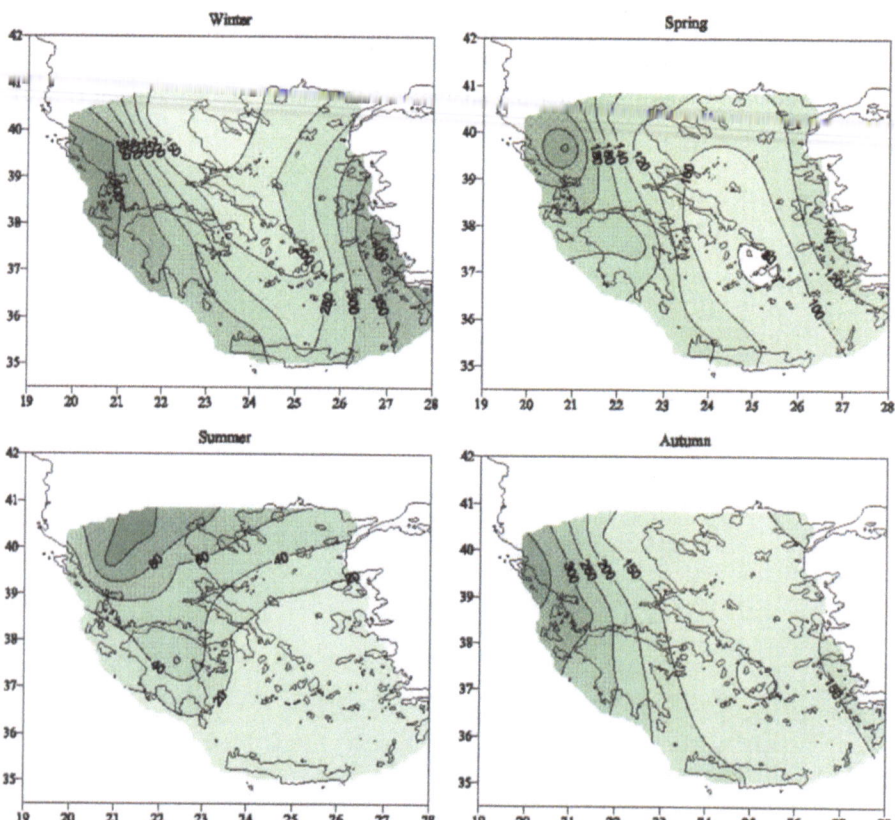

Fig. 3. Geograpical distribution of seasonal precipitation amounts

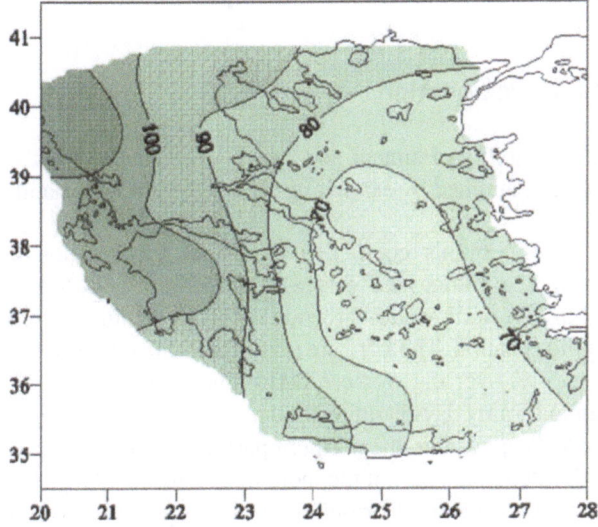

Fig. 4. Geographical distribution of rainy days per year

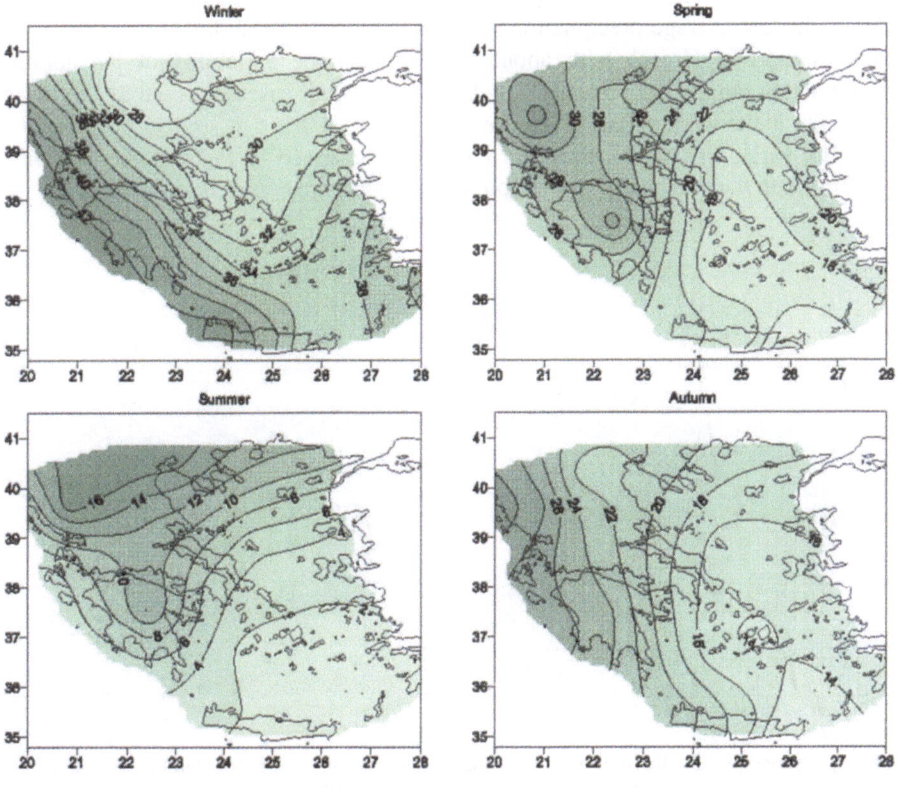

Fig. 5. Geographical distribution of the rainy days per season

3 Variability and Trends of Precipitation

3.1 Variability of precipitation

The interannual variability of annual and seasonal precipitation as well as of rainy days over Greece is examined by examining the coefficient of variability (CV) for 20 stations.

According to Fig. 6 this coefficient of annual precipitation decreases from Aegean to Ionian Sea (along the east-west direction) and from Macedonia to Peloponesus (along the north–south direction) showing a rather complex distribution pattern than simple zonation. The CVs are quite above 30% over the northern Aegean Sea and below to 20% over the western part of the country. The spatial distribution of the CVs of annual rainy days is quite similar with the previous distribution. Again the maximum of CVs appears in the northern part of the Aegean Sea (Fig.7) whereas the minimum is located in the part of the more humid areas of Greece.

Fig. 8 shows the spatial distribution of interannual variability of seasonal precipitation. According to this figure the CVs of winter precipitation decrease from north to south of the country showing a simple zonal distribution. The CVs are quite above 45% over northern Greece and about 37% over Crete. The CVs of the spring precipitation varies from 55% over Crete to below 35% over the Ionian Sea. Variability of the summer precipitation is very high in meridional Aegean Sea (240%) due to the low average precipitation whereas the lowest variability appears in north-western part of Greece with about 60%. Finally the variability of the autumn precipitation ranges from greater than 65% in the south-east of Aegean Sea to lower than 40% in the western part of Greece.

The variability of seasonal rain days is summarised in Fig. 9, where similar patterns with the seasonal CVs precipitation amounts are depicted.

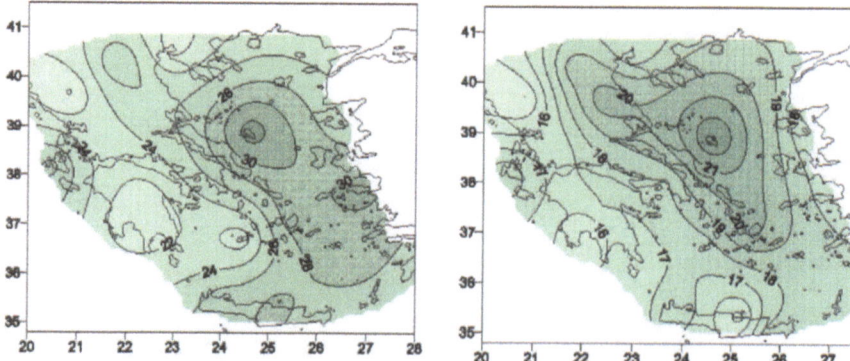

Fig. 6. Geographical distribution of annual precipitation CVs

Fig. 7. Geographical distribution of annual rain days' CVs

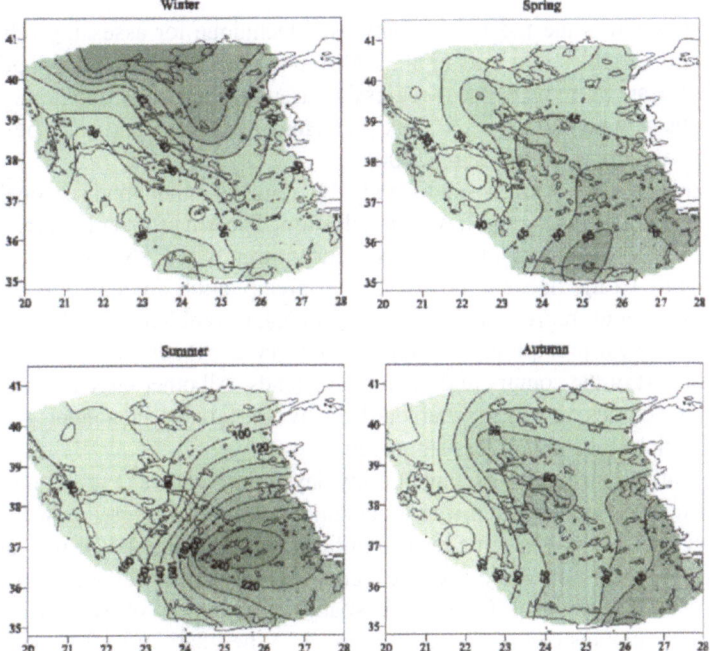

Fig. 8. Geographical distributiom of seasonal precipitation's CVs

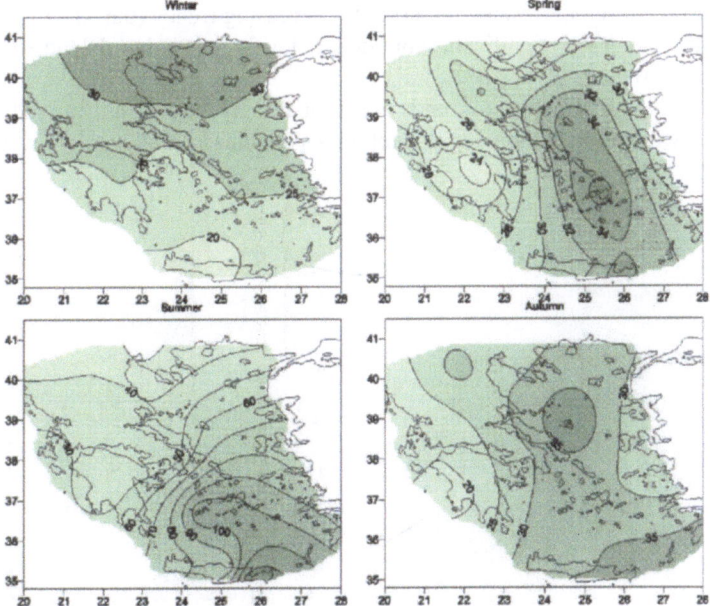

Fig. 9. Geographical distribution of seasonal rain days' CVs

3.2 Trends of Precipitation

Linear regression is the most commonly used technique for assessing relationships between two variables (De Luis et al., 2000), however, trends in precipitation series are rarely linear. Moreover, linear regression assumes normality and homogeneity of variance throughout the series (Clark and Hosking, 1986) and may be adversely affected by outliers (De Luis et al., 2000).

Linear regression was chosen to study the trends of annual and seasonal for both precipitation totals and rain days. In addition Spearman's rank correlation test (Sneyers, 1992) was chosen to detect significant trends at a 95% significance level.

In Figure 10 the linear trends of annual precipitation amounts are shown. The -3.0mm/year value represents the 95% significant confidence level. A significant decrease in annual precipitation amounts is observed over the mountainous regions in the west part of the country and the Ionian Islands. All other Greek regions (except a part of Cyclades) are also characterised by a decrease in annual precipitation amounts, which is, however, less significant.

The spatial distribution of annual rainy days trends shows (Fig.11) an important decrease in the Ionian Islands, the mountainous regions in west, as well as in the north–east part of the Aegean Sea. Insignificant negative trends are indicated in the Meridional Aegean Sea while positive trends (not significant as well) are observed in the central continental Greece. A spatial distribution of seasonal trends of precipitation amounts can be found in Figure 12. The -2.4mm/year value represents the 95% significant confidence level. The Ionian Islands, the mountainous region in west as well as the north, eastern and south – eastern of Aegean Sea show significant decrease of winter precipitation (3–6mm/year). Spring reveals decreasing trends in the major part of the country and summer only positive trends that are both insignificant. Autumn presents remarkable decreasing trends (the -1.1mm/year value represents the 95% significant confidence level) in Ionian Islands and in the mountainous regions in west.

The spatial distribution of trends in seasonal rainy days is summarised in Fig. 13, where the winter and autumn significant trends are clearly depicted.

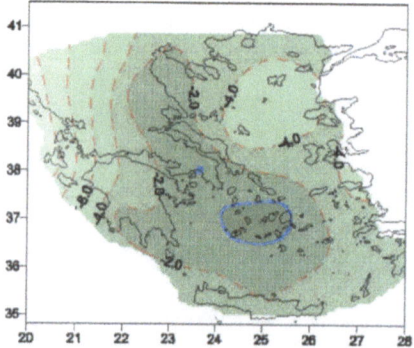 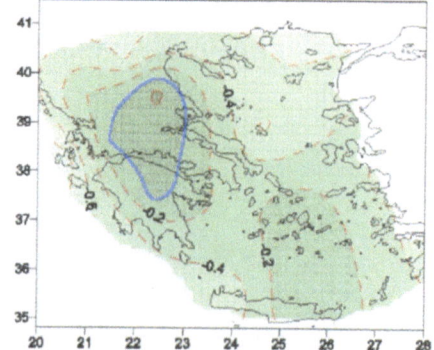

Fig. 10. Geographical distribution of the linear trend of annual precipitation amounts

Fig. 11. Geographical distribution of the linear trend of annual rain days

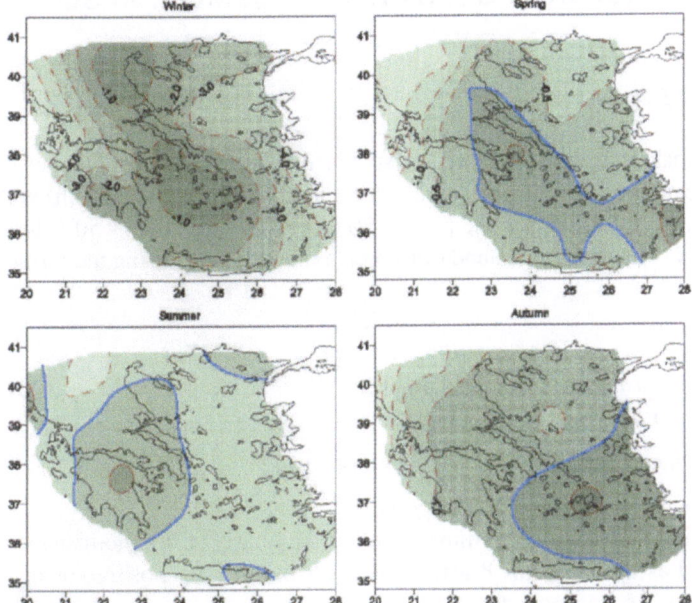

Fig. 12. Geographical sistribution of the linear trend of seasonal precipitation amounts

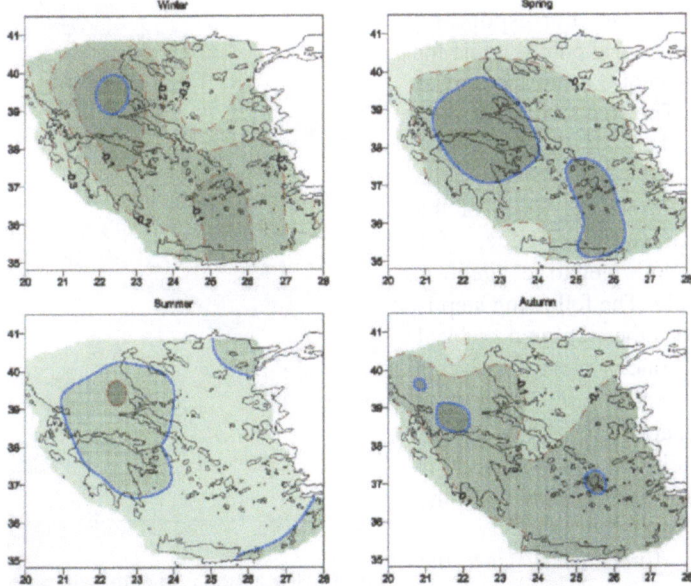

Fig. 13. Geographical distribution of the linear trend of seasonal rain days

4 The Automated Circulation Classification Scheme

4.1 Principle

The automated scheme is based on standardised 500hPa geopotential height data
(NCEP reanalysis data; Kalnay et al., 1996, at 00h, 06h, 12h and 18h) with a spatial
resolution of 2.5° x 2.5° for the region 20°N - 65°N and 20°W - 50°E for the period
1958-1997. The data were standardised on a monthly basis using the formula:

$$z_i = \frac{x_i - \overline{x_i}}{\sigma} \tag{8}$$

z_i is the standardised value at grid point i,

x_i is the observed value at grid point i,

$\overline{x_i}$ is the mean monthly value at grid point i, and

σ is the standard deviation at grid point i.

The first step is to determine whether the hourly (12h) anomalies over Greece
(calculated as the mean for 8 grid points, 2.5° x 2.5°) are positive or negative. The
second step is to look for the centre of the positive or negative anomalies. The centre is located in the region of the absolute highest positive or negative anomaly in the examined area. For an anticyclonic centre, the value of the anomalies at the centre must be greater or equal to the corresponding neighbouring grid points, while for a cyclonic centre it must be lower or equal. The next step is to check whether there is a continuous decrease or increase of the anomalies from the centre of the system towards Greece, for positive or negative centres respectively. The following step is to look for other regional or local centres. If one or more positive or negative centres are found with similar characteristics, the centre located closest to Greece is defined as the centre of the anomaly system.

The final step is the classification of the anticyclonic and cyclonic circulation types.

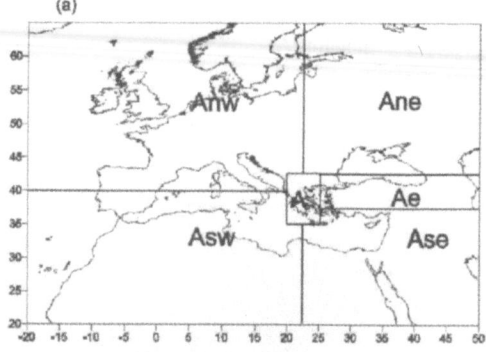

(a)

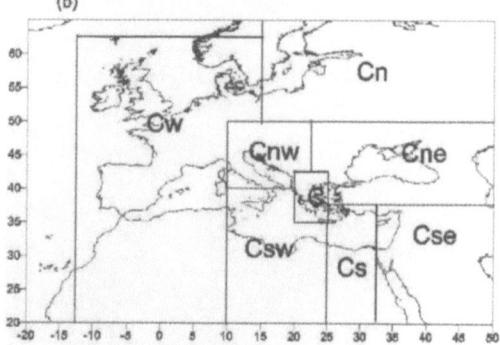

(b)

Fig. 14. The location (a) of the anticyclonic types and (b) of the cyclonic types for the 500 hPa level classification

Each anticyclonic circulation type is defined according to the location of the positive anomaly centre. Similarly, each cyclonic circulation type is defined according to the location of the negative anomaly centre. This classification produces 6 anticyclonic types (A_{NW}, A_{NE}, A, A_{SW}, A_{SE} and A_E, Fig. 14a) and 8 cyclonic types (C, C_s, C_{SW}, C_{NW}, C_{NE}, C_{SE}, C_N and C_w, Fig. 14b). The characterisation A, A_{NW} or C, C_{SW} refers to the location of the positive or negative anomaly centre in relation to the Greek area.

4.2 Short Description of the Circulation Types

It is expected that each of the 14 circulation types should have a characteristic synoptic pattern, which is physically distinct and produces the expected weather type direction of surface and upper-air flow and associated air-masses. In order to test this, and to describe the circulation types, composite maps showing the mean anomaly pattern were constructed from the 500hPa anomaly data field for each circulation type and each season. Only maps for winter are shown here (Fig. 15 and 16).

The anticyclonic types are:

A_{NW} The positive anomaly centre is located to the northwest of the Greek area, usually over western, central or northern Europe (France, Germany, Italy, Austria or Scandinavia, Fig. 14a). Positive anomalies are always prevalent over the Greek area (Fig. 15a).

A_{NE} The positive anomaly centre is located to the northeast of the Greek area (Fig. 14a). Positive anomalies are located over the Greek area (Fig. 15b).

A   The positive anomaly centre is located over the Balkans and Greek area (Figs. 14a and 15c).

A_{SW} The positive anomaly centre is located to the west or southwest of Greece, generally over the central or western Mediterranean, or north Africa (Figs. 14a and 15d).

A_{SE} The anticyclonic anomaly centre is located to the southeast of Greece (Figs. 14a and 15e).

A_E The anticyclonic anomaly centre is located to the east of the Greek area (Figs. 14a and 15f).

The Cyclonic types are the following:

C   The negative anomaly centre is located over the Greek area (Figs. 14b and 16a).

C_S The centre of the negative anomaly system is located to the south or southeast of the Greek area (Figs. 14b and 16b).

C_{SW} The centre of the negative anomaly system is located to the west or southwest of the Greek area (Fig. 16c).

C_{NW} The negative anomaly centre is located to the northwest of the Greek area (Fig. 16d), usually over the Adriatic Sea.

C_{NE} The centre of the negative anomaly system is located to the east or northeast of the Greek area (Fig. 16e), usually over the Black Sea.

C_{SE} The negative anomaly centre is located to the southeast of the Greek area (Fig. 16f), usually over or near Cyprus.

C_N The centre of the negative anomaly system is located to the north of the Greek area, usually north of 50° N (Fig. 16g).

C_W The negative anomaly centre is usually located far to the west of Greece in the western Mediterranean, western Europe or north Africa (Fig. 16h).

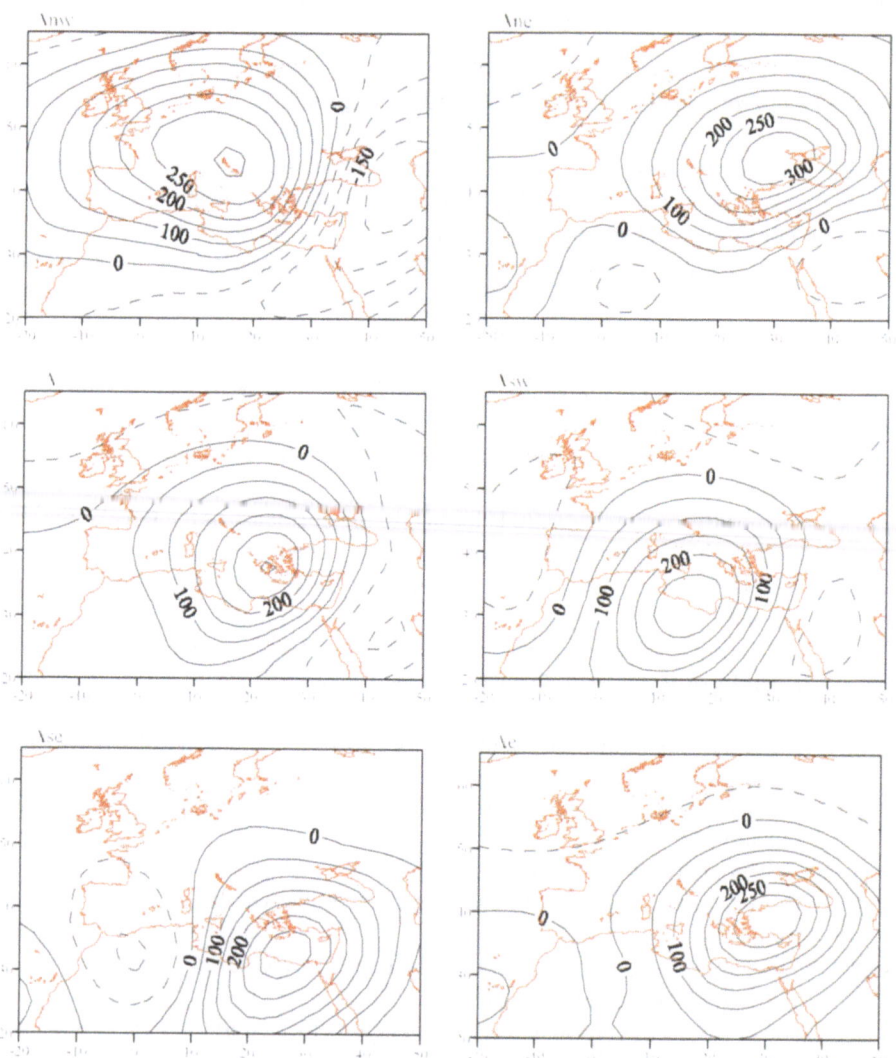

Fig. 15. Anticyclonic types: Mean winter anomalies (x100) calculated from 500 hPa level field data, 1958-1997. Figures run from (a) top left to (f) bottom right

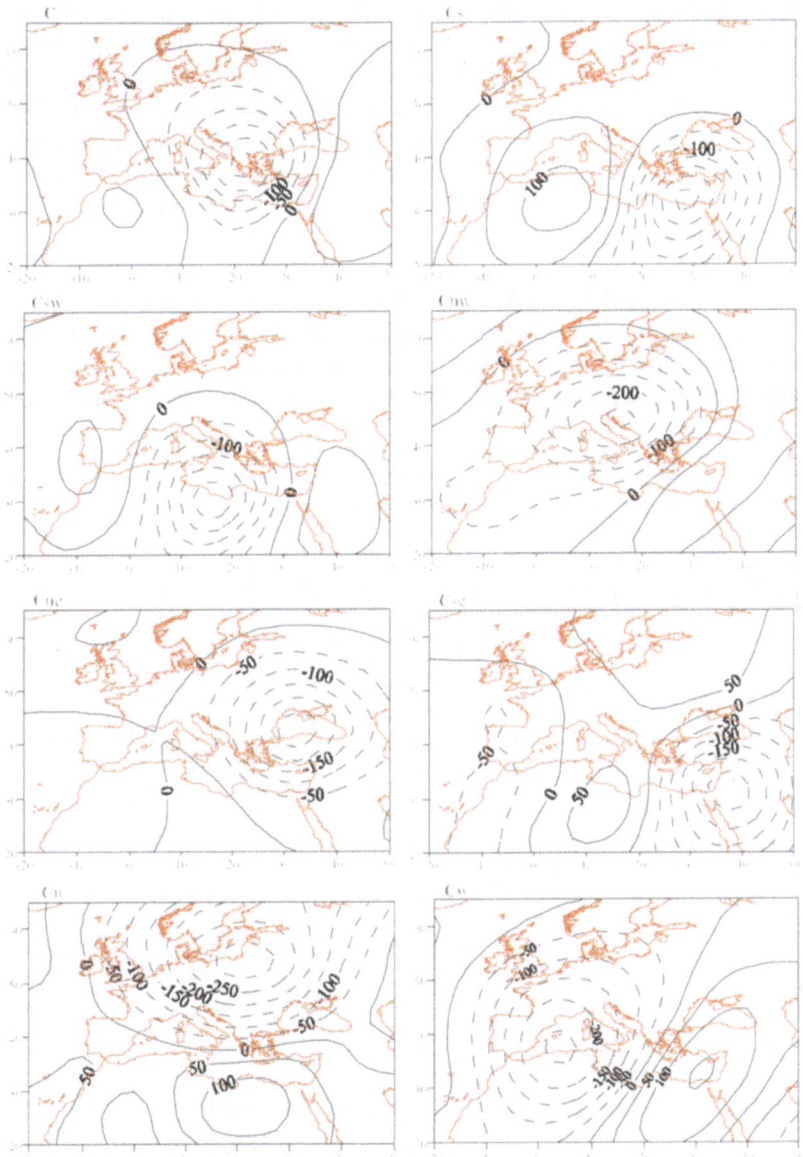

Fig. 16. Cyclonic types: Mean winter anomalies (x100) calculated from 500 hPa level field data, 1958 - 1997. Figures run from (a) top left to (h) bottom right

4.3 Frequencies of The Circulation Types and Their Relationship With Precipitation

The mean monthly and annual frequencies of the 14 circulation types at 12 UT are given in Figure 17 and in Table 1. The annual cumulative frequency of the six anticyclonic types is 47.8%. Similarly, the annual cumulative frequency of the eight cyclonic types is 52.2%. The cyclonic type C_{SW} appears the highest frequency (13.1%), with the next most frequent type the anticyclonic type A_{SW} (11.3%). For both types, the anomaly centres are located to the southwest of the Greek area. With respect to annual precipitation totals, the cyclonic types are responsible for 90.5% of the total precipitation over the Greek area, while the anticyclonic types are responsible for only 9.5%.

For the rainy period (October to March), the contribution of the anticyclonic types to total precipitation is 8.2%, while the cyclonic types contribute 91.8% (Table 2). The cyclonic type C_{SW} presents the highest individual percentage contribution to total precipitation (30.8%), followed by the cyclonic type C (28.8%). In terms of average precipitation per type of circulation the highest values (6.8mm/day) are

Table1. Monthly relative frequency of circulation types

M	Anw	Ane	A	Asw	Ase	Ae	C	Cs	Csw	Cnw	Cne	Cse	Cn	Cw
J	1.4	3.5	3.5	16	7.6	3.1	7.7	9.8	22.3	3.1	13	1.7	1	6.5
F	4.7	2.8	9.2	7.2	6.2	3	13.4	9.6	18.7	4.5	12.5	2	0.9	5.4
M	3.9	4.2	4.2	10.3	10.8	2.2	10.6	12.5	16.9	5.2	9.6	3.5	0.7	5.5
A	9.8	3.8	7.8	5.3	9.8	2	13.3	10.9	13.6	8.6	9.1	1.8	1.3	3
M	7.6	5.3	8.3	10.4	4.4	8.5	15.2	5.8	13.6	3.6	11.6	2.7	0.6	2.3
J	17.4	7.1	7.8	5.5	9.3	8.5	12.2	5.7	7.8	5.1	10.8	1.1	0.3	1.6
J	13	10	12.8	11	11.8	5.5	13.5	3.4	6.4	2.4	9	0.3	0.1	0.9
A	10.2	7.4	15.2	18.2	12.9	5	13.5	2.6	6.7	3.2	4.2	0	0.4	0.4
S	9.7	5.6	11.3	20.3	12.7	5.6	12.3	4	8.7	2.2	6	0.3	0.3	1.3
O	6.3	11.5	12.2	12.3	9.6	4.3	12.2	4	10.9	2.1	11	1.1	0.2	2.3
N	10.7	1.1	12.5	9.1	4.8	1.7	14.4	7.8	15	7.7	8.7	1.3	0.3	5.1
D	5.5	7.4	5.1	10.1	4.6	4.7	12.2	7.9	16.1	5	15	1.5	0.2	4.7
Y	8.3	5.8	9.2	11.3	8.7	4.5	12.5	7	13.1	4.4	10	1.4	0.5	3.2

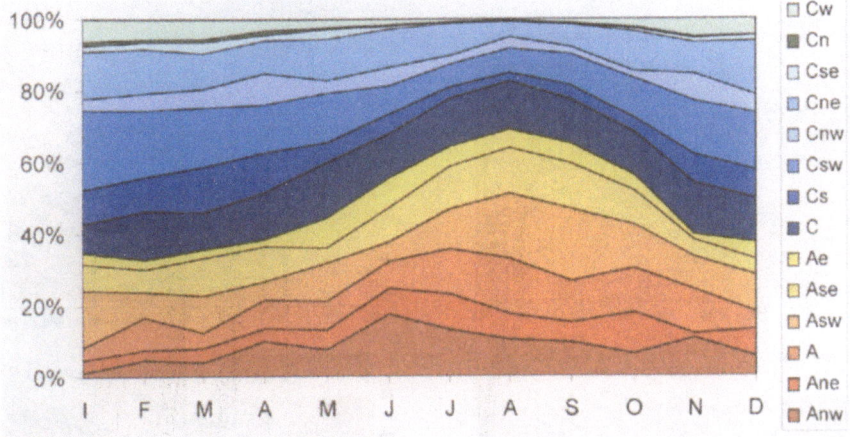

Fig. 17. Mean monthly percentage frequency of 14 circulation types

Table 2. Contribution from different circulation types (CT) to precipitation days and precipitation totals during the rainy period (October –March) in Greece (1958-1997)

Type	Absolute CT Frequency	Relative CT Frequency %	Percentage of precipitation per CT %	Average precipitation per CT for rainy days mm/day
A_{NW}	392	5.3	0.7	1.5
A_{NE}	375	5.1	1.24	2.3
A	563	7.7	0.67	1.3
A_{SW}	794	10.8	2.46	1.5
A_{SE}	531	7.2	2.7	1.9
A_E	230	3.1	1	2.5
C	853	11.7	28.8	6.8
C_S	626	8.6	8.5	3.2
C_{SW}	1212	16.6	30.8	5.7
C_{NW}	333	4.6	6.9	4.4
C_{NE}	847	11.6	8.7	2.5
C_{SE}	135	1.8	0.5	1.3
C_N	41	0.6	0.4	2.7
C_W	358	4.9	6.2	4.1

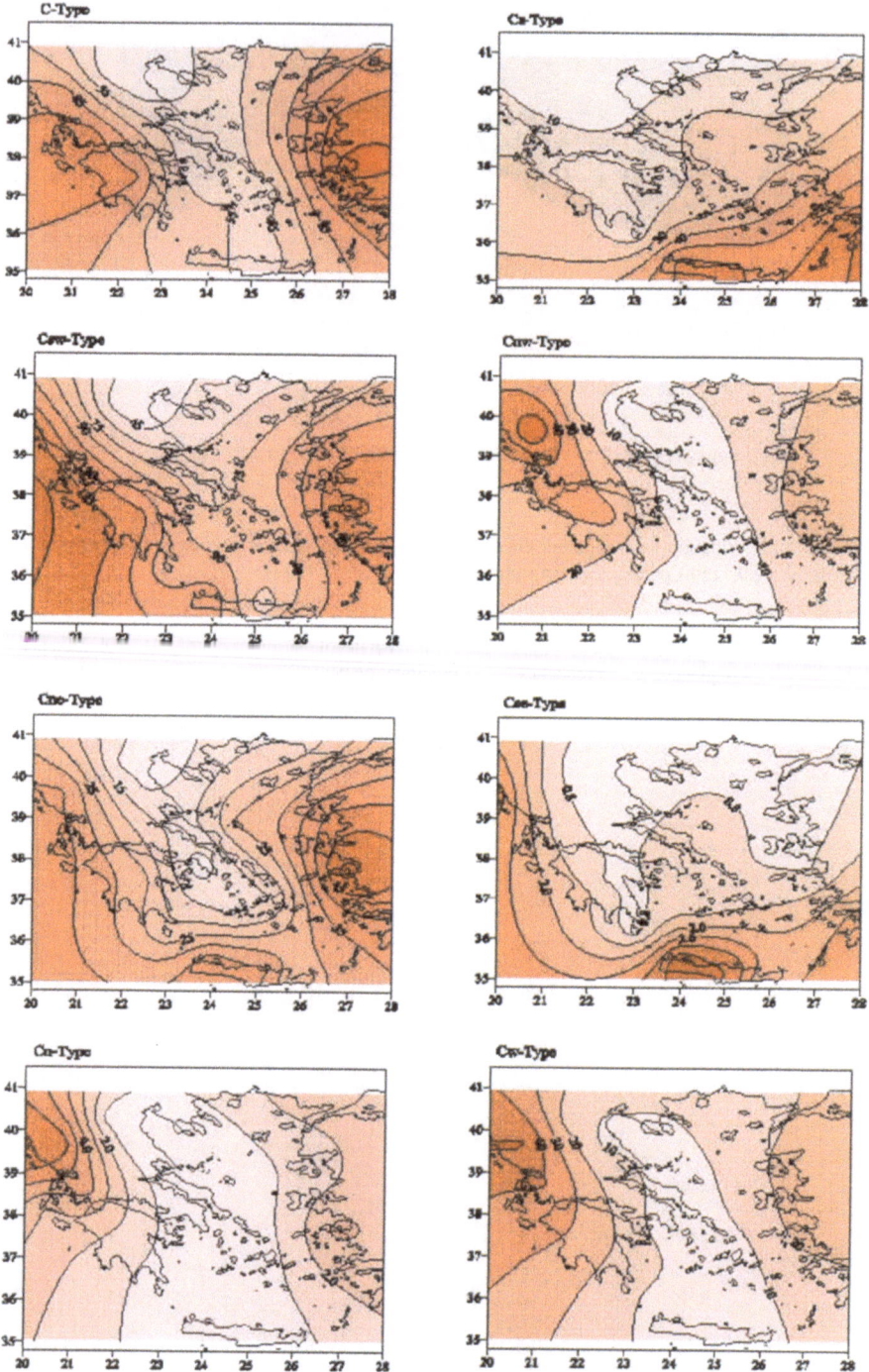

Fig. 19. Geographical distribution of winter precipitation amount per cyclonic type circulation

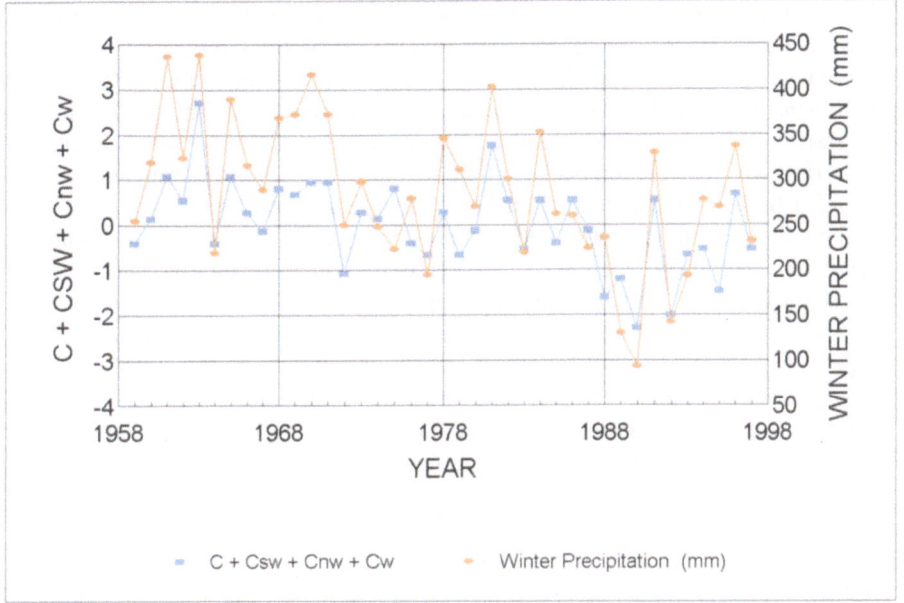

Fig. 20. Correlation between normalized frequencies of types circulation (C+Csw+Cnw+Cw) and winter precipitation for Greece (corr. coef.=0.84)

observed when the type C occurs, followed by C_{SW} (5.7mm/day), C_{NW} (4.4mm/day) and C_W (4.1mm/day).

Figure 18 shows the spatial distribution of winter probability of precipitation per cyclonic circulation type (Vafiadis et al., 2000). According to this figure, type C is associated with a fairly homogeneous high probability distribution of precipitation almost over the whole country.

The maximum of this probability appears to the west of Peloponesus and the Ionian Islands, whereas the minimum is found in the north part of Greece. Type C_S shows a rather zonal distribution with the maximum in the south of Greece (Crete) and the minimum in the north. The type C_{SW} reveals similar distribution with the type C but with lower values of probability. The type C_{NW} shows a relatively distinct distribution of probability, especially along the Aegean Sea, as compared to the other "rainy types" C and C_{SW}. Again the maximum (0.75) appears in the west but the minimum is located in Crete. The other cyclonic circulation types present lower values of probability than the previous types. The two types with eastern location of the cyclonic centre C_{NE} and C_{SE} show the maximum probability in the south and the minimum in the north, with the type C_{SE} showing the lowest values among the cyclonic types. Finally, for the two most marginal types the maximum of probability are found in the west and the minimum over Crete.

Figure 19 shows the spatial average distribution of the winter precipitation for all days within each cyclonic type circulation. A clear distinction could be made between high (C and C_{SW}), medium (C_S, C_{NW} and C_W) and low winter precipitation associated with the cyclonic types.

In order to further investigate the relationship between precipitation and circulation types on a seasonal basis, the correlation coefficients were calculated between seasonal occurrence rain days and seasonal total precipitation and the seasonal frequency of the "rainy" circulation types. The multiple correlation coefficients between the same parameters were also calculated. The results obtained are shown in Table 3 and reveal a high degree of association between the two timeseries, with correlation between 0.56 and 0.87 for the days of rain and between 0.51 and 0.84 for the precipitation amounts, all statistical significant at the 99% confidence level. It is worth noting that the values calculated for the multiple correlation coefficients are higher: between 0.7 and 0.91 for the rain days and between 0.7 and 0.9 for the amount of precipitation.

Figure 20 shows the long term variability of precipitation in Greece and the frequencies of the "rainy" cyclonic types for winter. The high degree of association between the two times series is evident. This figure shows that the significant decrease of winter precipitation amounts just after 1970, come along with an also significant fall in the frequency of occurrence of the "rainy" cyclonic circulation types.

Table 3. Correlation coefficients (1) and multiple correlation coefficients (2) between seasonal rain days and seasonal precipitation amounts and the frequencies of cyclonic circulation types (500hPa, period 1958-1997) for Greece.

Season		Winter	Spring	Summer	Autumn
1	Days of precipitation	0.87	0.74	0.69	0.56
	Precipitation	0.84	0.7	0.66	0.51
2	Days of precipitation	0.91	0.84	0.73	0.7
	Precipitation	0.9	0.85	0.7	0.7

5 Extreme Episodes

In order to define the driest and the wettest years the technique of Principal Component Analysis was applied on the annual data of precipitation amount for the hydrological year that was defined as the period from September to August for the time interval 1958 to 1997. The absolute higher negative score detected in the first principal component without rotation defines the driest seasons or years. The same definition may be drawn with respect to the wettest year, which also present the highest values of positive scores.

Figure 21 shows the spatial distribution of percentage of precipitation of the driest year 1989-1990 as compared to the average annual mean from September 1958 to August 1997. According to this figure the Aegean Sea and the Ionian Sea were characterized by much lower than the average precipitation amounts (~50%). Only the regions of south east Aegean Sea and Thessaly reveal precipitation amounts between 70 and 80% of the average.

The spatial distribution of the precipitation amounts for the wettest year (1962-1963, Fig. 22) reveals a very important enhancement of precipitation over Greece with a minimum over the meridional Aegean Sea (Cyclades ~110%) and two maxima, the first over Crete (~185%) and the second over Ionian Islands (~185%).

In order to analyse the seasonal evolution of extreme drought episode, the mean values of precipitation for every month of hydrological year and the corresponding values of the driest episode 1989-1990 are compared in Fig. 23 a and b. This extreme case presents a very pronounced dry winter period with close to normal precipitation during autumn, late spring and summer. An analysis of the corresponding values of frequency of C, C_{SW}, C_{NW} and C_W types circulation may be observed in the same figure (23 b). The agreement between the corresponding figures (monthly precipitation and monthly frequencies of circulation types) is obvious and in this case of extreme event shows the decisive role that is played by the identified links between anomalies monthly

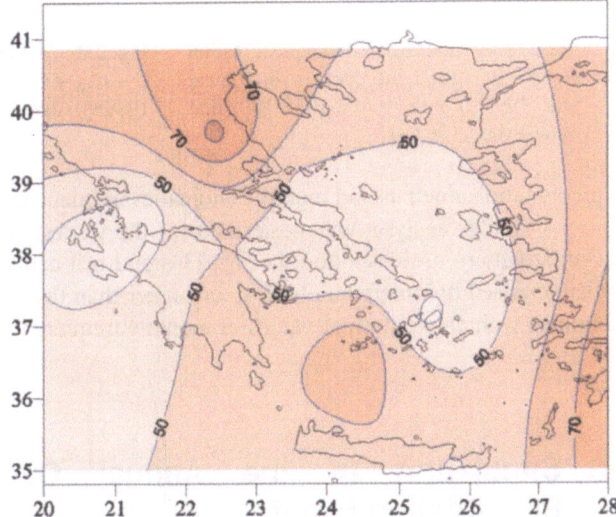

Fig. 21. Geographical distribution of percentage of precipitation amount of the driest year 1989/90 in relation to the annual mean for the period 1958-1997

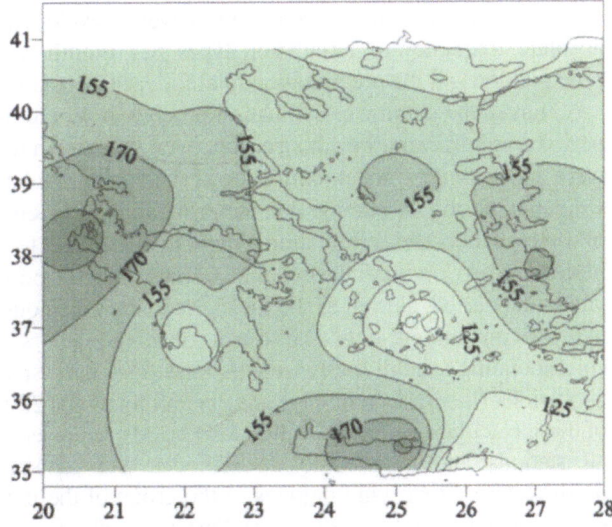

Fig. 22. Same as Fig. 21 but for the wettest year 1962/63

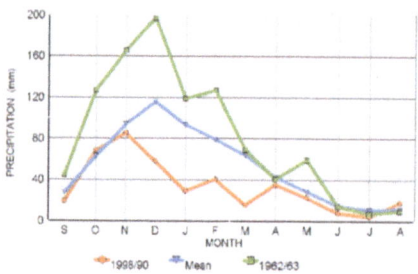 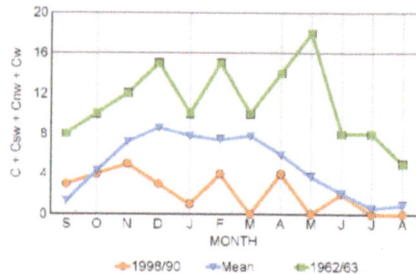

Fig. 23a. Annual course of the precipitation per month during the driest hydrological year (1998/9) of the period 1958-1997 and wettest year (1962/3)

Fig. 23b. Annual course of the circulation types $C + C_{SW} + C_{NW} + C_W$ for the hydrological years 1989/90 (dry) and 1962/63 (wet)

precipitation amounts and monthly anomalies frequencies of "rainy" circulation types.

A similar analysis was performed for the wettest year 1962-1963, revealing a similar pattern of association. In Fig. 23 a and b it can be seen that months, that are characterised by precipitation close or higher than the average value, correspond to months with above normal values of monthly frequency of "rainy" circulation types.

6 Simulation of the Annual Cycle and Extreme Precipitation Events

Despite the large variation in the amounts of the observed precipitation for each circulation type, we would like to assess the capacity for modelling the annual cycle of rain days and the corresponding precipitation amounts, as well as selected extreme precipitation events for a number of Greek stations.

Thus, multiple regression models were developed for each station using the absolute frequency of circulation types per month as well as the probability of precipitation and the precipitation total for each circulation type.

During the parameter estimation process, it was found that a 20-year calibration period was sufficient. For this reason, even years from the studied period 1958-1997 were used as the model calibration, while the 20 odd years were used for the validation period. Assuming that the probability of precipitation and mean daily totals of precipitation per circulation type are the same for validation and calibration periods, days of rainfall during annual, rainy and dry periods were calculated for the validation period, as well as the mean precipitation amounts.

The annual cycle of the mean observed and simulated rain days and the corresponding rainfall totals were estimated for each station and then averaged over all stations (i.e. mean rain days and rainfall totals for 20 Greek stations). The results of these calculations for the validation period are presented in Fig. 24, averaged over all stations and individually for Rhodes and Ioannina. Ioannina is located in northwestern Greece and belongs to the group of the most rainy continental stations (mean annual rainfall is 1138 mm). During the dry period it receives a relatively high amount of rainfall (238mm). On the other hand, Rhodes is located in the southeast Aegean Sea and is the wettest station in the south Aegean (mean annual rainfall is 787mm), while precipitation received during the dry period is much lower (59mm).

As expected, the simulated rainy days are more consistent with the observed values than the precipitation totals for all the Greek stations and the two individual ones. In some cases, such as December for all stations, August for Ioannina and February for Rhodes, the agreement is exact. The X^2 statistic was used to test the adequacy of the agreement between the observed and simulated rain days, with an acceptance level of $p(X^2 \geq X) \geq 0.05$. According to the results of this test, for the majority of the months (all stations, Ioannina and Rhodes) the agreement is acceptable except for June and October (all stations), April, June, July and November (Ioannina) and February, May, June, October and December (Rhodes).

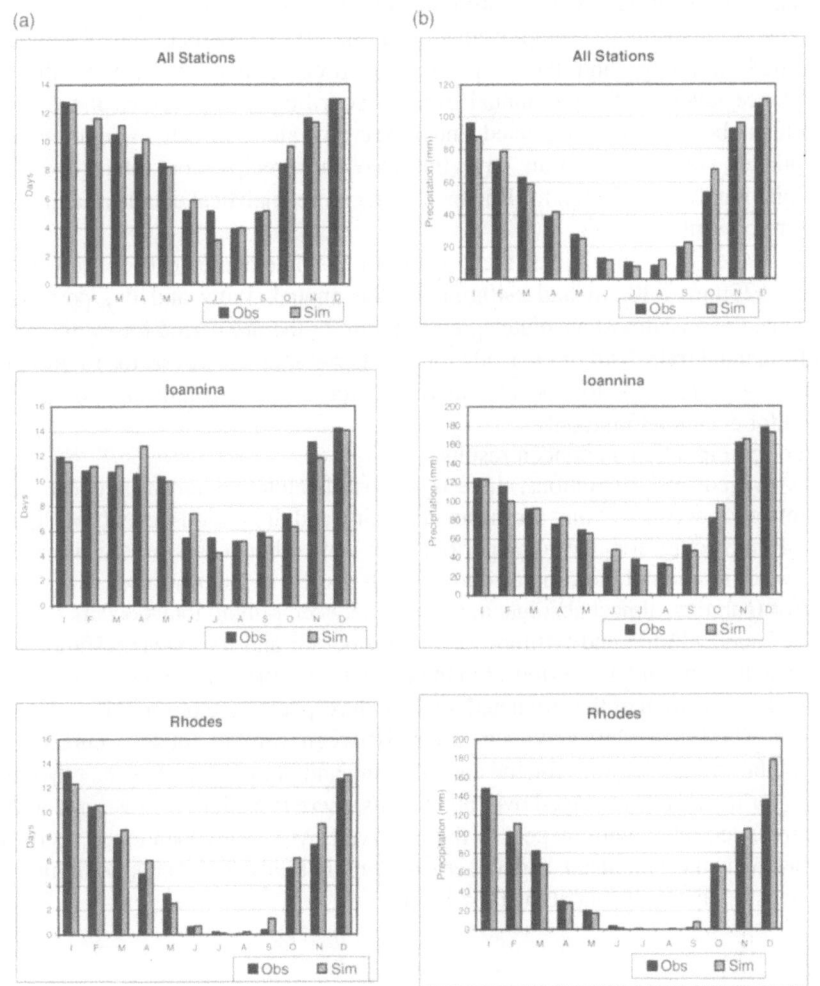

Fig. 24. Mean monthly observed and simulated days (a) and amounts (b) of rainfall for the validation period (odd years 1958 - 1997) averaged over all stations and separately for Ioannina and Rhodos

With respect to the simulated monthly precipitation totals, Ioannina shows the best results since the agreement between observed and simulated totals is acceptable for all months except February. For some months, there is almost exact agreement (January, March, August). The precipitation totals averaged over all stations show very good agreement, except for January (when observed totals are higher than simulated) and February (when the opposite occurs). Rhodes appears large discrepancies between observed and simulated totals for December, January, March and September, while agreement in other months is acceptable according to the X^2 test.

As for the annual cycle, the mean observed and simulated rainy days and the corresponding precipitation totals were calculated for selected extreme daily precipitation events. Daily amount thresholds of >25mm, >30mm, >40mm and >50mm were selected. The results of these calculations for the validation periods for annual, rainy and dry periods are presented in Fig. 25a and 25b.

The X^2 statistic was employed to test the agreement between observed and simulated rain days and the corresponding precipitation totals, with the same acceptance levels as for the annual cycle. According to the results of the X^2 test, agreement between the simulated and observed values for all extreme events is acceptable, not only for rain days but also for precipitation totals (except for precipitation events >50mm for Rhodes). Some more analytical details relating to rain days are presented below.

Averaged over all stations, there is almost absolute agreement for rainy days for events >25mm, >30mm and >40mm for the annual, rainy and dry periods. For Ioannina, there is almost absolute agreement during the dry period for events >30mm and for annual, rainy and dry periods for events >40mm. However, for Rhodes, there is absolute agreement for the annual period for events >25mm and for the rainy period for events >30mm only.

For precipitation amounts it results:

a) Averaged over all stations, there is almost absolute agreement between rainfall totals for events >25mm for the annual, rainy and dry periods, events >30mm and >40mm only for the dry period. For the other cases, there is a slight overestimation of the simulated totals with respect to the observed.

b) For Ioannina, almost absolute agreement between rainfall totals occurs for events >25mm, >30mm and >40mm for the dry period and for events >50mm for the annual, rainy and dry periods. On the contrary, for the other cases there is a slight underestimation of the simulated totals with respect to the observed.

c) For Rhodes, a very good agreement between rainfall totals occurs only for rainfall events >40mm for the annual, rainy and dry periods. For the other cases, as for the results averaged over all stations, there is a slight overestimation of the simulated totals with respect to the observed. As already mentioned, agreement between the simulated and observed rainfall totals for events >50mm is not statistically significant at the 0.05 level.

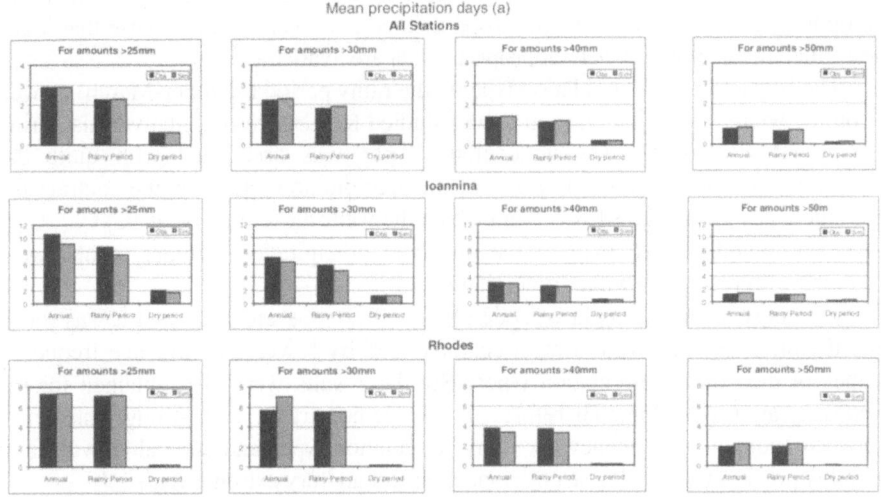

Fig. 25a. Mean observed (left column) and simulated (right column) days of rainfall for selected extreme precipitation events for the annual, wet and dry validation periods (odd years from 1959 to 1997) averaged over all stations and separately for Ioannina and Rhodes

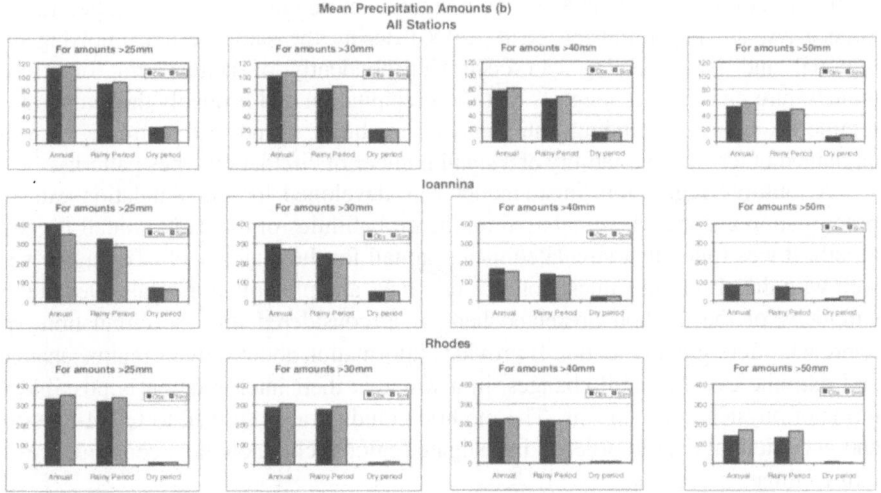

Fig. 25b. Mean observed (left column) and simulated (right column) amounts of rainfall for selected extreme precipitation events for the annual, wet, and dry validation periods (odd years 1959-1997) averaged over all stations and separately Ioannina and Rhodes

7 Conclusions

Daily atmospheric circulation affecting Greece is characterized using spatial methods of topology and geometry. The proposed method is universal in the sense that it can be applied in any region where appropriate data are available.

The impact of each circulation type on the rainy period (October-March) as well as on the winter precipitation regime was studied for Greece on a daily basis during the period 1958-1997. The anticyclonic types was associated with quite dry conditions (responsible for only 8.2% of the precipitation total). On the contrary the cyclonic types accounted 91.8%, from which the wettest cyclonic types (C, C_{SW}, C_{NW}, C_W) contributed almost 75% of the observed precipitation. Similar results were obtained with respect to winter precipitation.

The most intense drought episode (1989-1990) was analysed and it was found that the driest months were always characterized by lower than average frequency values of the "rainy" circulation types. Similar results were also obtained for the wettest year 1962-1963. Therefore, it was demonstrated that strong links exist between anomalous frequency of "rainy" circulation types and anomalous values of monthly precipitation with a strong impact on wet and dry extreme years.

Regional circulation types can thus be used as a mean to relate the broad pattern of atmospheric circulation around Greece with the regional pattern of weather and climatic trends. For example, it could be claimed that one possible reason for the negative significant trends in annual, winter and autumn precipitation amounts over Greece during the period 1958-1997 was an under average frequency of "rainy" cyclonic circulation types and in the contrary any over average frequency of anticyclonic circulation types (Maheras et al., 2000a). This negative significant trends of precipitation was accompanied by a significant negative trend of temperature in Greece (Arseni-Papadimitriou et al., 2000) during the same period, implying that regional and local cooling can occur independently of hemispheric or global temperature changes. It suggests that the precipitation and probably the temperature regime in Greece including interannual variability, trend and extremes can be explained in terms of variability of a small number of circulation types patterns.

A multiple regression linear model was developed for the simulation extreme precipitation events, which is based on the probabilities of the daily rainfall for each month. It was found that simulated mean rainfall for the majority of cases agree very well with the observed values, both for rain days and precipitation totals. Despite the simplicity of the method presented here, it is considered that the circulation-type approach offers great potential. It provides information about rainfall regime changes and extreme event rainfall changes, in a way that their interpretation is possible in terms of physical mechanisms. The method could be also applied to GCM pressure and geopotential data in order to investigate regional climatological consequences of future climate scenarios.

References

Arseni-Papadimitriou A, Maheras P, Patrikas I, Anagnostopoulou Ch (2000) Distribution géographique des températures maximales par type de circulation et leurs tendances, en Grèce, Publ. de l'AIC, 13:347-355

Clark WAV, Hosking PL (1986) Statistical methods for geographers, Wiley, New York

De Luis M, Raventos J, Gonzalez-Hidalgo JC, Sanchez JR, Cortina J (2000) Spatial analysis of rainfall trends in the region of Valencia (East Spain). Int. J. Climatol., 20:1451-1469

Kalnay E, Kanamitsou M, Kistler R, Collins W, Deaven D, Gandin L, Irebell M, Saha S, White G, Woollen J, Zhu Y, Leetmaa A, Reynolds R, Chelliah M, Ebisuzaki W, Huggins W, Janowiak J, Mo KC, Ropelewski C, Wang J, Jenne R and Joseph D (1996) The NCEP/NCAR 40-year Reanalysis project. Bulletin of the American Meteorological Society, 77:437-471

Kozuchowski K, Widig J, Maheras P (1992) Connections between air temperature and precipitation and the geopotential height of the 500hPa level in a meridional cross-section in Europe. Int. J. Climatol., 12:343-352

Kutiel H, Maheras P, Guika S (1996a) Circulation and extreme rainfall conditions in the eastern Mediterranean during the last century. Int. J. Climatol., 16:73-92.

Kutiel H, Maheras P, Guika S (1996b) Circulation indices over the Mediterranean and Europe and their relationship with rainfall conditions across the Mediterranean. Theor. Appl. Climatol., 54: 125-128

Maheras P, Patrikas I, Karacostas Th, Anagnostopoulou Ch (2000a) Automatic classification of circulation types in Greece: Methodology, Description, Frequency, Variability and Trend Analysis. Theor. Appl. Climatology, 67:205-223

Maheras P, Anagnostopoulou Ch. and Patrikas I (2000b) An objective classification method of circulation types, for Greece. 5th Hellenic Conference for Meteorology-Climatology and Atmospheric Physics. Thessaloniki 28-30 September 2000, in print.

Sneyers R (1992) Use and measure of statistical methods for detection of climatic change. In: climate change detection Project, Report on the Informal Planning meeting a statistical procedures for climate change detection. WCDMP, 20:176-181

Vafiadis M, Tolika K, Patrikas I, Anagnostopoulou Ch (2000) Distribution géographique de la probabilité d'apparition de précipitations d' hiver par type de circulation en Grèce. Publ. de l'AIC, 13:381-388

Xoplaki E, Luterbacher J, Burkard R, Patrikas I, Maheras P (2000) Connection between the large–scale 500hPa geopotential height fields and precipitation over Greece during wintertime. Clim Res, 14:129-146

References

Investigations on Global Climate Change in Bulgaria

I. Raev

In the context of "Global Change" studies investigations have been started to study specific climate effects at climate gradients in Bulgaria. Two stations have been set up on both sides of the mountain watershed of Rila-Rhodope mountains, South East Europe (Fig.1). This is where the dividing line between the influence of the two climatic zones in Bulgaria passes (M - Mediterranean climate, C - continental climate).

The *Vassil Serafimov Station* on the southern slopes of Rila Mountains is a representative ecosystem of Pinus sylvestris at an altitude of 1500 m a.s.l., a southern slope, and contains an experimental watershed basin from 1200 to 2400 m a.s.l.. Studies have been initiated in1961. There is a marked impact of the Mediterranean

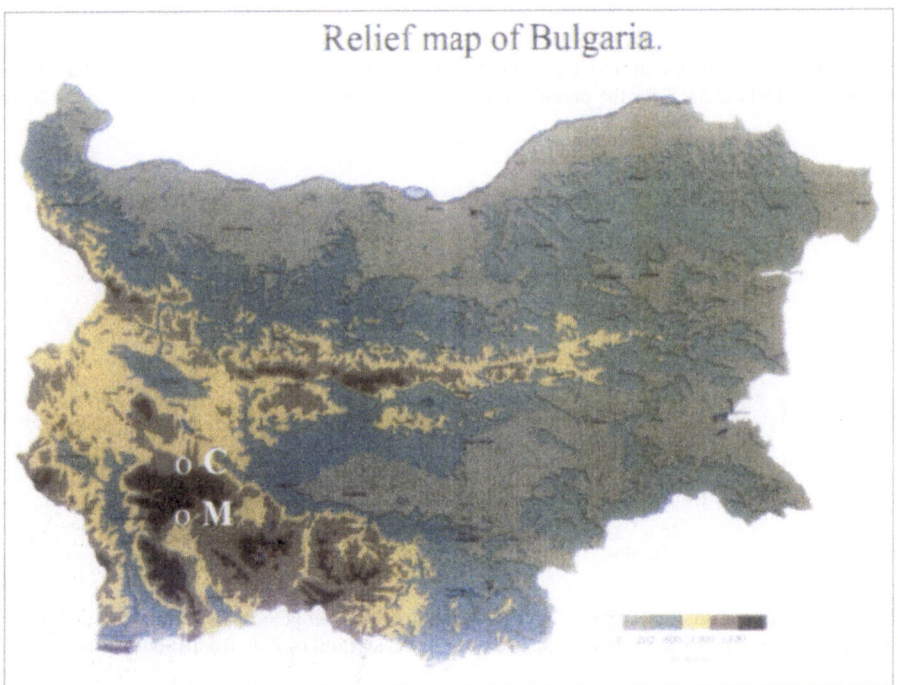

Fig. 1. Relief map of Bulgaria with the positions of the Govedartsi (C) and Vassil Serafimov (M) stations

climate.

The *Govedartsi Station* situated along the northern slopes of Rila Mountains is a representative ecosystem of *Picea abies, Abies alba, and Pinus sylvestris* at 1500 m a.s.l. and contains an experimental watershed from 1200 to 2500 m a.s.l. with a marked presence of a moderate-continental climate. Studies began in 1963.

Some data for air temperature and precipitation for the two stations at 1500 m above sea level are shown on Fig. 2, 3, 4, 5, 6, 7. Red = Station M, blue = Station C

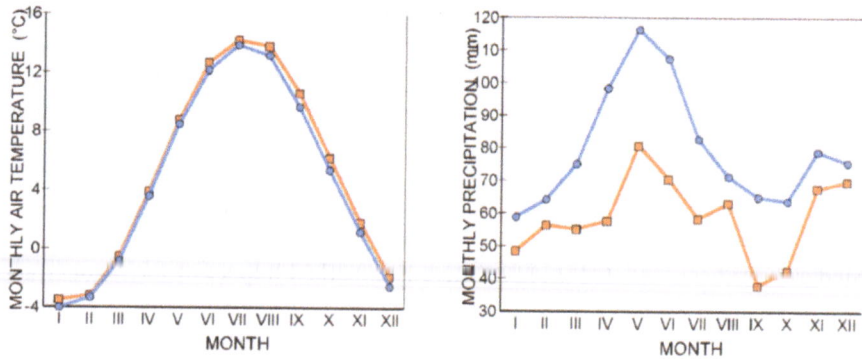

Fig. 2. Monthly air temperature (°C) at stations M and C at 1500 m.a.s.l. for the period 1964 to 2000

Fig. 3. Monthly precipitation (mm) at stations M and C at 1500 m.a.s.l. for the period 1964 to 2000

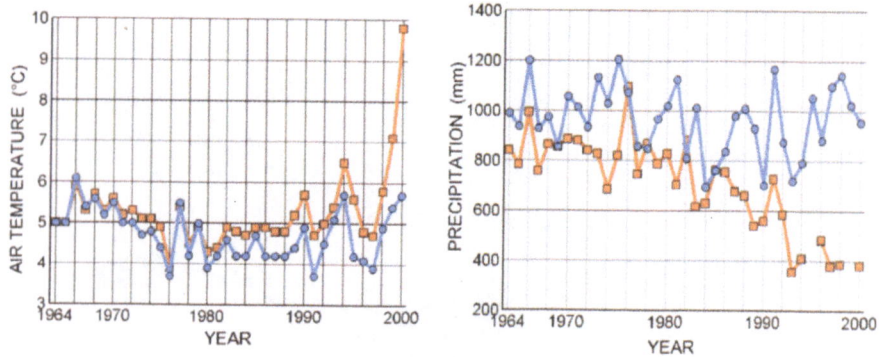

Fig. 4. Annual mean temperature (°C) at the stations M and C at 1500 m.a.s.l. from 1964 to 2000

Fig. 5. Annual precipitation (mm) at stations M and C at 1500 m.a.s.l. from 1964 to 2000

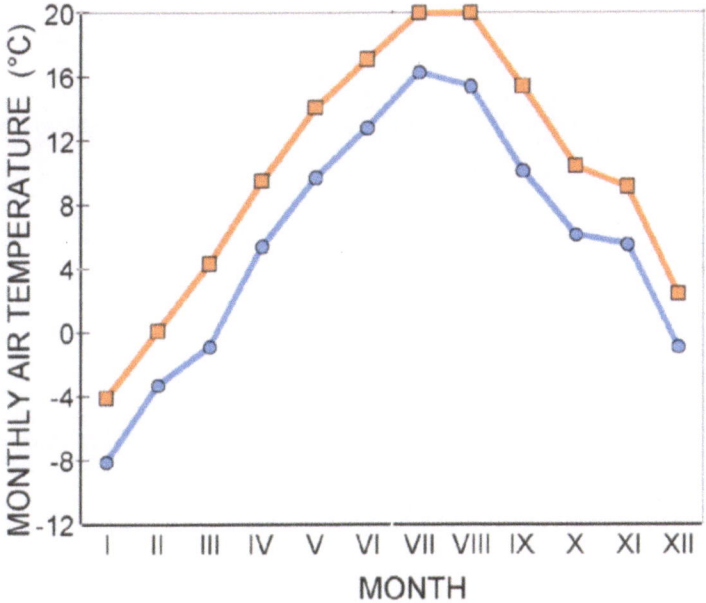

Fig. 6. Monthly air temperature (°C) at stations M and C at 1500 m.a.s.l. in 2000

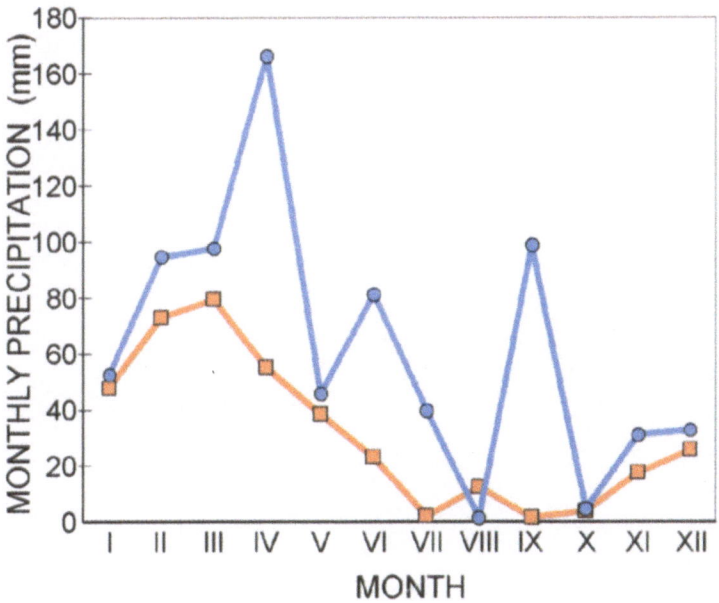

Fig. 7. Monthly precipitation (mm) at stations M and C at 1500 m.a.s.l. in 2000

Precipitation Scenarios in the Central-western Mediterranean Basin

E. Piervitali, M. Colacino

Abstract

The increase of greenhouse gases in the atmosphere could modify the hydrological cycle, in particular the rainfall regime. With reference to the Mediterranean basin, an analysis of precipitation in the period 1951-1995, indicates a rainfall reduction of 20%. On the basis of this trend a forecast of precipitation till 2030 has been performed using Winter's statistical model, which has been applied considering the Mediterranean Oscillation as forcing factor of climate. The obtained predictions show a rainfall decrease, particularly strong in the southern regions. In addition the soil moisture has been evaluated, using De Martonne's index. The results indicate a risk of desertification in some regions, due to the drying up of the soil in the future.

1 Introduction

Several investigations have been devoted to global change, in particular to study the variation in precipitation pattern, related to the important problem of the water resources availability.

Scenario models, based on the use of GCM, pay particular attention to this matter. They indicate an increase of the temperature between $1°C$ and $3.5°C$ in 2100 and the global mean precipitation should increase about 3% to 15% (IPCC-WG 1, 1995), however they do not give uniform variation on the globe about the trend of precipitations, because an increase is foreseen for high latitudes, while a reduction is expected in the low ones (Palutikof and Wigley, 1996).

In addition, the forecast at the regional scale presents great uncertainties, because the results obtained by using General Circulation Models (GCM) are largely different or contradictory, particularly in the assessment of precipitation trend. One problem is the relatively coarse spatial resolution of GCMs with respect to the fine resolution required by regional scenarios (Henderson-Sellers and Hansen, 1995).

Due to this large differences it is suitable to face the following two problems:

i)   to analyse long term series of observation data to verify if climatic variability is already detectable and

ii) to try to forecast the future pattern by statistical models.

With reference to the long term data series different authors agree on a reduction of rainfall in the Northern Hemisphere for latitudes lower than $50°$: however the values are in disagreement (Bradley and Groisman, 1989; Diaz et al., 1989; Vinnikov et al., 1990). Also regional studies have been carried out by using different methodologies and different samples areas (Groisman and Easterling, 1994; Ben-Gai et al., 1994; Norsallah and Balling, 1996). Since in the Mediterranean region only few authors give quantitative results (Palutikof et al., 1994), a systematic analysis of yearly and monthly precipitation trend has been performed in the Central–Western Mediterranean basin (Piervitali et al., 1998).

With reference to the forecast, the statistical approach widely used consists in trying climate prediction assuming a well defined external forcing. In particular, many works have been developed connecting the precipitation trend to ENSO and assessing the future evolution of rainfall by the forecast of the forcing phenomenon (Price et al., 1998; Rodò, et al. 1997).

In the present work the analysis of yearly precipitation over the Central-Western Mediterranean basin is reported. It has been carried out examining data collected in the period 1951 1995 in a network of 59 stations lying along the rim of the basin. It generally indicates a rainfall reduction in the whole region, particularly strong in the southern part.

Since in a study developed on the teleconnections between ENSO and precipitation in Central-Western Mediterranean basin (Colocino et al., 2001) no correlation has been found, we have analysed the role of the North Atlantic Oscillation (hereafter NAO) and of the Mediterranean Oscillation (hereafter MO). The last one seems to be anticorrelated with the precipitation values. On the basis of the correlation analysis we have assumed the MO as climatic forcing factor. We have used the statistical model of Winter to forecast the future values of the MO index and a prediction up to 2030 has been performed. Using, then, the anticorrelation between the MO index and the precipitation rate, rainfall until 2030 has been predicted for the different stations of the Central-Western Mediterranean basin.

Since also the evapotranspiration is affected by climatic change, because higher temperature imply high evapotranspiration and reduced run-off (Henderson-Sellers and Hansen, 1995) soil moisture has been evaluated, using De Martonne's index, which depends on precipitation and temperature. Scenario data for 2010 and 2030 show a soil aridity in progress with a possible risk of desertification in some areas.

2 Data Set

Precipitation data used in the present work have been gathered as yearly values for the 59 stations listed in Table 1. Since these series do not present missing values, no interpolation was necessary. For Italy, Spain, Portugal, Algeria, Tunisia, data quality control has been performed by the respective Meteorological Services, using a program of reduction or elimination of the error. As confirmed by the Italian Meteorological Service, it is based on the control of the physical internal coherence and the spatial inter-comparison. Data of other countries have been gathered from the World Climatic Disc, produced by East Anglia University (World Climate Disc, 1992).

The sample area has been then subdivided into three latitude belts: northern (> 42°N), that includes 23 stations, central (38°N-42°N), with 21 stations, and southern (<38°N) with 15 stations. Both for the whole area and each belt the statistical analysis has been effected.

Pressure field has been also examined, using data of 500 hPa height in Algiers and Cairo, received from the respective Meteorological Services.

Table 1. Stations used for yearly precipitation analysis

Ajaccio (France)	Campobasso (Italy)	Marseille (France)	Portaggr (Portugal)
Algeri (Algeria)	Catania (Italy)	Mertola (Portugal)	Potenza (Italy)
Alghero (Italy)	Coimbra	Messina (Italy)	Prizzi (Italy)
Alicante (Spain)	(Portugal)	Milano (Italy)	Ravenna (Italy)
Almeria (Spain)	Cozzo (Italy)	Napoli (Italy)	Roma (Italy)
Amendola (Italy)	Crotone (Italy)	Nimes (France)	Sevilla (Spain)
Ancona (Italy)	Evora (Portugal)	Nice (France)	Toulouse (France)
Arezzo (Italy)	Firenze (Italy)	Oran (Algeria)	Torino (Italy)
Barcelona (Spain)	Genova (Italy)	Palinuro (Italy)	Trapani (Italy)
Beia (Portugal)	Gibilterra (Spain)	Palma (Spain)	Trieste (Italy)
Bellavista (Italy)	Ginosa (Italy)	Perpignan (France)	Tripoli (Libya)
Bologna (Italy)	Grosseto (Italy)	Perugia (Italy)	Tunisi (Tunisia)
Bragan (Portugal)	Leuca (Italy)	Pescara (Italy)	Venezia (Italy)
Brindisi (Italy)	Lisboa (Portugal)	Piacenza (Italy)	Verona (Italy)
Cagliari (Italy)	Malta (Malta)	Pisa (Italy)	Viterbo (Italy)

3 Analysis of Precipitation Trend

To analyse precipitation pattern over the Central-Western basin a regional series has been realised, from the available data, by Thiessen's technique (Chow, 1964). This is used when, as in our study, the observations are not uniformly spaced over the examined region and consists in weighting the datum of each station by a suitable coefficient.

The obtained results indicate a decrease by a rate of -3.2 mm/year, that for the whole period gives a reduction by 142 mm that is a reduction by about 21% (Piervitali et al., 1998).

In order to obtain further confirmation of this results the Standardized Anomaly Index (SAI) has been computed which is defined by

$$I_j = \frac{1}{N} \sum_{i=1}^{N} \frac{(P_{ij} - \overline{P_i})}{\sigma_i} \qquad (1)$$

where:

I_j is the index for the jth year,

$\overline{P_i}$ is the mean precipitation in the ith station,

σ_i is standard deviation of the precipitation in the ith station,

P_{ij} is the precipitation in the ith station for the jth year,

N is the number of stations.

This index can be evaluated since the yearly values of the precipitation can be assumed to follow the Gaussian distribution (Landsberg, 1986). Nicholson (1983) recognized that the scale of this parameter, shown in Table 2, establishes a correspondence between these values and the quintiles of the distribution.

Table 2. Scale of SAI values

	$I_j \leq$	-0.9	much lower than mean
-0.85	$\leq I_j \leq$	-0.25	lower than mean
-0.25	$\leq I_j \leq$	0.25	in the mean
0.25	$\leq I_j \leq$	0.85	higher than mean
	$I_j \geq$	0.85	much higher than mean

The results are shown in Fig.1. The SAI in the examined period extends from a value higher than 0.25 to values lower than -0.25, with a statistically significant variation. This confirms what was found by the above mentioned scientists, who claim for a reduction of the rainfall for latitudes lower than 50°.

After the analysis relative to the whole region, the trend in each latitude belt has been computed by the before mentioned technique. Table 3 shows the results relative to the three sub-regions. A clear negative trend is found and the highest decrease occurs in the southern belt, where a reduction by 26% is recorded.

In Fig.2 a, b, and c the SAI indices are reported. They show that the decreasing trend is statistically significant and confirm the above mentioned results.

The strongest reduction in the Southern belt could be sign of an expansion towards the north of the desert drought and of a possible desertification, according to the self induction mechanism proposed by Charney (1975), which could be enhanced by the anthropic activity.

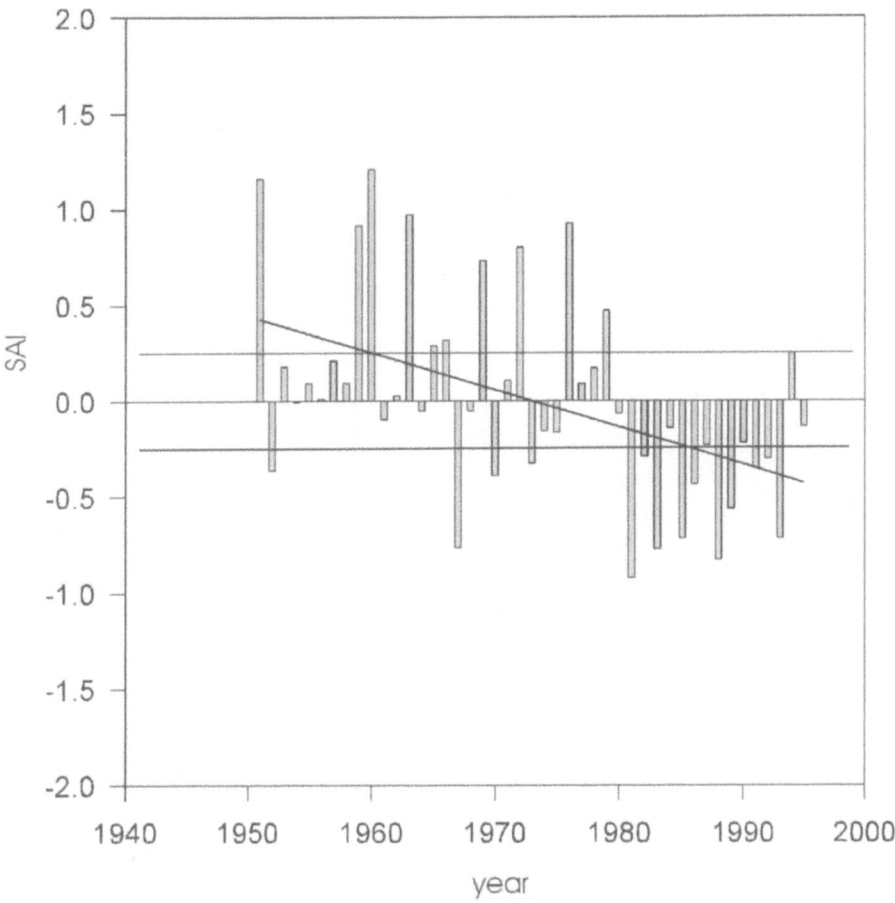

Fig.1. SAI precipitation trend over the Central-Western Mediterranean basin

Table 3. Variations of the annual rainfall in the Central-Western Mediterranean basin and in the three latitudinal belts, in the period 1951-1995

Area	ΔP (mm)	ΔP (%)	Trend (mm/y)
Whole basin	-142	-21	-3.2
Northern belt	-107	-13	-2.4
Central belt	-148	-20	-3.3
Southern belt	-157	-27	-3.5

E. Piervitali, M. Colacino

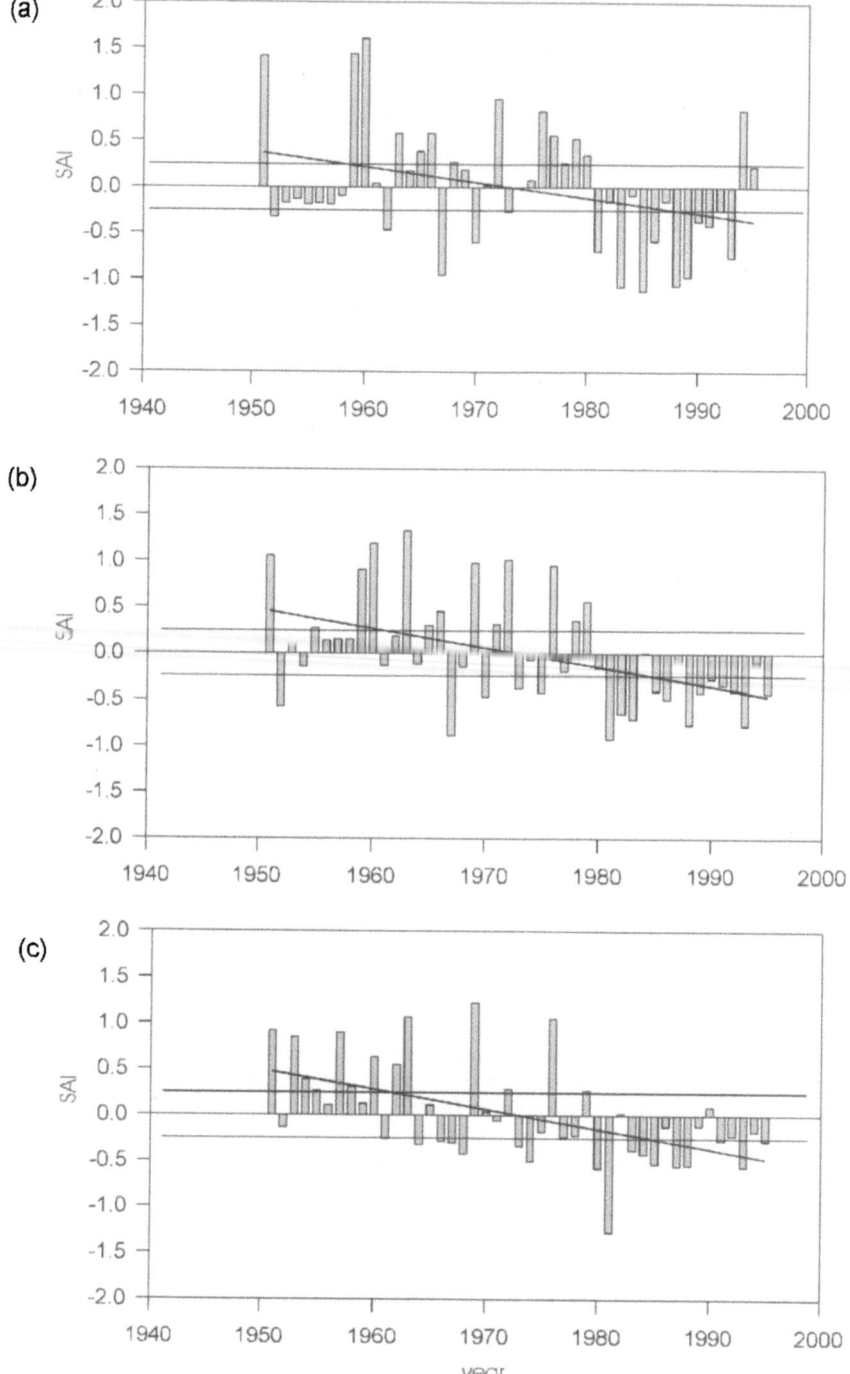

Fig.2. SAI precipitation trend in the Northern latitudinal belt (a), in the Central belt (b) and in the Southern belt (c)

4 Analysis of Correlation Between Precipitation, NAO and MO

The correlation analysis between precipitation and both NAO and MO has been performed in order to investigate about a climatic forcing of Mediterranean rainfall.

The NAO is defined as the normalized pressure difference between two stations, one lying in Iceland and other in the Azores (Van Loon and Rogers, 1978; Jones et al., 1997). In the present work we have used the index found on the home page of the Climatic Research Unit (http://www.cru.uea.ac. uk).

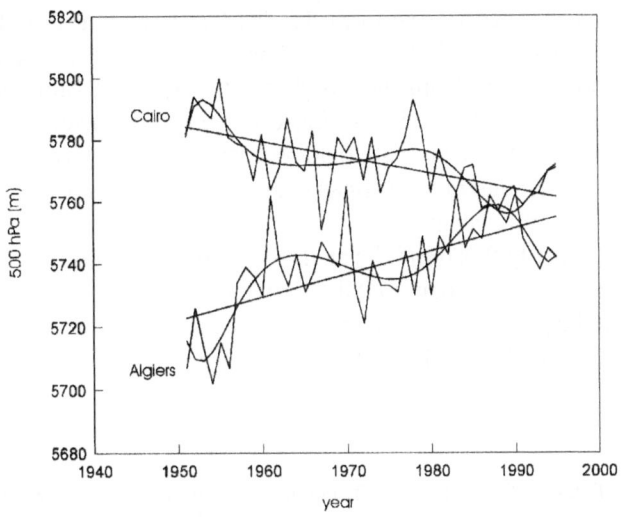

Fig.3. Pattern of the 500 hPa height in the stations of Algiers and Cairo, with the superimposed linear trend and the 8th order polynomial smoothing

The MO, less known than NAO, has been investigated firstly by Conte et al. (Conte et al., 1990). It concerns the pattern of the 500 hPa surface height monitored by the soundings in the stations of Algiers and Cairo, assumed as representative of the Western and Eastern basin respectively. Fig.3 summarizes the obtained trends. The first one is increasing and the other one is decreasing. The 8th order polynomial smoothing shows a 22-year wave in both series of data. The presence of this wave is confirmed by the spectral analysis, that presents one peak corresponding to 22 years

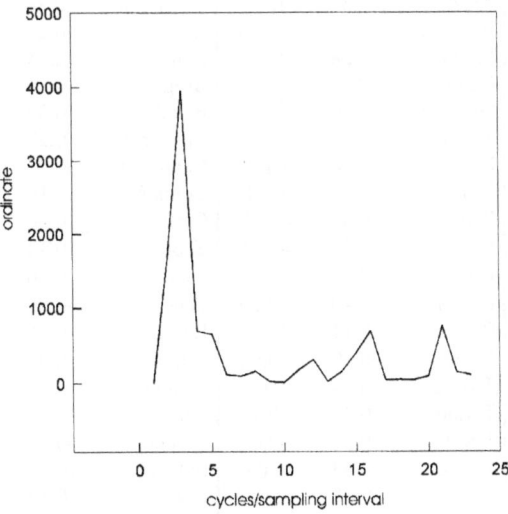

Fig.4. Spectral analysis for the 500 hPa series of Algiers. It is clear a 22 year cycle

(Fig.4). The wave of Algiers is in phase opposition with that of Cairo. When the pressure increases in the Western basin it decreases in the Eastern one and vice versa. This seesaw characterizes the MO. The index used to define the MO is the height of 500 hPa surface recorded in Algiers.

The analyses of regression between precipitation series and both NAO and MO indexes have been performed for each station of the three belts. The obtained results are reported in Table 4. As it appears evident, the observatories lying in Portugal and Western Spain show higher anticorrelation values with NAO than with MO, while the stations in the Eastern Iberian Peninsula and Italy seem to be more sensitive to the pressure field oscillation recorded over the Mediterranean (MO). In Gibraltar and in the African coasts the influences of NAO and MO seem to be more or less equivalent.

Due to this trend the analysis has been performed considering only the stations from Eastern Spain to Italy, for which we can assume the MO as a climatic forcing of the precipitation. On this basis we have tried to forecast the future patterns of the pressure field and then to predict the precipitation.

Table 4. Correlation coefficient between the precipitation series in each station of the three latitudinal belts and both the NAO and the MO indexes

Northern belt	Correlation coefficient		Central belt	Correlation coefficient		Southern belt	Correlation coefficient	
	MO	NAO		MO	NAO		MO	NAO
Ajaccio	-0.25	-0.15	Alghero	-0.26	-0.03	Algiers	-0.43	+0.22
Ancona	-0.40	$-2 \cdot 10^{-3}$	Alicante	-0.33	+0.08	Almeria	-0.06	-0.16
Arezzo	-0.31	-0.12	Amendola	-0.38	+0.05	Beia	-0.10	-0.38
Bologna	-0.48	+0.13	Barcellona	-0.28	$-6 \cdot 10^{-4}$	Catania	-0.40	+0.19
Firenze	-0.28	-0.06	Bellavista	-0.34	+0.18	Cozzo	-0.46	+0.36
Genova	-0.22	-0.02	Bragan	-0.10	-0.28	Gibraltar	-0.22	-0.24
Grosseto	-0.30	-0.09	Brindisi	-0.58	+0.06	Malta	-0.38	+0.27
Milano	-0.30	-0.19	Cagliari	-0.22	-0.05	Mertola	-0.10	-0.39
Nice	-0.38	+0.01	Campobasso	-0.23	-0.08	Messina	-0.48	+0.07
Nimes	-0.30	-0.01	Coimbra	-0.16	-0.27	Oran	-0.26	+0.38
Perpignan	-0.06	-0.09	Crotone	-0.36	+0.02	Prizzi	-0.02	-0.26
Perugia	-0.23	-0.10	Evora	-0.14	-0.38	Sevilla	+0.05	-0.46
Pescara	-0.39	+0.24	Ginosa	-0.46	+0.04	Trapani	-0.19	+0.08
Piacenza	-0.40	+0.26	Leuca	-0.66	+0.04	Tripoli	-0.19	+0.13
Pisa	-0.11	-0.13	Lisboa	-0.06	-0.49	Tunisi	-0.43	+0.39
Ravenna	-0.39	+0.02	Napoli	-0.03	-0.11			
Toulouse	-0.14	-0.04	Palinuro	-0.22	-0.03			
Torino	-0.43	+0.17	Palma	-0.27	-0.16			
Trieste	-0.23	+0.03	Portalegre	-0.03	-0.48			
Venezia	-0.43	-0.08	Potenza	-0.29	-0.09			
Verona	-0.35	-0.08	Roma	-0.17	-0.10			
Viterbo	-0.31	+0.16						

5 Forecast of Rainfall Using Winter's Model

A forecast of precipitation over the Central-Western Mediterranean has been performed until 2030, by using the Winter's statistical model (Granger and Engle, 1987).

This model has been applied on the series of the pressure field (500 hPa height) in Algiers, the representative station of the Western part of the basin. Then, on the basis of the connection between the Mediterranean Oscillation and the rainfall series in the Central-Western Mediterranean, also a precipitation prediction has been performed.

The Winter's multiplicative model, in fact, is useful to make predictions of time series which present the tendency to a cyclical pattern. This tendency is called seasonality, while the length of the cycle L is the seasonal period. As found by the spectral analysis, the Mediterranean Oscillation shows a 22 years periodicity. In this case L = 22.

Winter's multiplicative model assumes that the given series can be decomposed into three components, according to the following form:

$$y_t = T_t \times S_t + e_t \tag{2}$$

where:

y_t is the observational data at time t,

T_t is the trend component or level of the series, modeled by the linear trend $T_t = \mu_t + \beta t$,

S_t is he seasonal component, modelled by seasonal indicators $I_i = I_{i+L} = I_{i+2L}$ where

$$\sum_{i=1}^{L} I_i = L \qquad\qquad i = 1.......L,$$

e_t is a random or error component.

Starting values for T, μ, I, have been selected according to Abraham and Ledolter (1983).

Winter's model is based on the following equations:

$$T_t = \frac{a \cdot y_t}{I_{t-L}} + (1-a)\big[T_{t-1} + \beta_{t-1}\big]$$

$$I_t = \frac{b \cdot y}{T_t} + (1-b) \cdot I_{t-L} \tag{3}$$

$$\beta_t = c \cdot \big[T_t - T_{t-1}\big] + (1-c) \cdot \beta_{t-1}$$

The prediction at the time t+τ is given by:

$$P = \left[T_t + \tau \cdot \beta_t \right] \cdot I_{t+\tau-L} \qquad\qquad if \, \tau \leq L$$

$$P = \left[T_t + \tau \cdot \beta_t \right] \cdot I_{t+\tau-integer\left(\frac{\tau+L}{L}\right) \cdot L} \qquad if \, \tau \geq L \qquad (4)$$

and a, b, c are three smoothing constants between 0 and 1, calculated by the minimization of the square difference between the theoretic series, defined by the above equations, and the observed series.

The obtained prediction, for the 500 hPa surface, is shown in Fig.5. Values until 1995 are observed data, while values from 1996 to 2030 are the predictions. In Fig.5 also the linear trend is shown, that is a positive trend, and the simple 7 term weighted average, that indicate an oscillation of about 22 years.

The prediction of the mean precipitation over the Central Western Mediterranean
has been obtained using the anticorrelation between the Mediterranean Oscillation

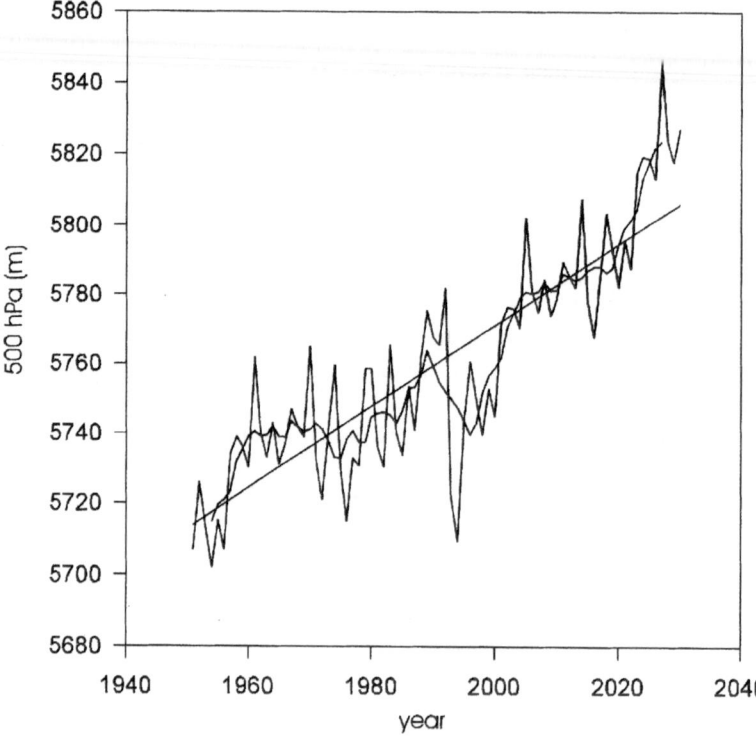

Fig.5. 500 hPa pressure pattern in Algiers station with the relative linear trend and the simple 7-term weighted moving average: observations until 1995 and forecast from 1996 until 2030, using Winter's model

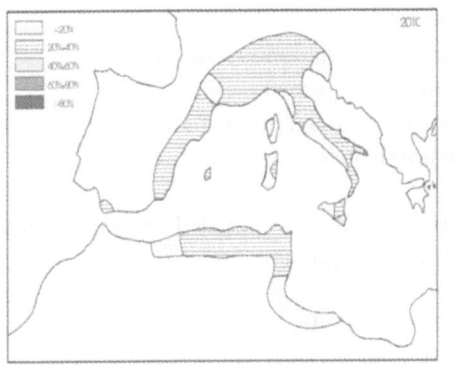 

Fig.6. Geographical distribution of the percent rainfall variation from the reference mean 1951-1980, predicted by Winter's model for 2010

Fig.7. As Fig.6 but for 2030

and the average rainfall series. It has been evaluated a rainfall decrease of -127.0 mm ≅ -20 % in 2010 and -173.8 mm ≅ -28% in 2030 from the 1951-1980 reference mean.

The same described process has been applied to forecast rainfall in each of the stations used to construct the mean Central-Western Mediterranean series, in which the above mentioned anticorrelation between precipitation and 500 hPa height was found.

The obtained results, reported as percentage variations from the reference mean 1951-1980, are shown in Figs. 6 and 7, respectively for 2010 and 2030. In particular in 2030, a strong decrease over Southern Italy (Puglia, Sicily, Sardinia) and over Northern Africa (Tunisia) is found.

6 Forecast of Soil Moisture

Scenarios of soil moisture over the Central-Western Mediterranean basin has been calculated until 2030, using De Martonne's index (Burgos et al.), depending on temperature and precipitation. It is defined by

$$I = \frac{P}{T+10} \tag{5}$$

where:
P = total annual precipitation (mm) and
T = mean annual temperature (°C).

On the basis of the index values, the following classification has been proposed:

I ≤ 10   desert or almost desert
10 < I ≤ 15   almost-arid; continuous irrigation is requested
15 < I ≤ 20   sub-humid; irrigation is requested
20 < I ≤ 30   almost-humid; well-timed irrigation is requested
30 < I   moist soil; seasonal irrigation

The GISS model of NASA (Hansen et al. 1988) was adopted to evaluate the future evolution of the temperature. It shows three possible future scenarios:

a. pessimistic scenario - business as usual,
b. realistic scenario - global emissions partly limited,
c. optimistic scenario - global emissions limited at the actual level.

In De Martonne's index calculation P is the precipitation value predicted by Winter's statistical model and T is the temperature value of the GISS realistic scenario b), which forecasts an increase of the global temperature of 0.9°C for 2010 and of 1.4 °C for 2030, with respect to the reference mean 1951-1980.

The moist soil scenario relative to the 1951-1980 period (Fig. 8) is reported together with the predicted maps for 2010 and 2030 (Figs. 9 and 10). It shows an expansion of the soil aridity, with index values lower than 10 in some southern regions, indicating almost desert areas.

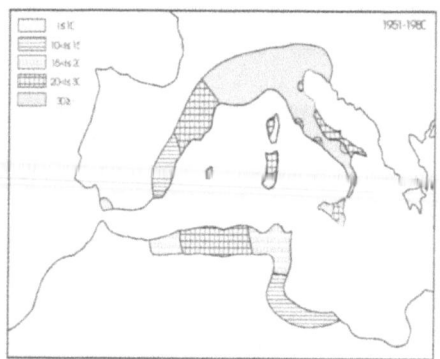

Fig.8. De Martonne's index over the Central Western Mediterranean basin, calculated for the period 1951-1980

7 Conclusions

From the above analysis the following conclusions can be drawn:

i) The precipitation in the period 1951-1995 decreased at a rate of 3.2 mm/year and this trend is statistically significant.

ii) The analysis of regression between rainfall series and both MO and NAO shows significant correlation coefficients only in the first case.

iii) The application of Winter's model foresees an increase of the pressure field and as consequence a reduction of

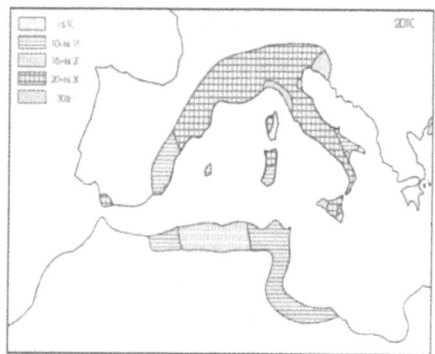

Fig.9. De Martonne's index over the Central Western Mediterranean basin, predicted for 2010

the precipitation amount and this reduction is stronger at lower latitudes.

iv) The evaluation of De Martonne's index suggests an expansion of soil aridity in the future, above all in the southern areas of the basin where there is a serious risk of desertification.

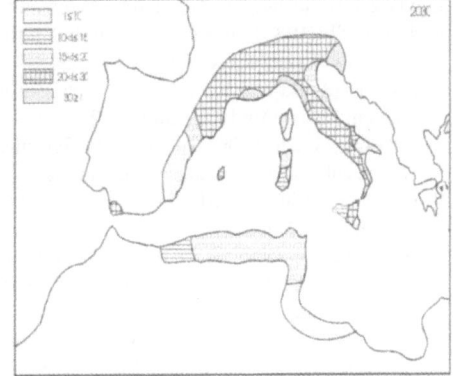

Fig. 10. As Fig. 8 but for 2030

References

Abraham B, Ledolter J (1983) Statistical methods for forecasting. Wiley & Sons, pp.445

Ben-Gai T, Bitan A, Manes A, Alpert P (1994) Long term changes in annual rainfall patterns in Southern Israel. Theor. Appl. Climatol., 49:59-67

Bradley RS, Groisman PYa (1989) Continental scale precipitation variations in the 20th century. In: Proceedings of the International Conference on "Precipitation Measurements", WMO, Geneva, pp.168-184

Burgos JJ et al. Land use and agrosystem management under severe climatic conditions WMO Tech Note n.184

Charney JG (1975) Dynamics of desert and drought in the Sahel. Quart. J. R. Met. Soc., 101:193-202

Chow VT (1964) Handbook of Applied Hydrology a compendium of water resources technology. McGraw Hill Book Company

Colacino M, Diodato L, Malvestuto V, in progress

Conte M, Colombo F, Agostinone F (1990) Variazioni climatiche recenti nel Mediterraneo centrale e scenari climatici possibili nella prima parte del 21mo secolo,. In: Tecnagro E (ed.) Proceedings of the international seminar on "La modifica del tempo: aspetti scientifici ed operativi". University of Palermo

Diaz HF, Bradley RS, Eischeid JK (1989) Precipitation fluctuations over global land areas since the late 1800s. J. Geoph. Res., 94:1195-1201

Granger CWJ, Engle RF (1987) Econometric forecasting: A brief survey of current and future techniques. In: Land KC, Schneider SH (eds.) Forecasting in the Social and Natural Sciences. Reidel, Dordrecht, pp.117-139

Groisman PY, Easterling DR (1994) Variability and trends of total precipitation and snowfall over the United States and Canada. J. of Climate, 7:184-205

Hansen J, Fung I, Lacis A, Rind D, Lebedeff S, Ruedy R, Russel G (1988) Global climate changes as forecast by Goddard Institute for Space Studies Three-dimensional model. Jour. of Geoph. Research, 93:9341-9364

Henderson-Sellers A, Hansen AM (1995) Climate Change Atlas. Kluwer Academic Publishers, Dodrecht, pp.160

IPCC-WG 1 (1995) Scientific Assessment of Climatic Change. WMO/UNEP, Cambridge University Press

Jones PD, Jonsson T, Wheeler D (1997) Extension of the North Atlantic Oscillation using early instrumental pressure observations from Gibraltar and South-West Iceland. Int. Jour. of Climatology, 17:1433-1446

Landsberg HE (1986) World Survey of Climatology, Vol.1B. Elsevier, Amsterdam

Nicholson SE (1983) Subsaharan rainfall and the years 1976-80: Evidence of continued drought. Monthly Weather Rev., 111:1646-1654

Norsallah HA, Balling RC (1996) Analysis of recent climatic changes in the Arabian peninsula region. Theor. Appl. Climatol, 53:245-252

Palutikof JP, Goodess CM, Guo X (1994) Seasonal scenarios of the change in evapotranspiration due to the enhanced greenhouse effect in the Mediterranean basin, Intern. J. Climat., 14:853-868

Palutikof JP, Wigley TML (1996) In: Climatic Change and the Mediterranean, Vol.2. Arnold, London, pp.27-56

Piervitali E, Colacino M, Conte M (1998) Precipitation pattern in the Central-Western Mediterraneanbasin in the period 1951-1995: Part I: precipitation trend. Il Nuovo Cimento, 21C: 331-344

Price C, Stone L, Huppert A, Rajagopalau B, Alpert P (1998) A possible link between "El Niño" and precipitation in Israel Geoph. Res. Letters, 25:39663-39675

Rodò X, Beert E, Comin FA (1997) Variations in seasonal rainfall in Southern Europe during the present century: relationship with the North Atlantic Oscillation and El Niño Southern Oscillation. Climate Dynamics, 13: 275-284

Van Loon H, Rogers J (1978) The seesaw in winter temperatures between Greenland and Northern Europe. Part I: General description. Mon. Wea. Rev., 106:296-310

Vinnikov KYa, Groismann PYa, Lugina, KM (1990) Empirical data of contemporary global climate changes (temperature and precipitation). Jour. of Climate, 3:662-672

World Climate Disc (1992) Global Climatic Change Data on CD-ROM. Chadwyck and Healey Ltd, Cambridge

Temperature and Precipitation Variability and Trends in Northern Spain in the Context of the Iberian Peninsula Climate

M.J. Esteban-Parra, D. Pozo-Vázquez, F.S. Rodrigo, Y. Castro-Díez

Abstract

The longest precipitation and temperature series of the peninsular Spain and Balearic Islands are analysed covering the periods 1880-1995 and 1900-1997 respectively. To identify coherent behaviour in these series, the PCA method has been used, resulting in several major noise-filtered time series. For temperature data, although two statistically significant EOFs have been obtained for the western and eastern parts of Spain, the entire region can be described by one EOF and its time evolution by the first PC series. These time series show a general upward trend and some abrupt changes. On the other hand, for precipitation, we have obtained three significant EOFs, except for winter, when we find four EOFs. The first EOF can be associated with Andalusia and Spanish Interior, the second and third EOFs with Mediterranean and Cantabric coasts, alternatively, depending on the season. The analysis of the principal components series using moving average and the Mann-Kendall test, show significant long term decreases in precipitation for Mediterranean region and Interior (at least in some seasons) and an increase of the precipitation for the Northern coastal region. The changes found have been related to changes in the circulation patterns and discussed from the perspective of climatic change.

KEYWORDS: Mean temperature, precipitation Spain, PCA, climatic change.

1 Introduction

The most common variable analyzed to assess climatic change due to increased greenhouse gases in the atmosphere is the average global temperature. The series of this variable present a significant rise during the period of instrumental records, by more than 0.6 °C over the past 135 years (Jones et al., 1994). This variable, the average temperature, is an agent in environmental changes (Hansen et al., 1995). Predicted increase in global temperature based on numerical models, is between 0.6 and 0.4° C over the next 25 years (IPCC, 1995), together with effects, which are not very well understood, so, for example, a most critical need is to investigate whether global warming will cause changes in regional precipitation, and some authors have underlined the need for researching the precipitation variability (e.g., Shuttleworth, 1996). This implies the need to examine local changes and to relate then to different possible environmental impacts of the climatic change in each region. Furthermore, to improve our knowledge of the climate, research is needed on local and regional scales (Karl et al., 1989). The use of empirical methods can reveal major patterns and provide qualitative estimates of the form and range of the regional climate change useful to validate climatic models with a physical basis (Giorgi and Mearns, 1991).

In the present work, the mean temperature and precipitation records throughout Spain are analysed, with an special emphasis in the Northern Spain. The climatic studies over this region have particular interest, due to the geographical position of the Iberian Peninsula, between the Atlantic Ocean and the Mediterranean Sea, and between the continents Africa and Europe. In addition, this region is located in the circumpolar vortex border (Capel Molina, 1981) and its climate is closely linked with the atmospheric circulation. The climatic variability of this area is also of interest due to the current problems provoked by droughts and other extreme events, which have been discussed by some authors (Katz and Brown, 1992) as good indicators of the climatic change and by the adverse predictions of the models in the Mediterranean area (IPCC, 1995). It is important to identify regions with similar behaviour, detect and quantify possible changes and relate them with possible causal mechanics .The present paper undertakes this task for the case of Spain.

2 Data

The database used in this study includes annual and seasonal mean, maximum and minimum air temperature and precipitation for 40 Spanish localities, covering the Iberian Peninsular area and Baleares Islands (Fig. 1.). They were selected from 65 series supplied by the Instituto Nacional de Meteorología (INM), attending quality criteria. Precipitation database is completed with some grid points of Portugal. The temporal coverage for temperature is 1880-1995 and for precipitation is 1900-1997.

Most of the stations have not changed their position, but only for a few of them do we know the metadata related to methods and instruments. The absolute homogeneity of the records was evaluated using the Thom and Bartlett tests (Thom, 1966; Mitchell, 1966). These methods were also used to check the relative homogeneity, analysing the difference (for temperature) or ratio (for precipitation) series obtained for neighbouring locations (Bradley et al., 1985, Rhoades and Salinger, 1993).

Homogeneity problems are most common in temperature records. Some of the

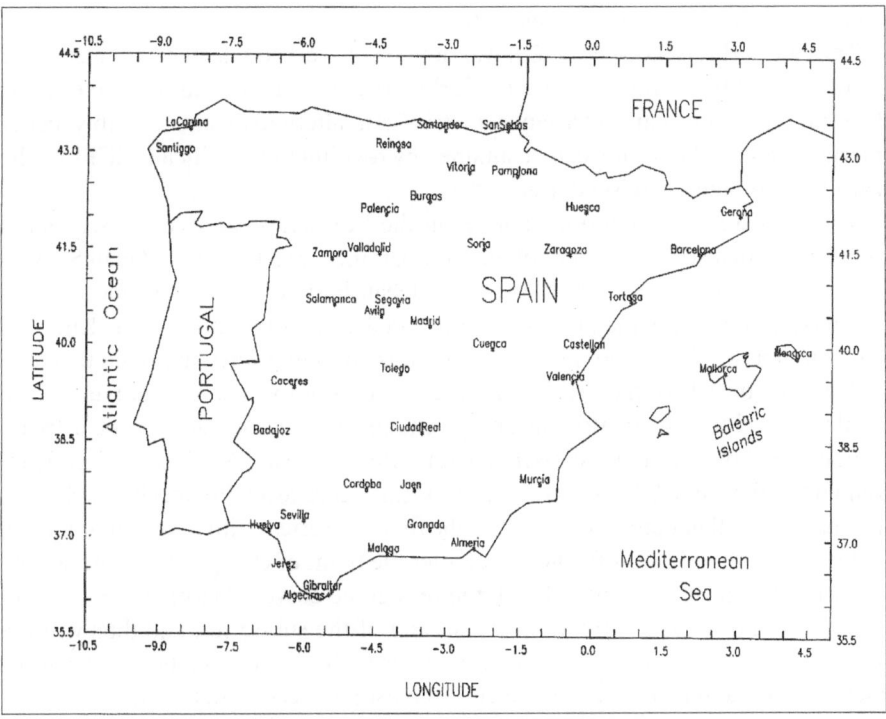

Fig. 1. Locations of the stations the data of which are used

annual and seasonal series present homogeneity problems that were corrected using the method proposed by Bradley et al. (1985), taking into account the difference between the mean values before and after the inhomogeneity, and when possible, nearby stations. In the correction, periods without trends or other effects were used, following the recommendations of Rhoades and Salinger (1993). The details of these corrections can be found in Esteban-Parra(1995) and Esteban-Parra and Castro-Díez (1996). For precipitation series, only Avila showed problems that were corrected using data form Segovia, a locality very closed to Avila and with a similar altitude.

Series with few missing values have been used (less than 3 consecutive years). The missing data have been filled in with the corresponding monthly mean value of the missing month calculated for the entire period.

3 Method

To identify the general behaviour of all the series, the Principal Component Analysis, PCA was applied (Preisendorfer, 1988). Briefly, given p variables (anomalies) of length n corresponding for example to p locations, the method consists of the diagonalization of the covariance/correlation matrix. The EOFs

(Empirical Orthogonal Functions)are the eigenvectors with an associated variance equal to their corresponding eigenvalue.

The rule N (Preisendorfer, 1988) was applied in order to select the number of the significant EOFs. This is a MonteCarlo procedure to obtain the eigenvalue distribution of random covariance matrix. An alternative and possibly better interpretation of the results can be attained by rotating the significant EOFs by the Varimax procedure (Preisendorfer, 1988).

To detect trends and abrupt changes in the PCs series we used the sequential Mann-Kendall test. Descriptions of this non-parametric test can be found in Sneyers (1975), Goossens and Berger (1986) and Esteban-Parra et al. (1995). One problem in applying this test is the presence of the serial correlation in the data (Kulkarni and von Storch, 1995), and therefore caution is advisable when applying this test.

The test consists of the graphical representation of two curves, computed in a similar way. Mathematically, an abrupt change is a particular case of a trend, characterized by two stable sub-series with different means. We have a significant trend when the curve C1 surpasses the 5% significance level. For an abrupt change, the curve C1 will not present a trend for the first sub-series (with a particular mean), and will pass the 5% significance level after the point of change, when the second sub-series begins. On the other hand, the retrograde curve C2 (which is equivalent to C1), will not present a trend for the second of the sub-series, and thus will not pass the significance level. However, both curves have the same behaviour at the change point, and therefore the two curves must intersect at this point.

4 Temperature Analysis

4.1 EOF Patterns

Except for winter, which has only one significant EOF, two significant EOFs were found for the rest of the seasonal and annual series. At this point we should note that the first unrotated EOF explains most of the variance (more than the 50%) with high correlations for all stations, and the second one less than the 10%, with significant correlations for stations on the Mediterranean and East Cantabric coasts. When we rotate these EOFs, the second has the effect of pulling the first, in such a way that in some cases the rotation divides the first unrotated EOF into two. Both have significant correlations for almost all the stations, although the first rotated EOF represents slightly better the western part and the second the eastern part. This is the case for example of the annual data and, to a lesser extend, for summer data that is, the regions most influenced by Atlantic weather types, and by the Mediterranean climate. In any case, the regionalisation drawn by these two rotated EOFs is quite limited and the first unrotated EOF can be considered representative of the entire area under study. Fig. 2 shows the loading factors associated with the first unrotated EOF for the annual and seasonal data.

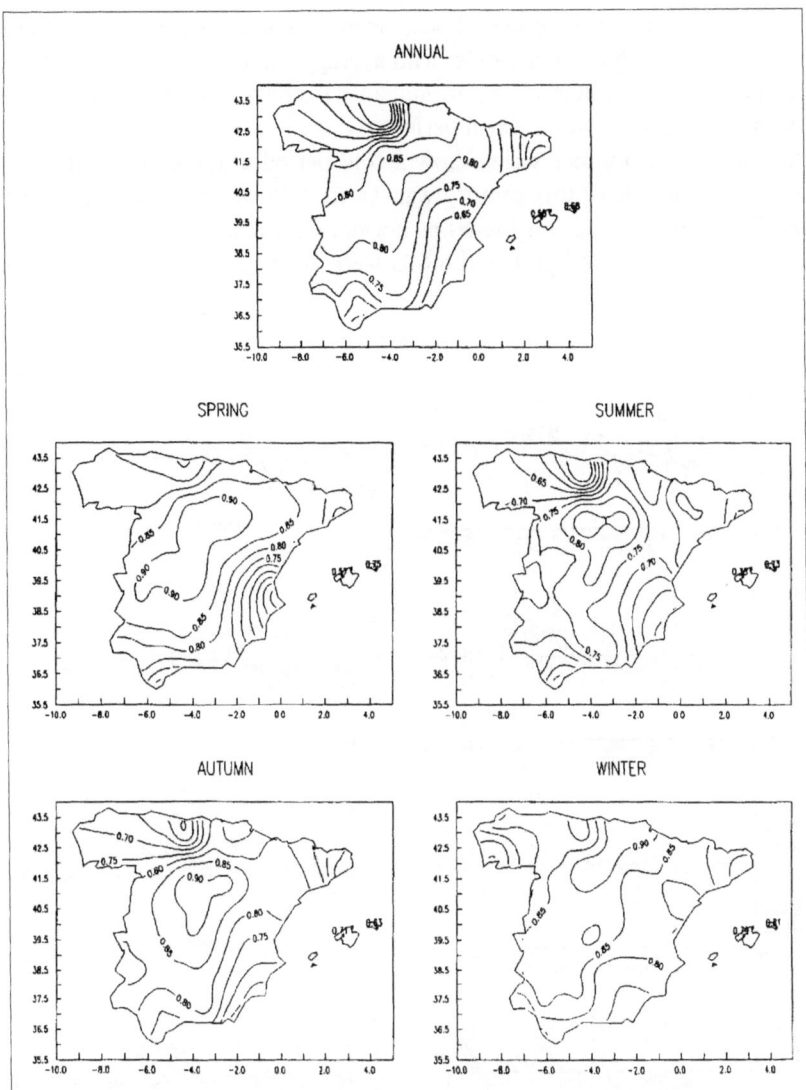

Fig. 2. Loading factors associated with the first unrotated EOF for the annual and the seasonal data

The same results are obtained for maximum and minimum temperatures.

4.2 PCA Series

The left-hand sides of Figures 3, 4, and 5 show the annual and seasonal mean, minimum and maximum series with the corresponding smoothed series, identifying the longer-term variability.

For the winter, a very cold period ocurred at the end of last century, followed by a long period (until the end of 1950s) with average values slightly below the mean. The 1960s presented values above normal and, from the end of the 1970s until the end of the series, there was a warm period.

The spring and summer series show a cold period at the end of 19th century, while the first decade of this century was warm, followed by cold years until the 1920s. The 1930s can be considered as normal with respect to the entire period, whereas, the 40s, 50s and part of the 60s were all very warm. The decade of the

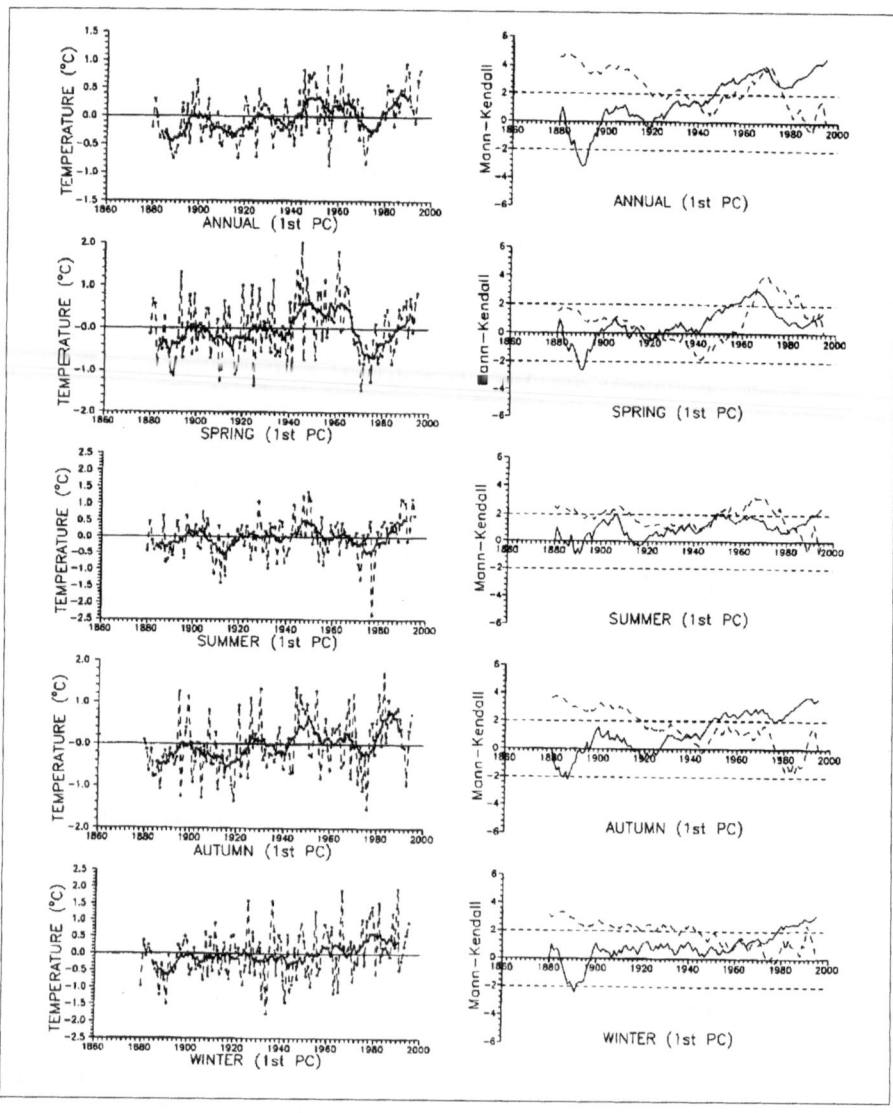

Fig. 3. Annual and seasonal mean temperature data series with smoothed data series identifying the longer termvariability.

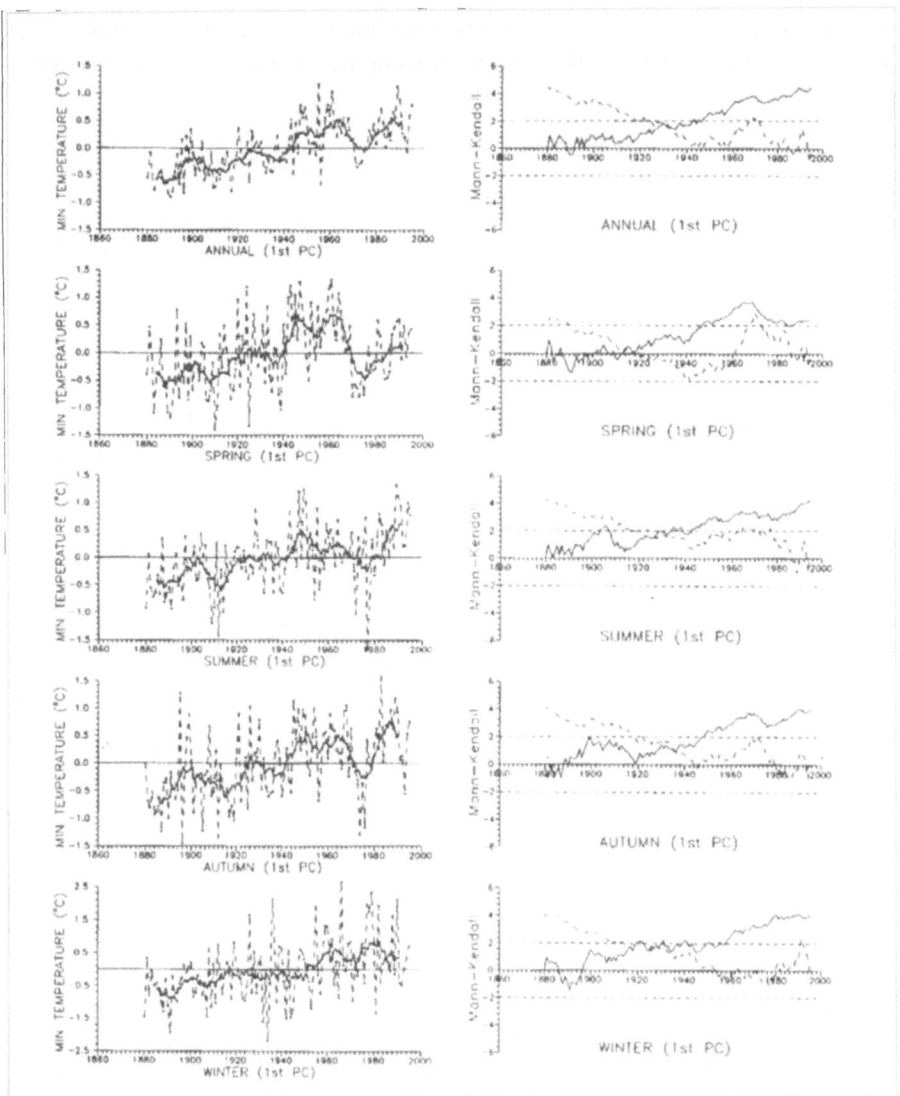

Fig. 4. Annual and seasonal minimum temperature data series with smoothed data series identifying the longer term variability.

1970s was very cold, followed by the relatively warmer 1980s.

Autumn shows similar fluctuations, with a remarkable intensity of warm period during the decades of the 1940s and 1950s. The cold period of the 1970s was shorter and less intense, there was a strong increase in the temperatures during the 1980s.

The annual series summarizes this behaviour, showing an increasing trend until 1970, superimposed over the cold/warm periods. There were remarkable cold conditions at the end of the last century and during the 1970s, followed by the strong warming in the final years of the series. Imaximum and minimum temperatures show similar fluctuations, with a very significant increased trends in the minimum temperatures.

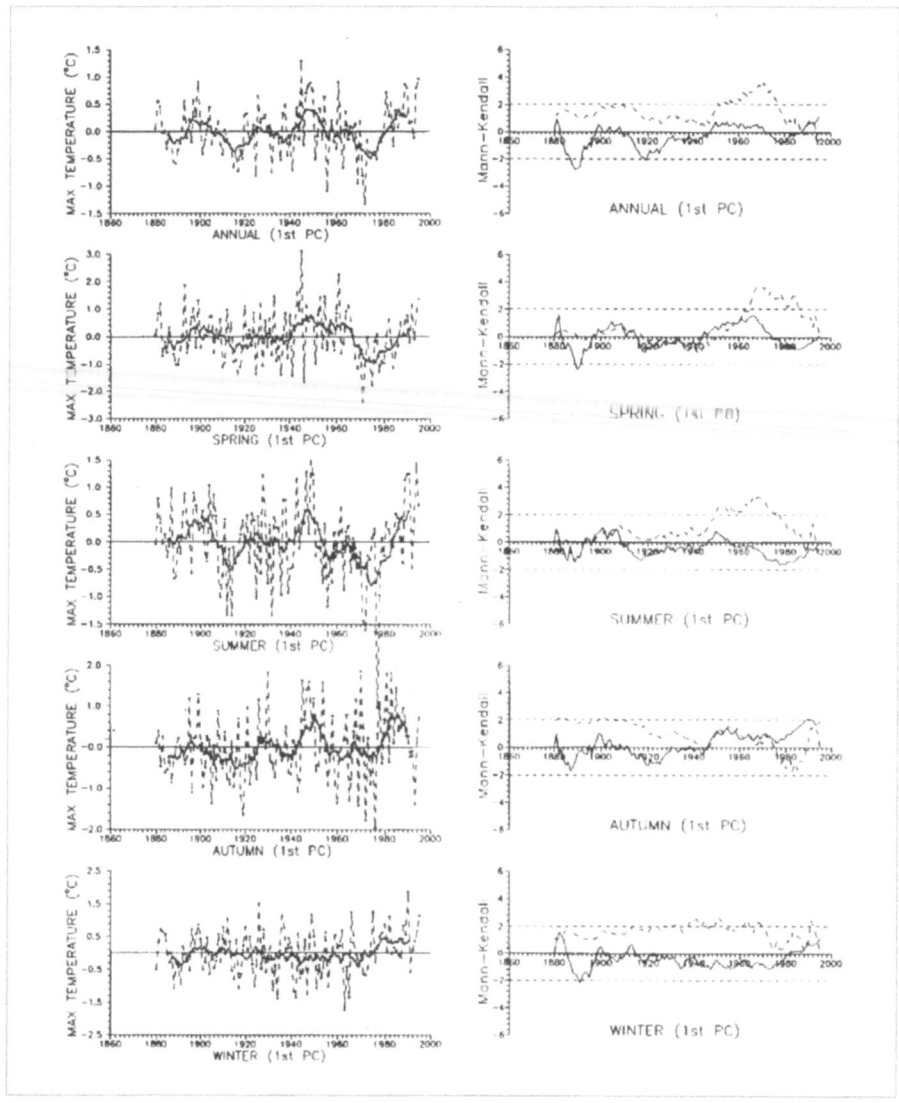

Fig. 5. Annual and seasonal maximum temperature data series with smoothed data series identifying the longer term variability

4.3 Trends and Abrupt Changes

This section deals with identifying the presence of trends and abrupt changes in the PC time series. For these series, the lag-one correlation is very small (no significant at 95% level). The right-hand side of Figures 3 - 5 also show the result of the Mann-Kendall test applied to the first principal component series.

All the series except the spring ones show a significant increase. Winter shows a smooth trend, while autumn presents an abrupt change at the end of 1930s. Summer series presents and increase, although without significance during the 1970s. For the annual series, the first series shows an abrupt change in 1936 (this point change is near the significant level and thus perhaps could be described as a trend).

For the Northern Plateau, the increase appears for the minimum series, with values that rise more than 2° C between the first and the last years of the records. Maximum temperatures present a high increase in the last years. (Figure 6).

4.4 Discussion of Temperature Results

The annual Spanish series have been compared with the global and Northern Hemisphere series (Jones et al., 1994). The correlations between this first PC annual series and the global and Northern Hemisphere series are 0.46 and 0.49 respectvely, (significant at 95% level) In general, the Spanish series show greater variability. For the Northern Hemisphere series (and also for the global one), the agreement with the Spanish series appears in the cold period at the end of 19th century, the slight increase at the beginning of this century and the strong decrease before 1910; however between 1910 and 1945, these global and hemisphere series do not show great fluctuations, but rather clearly increasing trends that stabilize during the 1950s (this last point agrees with Spanish series). The decade of the 1970s was cold throughout the world, although apparently that it was colder in Spain, which also had more intense the temperature increase in the 1980s.

With regard to the relationships between temperature and NAO and pressure conditions, Table 4 shows the correlations between the NAO index proposed by Jones et al. (1997) (using Gibraltar pressure data as the southern location) and also with the mean sea level pressure data of Gibraltar as representative of the circulation conditions over the Iberian Peninsula. It is surprising the zero correlation for winter between pressure or circulation index, bearing in mind that circulation conditions have been proposed as an important temperature variability agent during the cold season (Wallace et al., 1995). In fact, precipitation over the area is controlled mainly by the NAO (Esteban-Parra et al., 1998).This situation may be due to different effects on maximum and minimum winter temperatures: during high pressure and high NAO conditions, maximum temperatures increases due to solar heating, whereas minima decrease due to radiative cooling, and thus, for low NAO and low pressure conditions, the reverse is true with regard the temperature behaviour. This

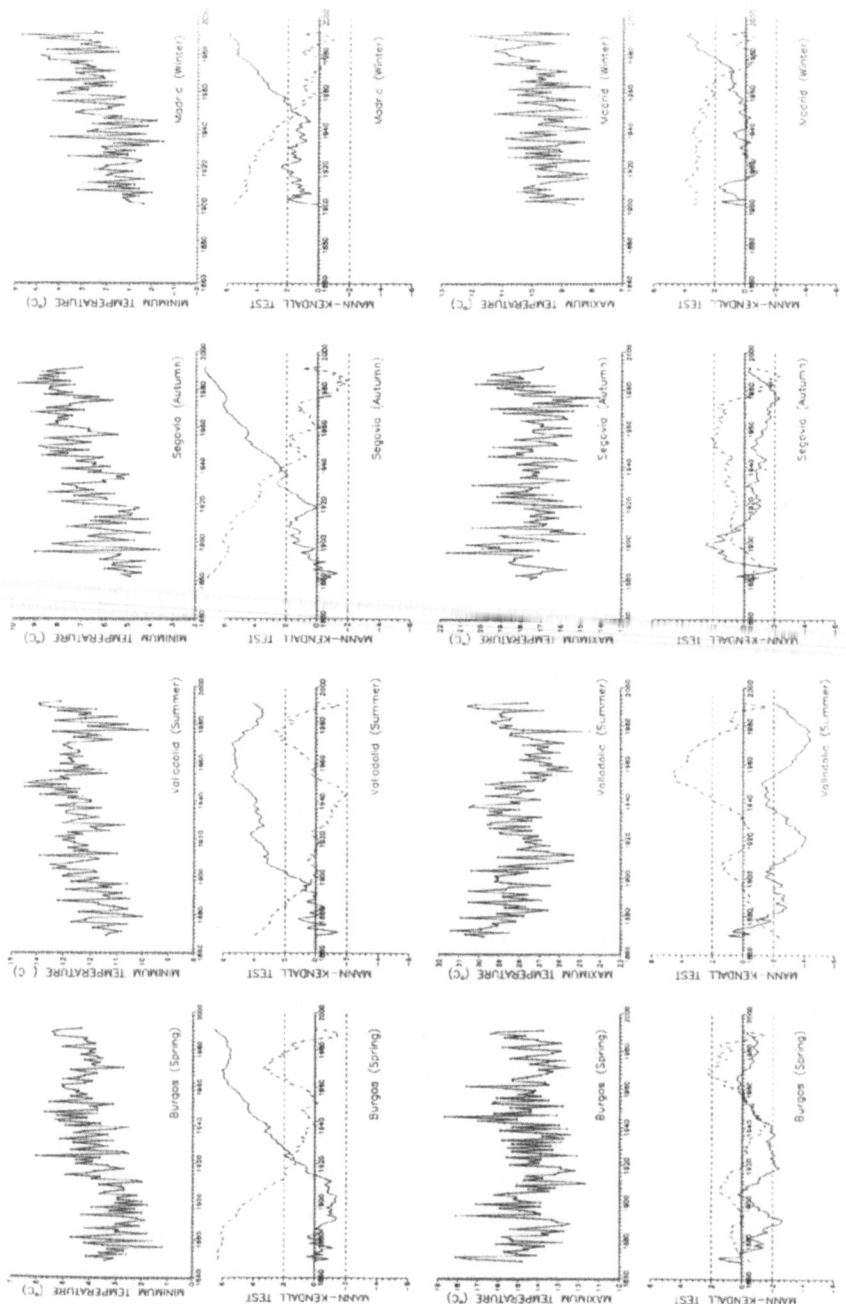

Fig. 6. Maximum and minimum temperature series and Kendall test results for selected stations and selected seasons in the northern plateau

low correlation between NAO and winter temperature for Spain has been reported by other researchers (Pozo et al., 1999). As shown above spring temperature is highly correlated (positive) with Gibraltar pressure, and can also be explained in terms of the high solar heating under clear conditions during the anti-cyclone patterns and a cooling during cyclonic activity, but the radiative mechanism is not enough to provoke a low minima during the clear nights. This is also applicable to autumn and annual data. Meanwhile low negative correlations in summer would be unexpected. This maybe due to the stable general high-pressure conditions and the weakening circulation. As some authors argue (Wallace et al., 1995, 1996), the temperature variability during summer regions would be caused by radiative forcing instead of circulation because this factor is very weak during the warm season. But as Figure 4 shows, these relationships have changed markedly over time, reflected in an apparent increase in the influence of pressure.

The Azores high appears to have been more intense and to have been desplaced towards the east and the north since 1980 (Beniston et al., 1994), and it could be responsible for the increases in temperatures during these years. In any case, such pressure or Atlantic circulation conditions are not enough to explain the temperature variability on scales of decades or years. For example, there is a good agreement with the pressure values over the Mediterranean sea at 40°N; high-pressure values correspond to cooler periods, and low-pressure values to warmer periods (Makrogiannis and Sashamanogluo, 1990).

A partial explanation of the increase found is the so-called urban island effect.In the present work the most part of the localities correspond to urban sites, from large cities such as Madrid and Barcelona, to rural localities such as Monflorite and Tablada. Most are small cities (41.5% of these localities have less than 100.000 inhabitants, and 22% less than 50.000 inhabitants). It is true that urban heating is also present in small localities as demonstrated by Karl et al.(1988), but this conclusion loses validity for cities with a very long history, as it is in this case. Urban growth has not had a constant rate, and thus, it is not possible to consider a linear trend. At least in Spain, growth has been greater during this century, especially since 1960, when the cities began to develop modern features, such as the presence of asphalt and industry. In addition, the number of cities with more than 100.000 inhabitants rose in the1970s, although the warming began before 1920.Also, the cooling in the 1970s does not seem masked by the urban heating. According to Jones et al. (1989), at least at hemispheric scale, the effect of the urbanism is less than 0.1 °C. Given the general agreement between their series and ours, the urban island effect is perhaps responsible for the temperature increase in Spain at a similar rate.

The magnitude of the urban heating effect depends of several variables: size and behaviour of the urban nucleus, topography, local industry and the meteorological conditions (Landsberg, 1981). Kozuschoswki et al., (1994) estimate the magnitude of this heating for the industrial city of Krakow in 0.5 °C during the last 40 years. Beniston et al. (1994) consider that the urban island effect is very small for cities like Zurich and Lugano, taking into account the global increase of the temperature in this case. Other researchers also argue that the temperature rise cannot be attributed solely to urbanization processes, due to the presence of this warming in sites with different development of the urbanization and land use (Palmieri et al., 1991).In fact, some localities, such as Segovia, with only 53.000 inhabitants, have increases higher than in Madrid, with more than 3.000.000 inhabitants.

Currently, one of the leading aims in the research on climatic change is not only to detect the anthropogenic effect, but also to quantify this effect and also to determine how much of the change is due to the natural climatic variability. One behavior. possible agent of the natural variability is the volcanism. In general, volcanic eruptions bring about short periods of cooling, by expelling sulphur gases and aerosols into the atmophere. At the end of the last century, volcanic activity was quite intense, exemplified by the stupendous Krakatoa (1883)and Tarawera (1886) eruptions, which influenced meteorological records of continental Europe (Bradley, 1988). Another major eruption was Pinatubo, in Philippines in June of 1991, which may be associated with the relative cold years in the last warm period (1991 began to be colder in autumn, when the Pinatubo effect arrived Spain (Olmo and Alados, 1995).

We cannot attribute all the warming detected to the greenhouse effect, but the comparative study of some of the fluctuations can help to define the magnitude of the effect, but following Wallace et al. (1995) arguments, for Spain, radiative forcing of the temperature climate could be the main responsible of the behaviour found. The difference between the last century and the beginning of this one can be considered natural fluctuation, perhaps related to the volcanic activity and the end of the Little Ice Age at the end of last century, provided the greenhouse effect was minor during this epoch (Bradley and Jones, 1995). From this point of view lower temperatures in the 1970s may reflect the high magnitude of the natural variability, and nullify (at least for local areas such as the Iberian Peninsula and Balearic Islands) the greenhouse effect during certain periods. The strong increase in the 1980 could be provoked by anthropogenic causes, natural variability or both. The first possibility could be related to an abrupt response of the climatic system to the human forcing, although it is the cold decade of the 1970 s which breaks the trends. Despite of the powerful eruption of El Chichón in 1982 and its cooling effects, the 1980s generally had the highest temperatures, the maximum temperature being for 1989 (prior to that period only isolated years present very high values). Thus, if the increase was due only to natural variability, it would be the first instance of this strong value and persistence. Given the relative good agreement with the global and hemispheric series, and that these series are relatively well represented by the climate models (IPCC, 1995), this increase may in fact be partially due to the increased greenhouse effect, as some researchers believe (IPCC, 1995). This hypothesis includes that the increasing trend will continue in the future.

5 Precipitation Analysis

5.1 EOF Patterns

Figs. 7a and b show the significant seasonal and patterns of the (Varimax) rotated EOFs of Spanish precipitation anomalies by drawing the isolines of the loading factors values. Table 2 shows the percentage of variance explained by each EOF. The first rotated EOF pattern is centred on principally Andalusia and the interior part of Spain. In this region, precipitation is associated with the westerly circulation, common to polar maritime (mP) and tropical maritime (mT) air masses. According to Capel Molina (1981), the most frequent flows are from the SW, followed by W

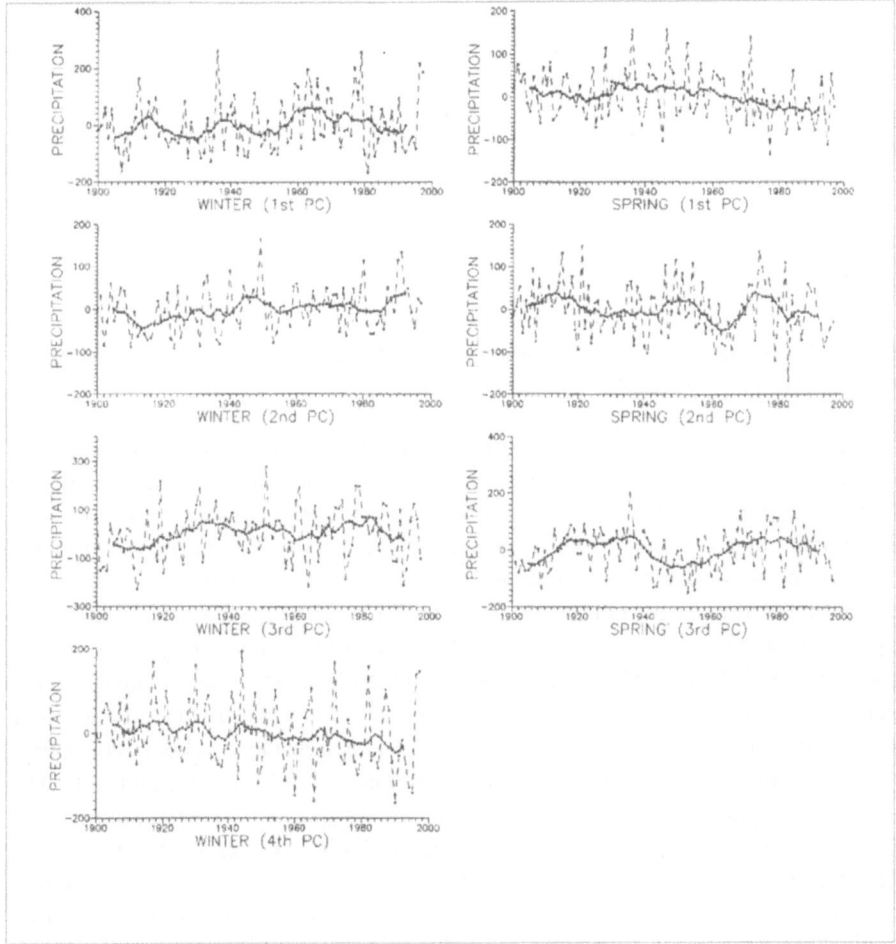

Fig. 7a. The significant principal components of the winter and spring precipitation anomalies in mm for the twentieth century

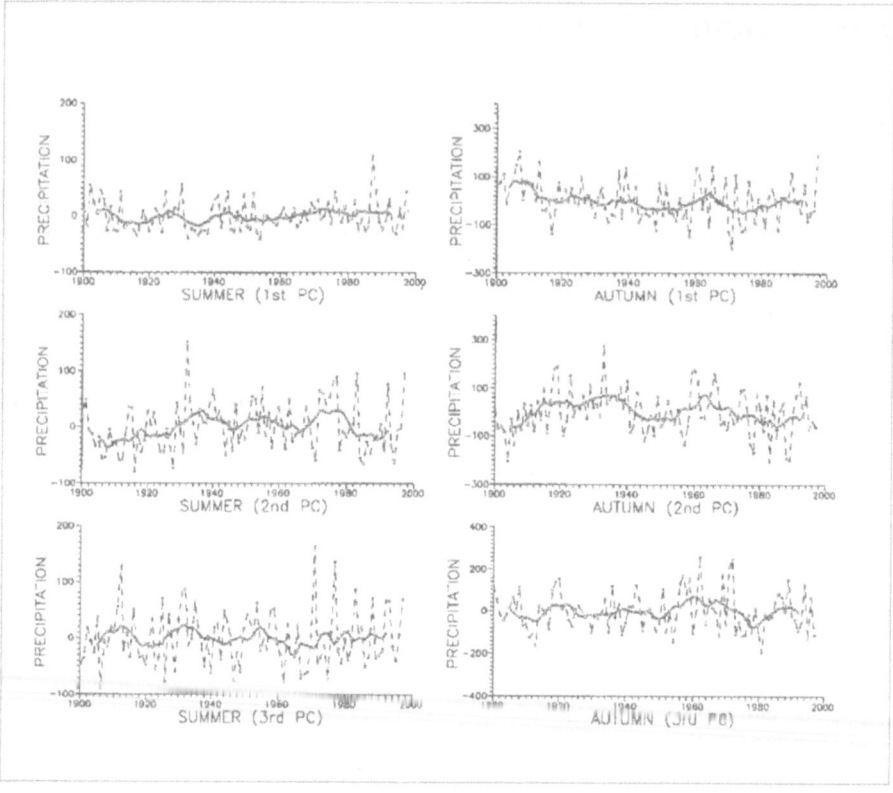

Fig. 7b. The significant principal components of the summer and autumn precipitation anomalies in mm for the twentieth century

and then NW. It can be observed that most of the mountain ranges in Spain (except for the Iberian and Penibetic ranges) have a west to east orientation. They do not impede the dominant westerly circulation regime prevailing from October to May. The second and the third EOFs are associated alternatively with the precipitation regimes in the Mediterranean and the Cantabric Coast (northern coastal stations), depending on the season. For the Mediterranean Coast, the precipitation is produced by easterly flows, associated with meridional incursions of maritime polar air which come from the Atlantic Ocean and flows into the Mediterranean sea after rotating in a clyclonic way (Capel Molina, 1981) or from a tropical continental air mass. The warm humid E and SE winds from the Mediterranean Sea lead to heavy convective precipitations over the region, specially when there is colder air at high levels. The northern Cantabric Coast is characterized by big amounts of precipitation. An important factor for precipitation in this area is the confluence of coastal-mountains. The precipitation originates in meridional N or NW circulation, associated with polar maritime and arctic maritime air masses. Cold air that arrives in the Iberian Peninsula with a high vertical instability due to thermal subversion (Capel Molina, 1981).

Only for winter, there is a fourth significant EOF, representing the northern Mediterranean coast.

5.2 Principal Trends and Abrupt Changes

This section deals with identifying the presence of trends and abrupt changes in the PC time series. For these series, the lag-one correlation is very small (no significant at 95% level). The right-hand side of Figures 3 - 5 also show the result of the Mann-Kendall test applied to the first principal component series.

The characterisation of the periods as wet or dry has been done using 10 year moving averages. Figs. 7a and b show the seasonal series with the corresponding smoothed series, identifying the longer term variability. The first winter series, associated with Andalusia and "Peninsular Interior" begins with a slightly dry period between 1900 and 1930.There is a relatively rainy period from then until 1945. Then, and during the 1950s, it follows a drier than normal period. The 1960s and, to a lesser extend, the 1970s were very rainy. The 1980s are characterized as a dry period. For the spring there is a significant decreasing trend in this region. For summer, the first years of the 20th century were quite humid. For autumn the most significant fluctuation ocurred during the dry 30s.

For the Mediterranean area the significant decreasing trend is remarkable which occurred during winter in the Northeastern area and for spring in the last years. Also for spring, we find important fluctuations in the dry 60s and the rainy 70s.

At the Northern coast, there is a slight increasing trend of winter precipitation. During spring, it is remarkable the low precipitation values during the decades of 40s and 50s, and in the 90s. Autumn precipitation shows low values from 1900 until 1930 and in the decade of the 80s.

5.3 Discussion of Precipitation Results

Annual precipitation basically is related to the precipitation from September to May as the correlation between the annual and seasonal PC series shows. The correlation ranges around 0.55 for the Interior and Andalusia and Cantabric coast. All correlation coefficients between annual and seasonal PC series for Andalucia, the interior stations, and the Cantabric region are significant, although the correlation between summer and annual series of these areas is smaller than the other seasonal correlations. For these areas, the winter is the most important season on the annual totals, whereas, for the Mediterranean region, the autumn (correlation with annual data of 0.67) is the most influential one.

The results agree partially with conclusions drawn by other researchers. There were relatively wet periods in the Mediterranean until 1914, in 1930s, and in 1960s and 70s for the Mediterranean area (Maheras, 1988; Maheras and Koliva-Machera, 1990; Maheras et al., 1992). The humid 1930sin the Spanish region seems to be widespread over Europe, at least for the latitudes below 50°N (Brazdil et al., 1985), although it was probably less pronounced for the Western and Central Mediterranean areas (Maheras, 1988; Maheras et al., 1992). There is also a good concordance between dry periods. The dry period around 1920 appears also in a general way in the Mediterranean and Southern Central Europe (Brázdil et al. 1985; Maheras 1990). The 1st annual series centred on the Andalusia and Interior region of Spain is similar to the Central Europe series, particularly those located in the south (Brázdil et al., 1985) from 1920 until the end of the records. The decreasing trends found for the 1st and 2nd PC series (Andalusia and Interior, Mediterranean Coast) are common for Southern Europe. The annual series corresponding to the Cantabric are more similar with series corresponding to Northern Europe (Brázdil et al., 1985).

The presence and intensity of the Azores High is a determinant in the Iberian precipitation regime. Effectively, the coincidence between the annual average central pressure value of the Azores High and its mean position (Sahsamanogluo, 1990) and the 1st annual PC series is very clear; low precipitation are related to high values of Azores central pressure or an eastwards shift of its mean position. There is also a concordance with the evolution of the other two annual series, specially during the relatively dry years between 1940 and 1960, when the Azores High shifted to the east, and during the 1980s when the intensification of the Azores high coincided with a new period of low precipitation. As a proof of this relation, Table 5 shows the correlation coefficients between the PC series and the seasonal and annual series of MSLP at Ponta Delgada (Azores) for which there is data from 1894 to 1981. The use of these data as representative of the Azores High is necessarily limited because this action centre has not a fixed position from year to year and from season to season (Sahsamanogluo, 1990). The relation is quite good for the 1st PC time series, where the Azores High influence is particularly significant, and for the winter season, when the position of the Azores High is to the south east of Ponta Delgada and in an optimum position for controlling the precipitation in Western Iberia (Capel Molina, 1981). It is worth mentioning the positive correlation between the autumn series for Cantabric and Azores, similar to the situation for Western Northern Europe.

There is also a good concordance between the evolution of the precipitation over the Mediterranean Coast and the mean sea level pressure over the Mediterranean (Makrogiannis and Sahsamanogluo, 1990), so the negative anomalies of the pressure in the 40^{0}N from 1880 to 1900 and from 1930 to 70 coincide with the rainy period in the beginning of the series and during the 60s and 70s, and positive anomalies in 1900-1930 and 1970-80, correspond to the dry periods of the decades 1910 and 1920 and 1980. There is not concordance between the positive anomalies of pressure on the 40°N and the precipitation during the first decade of this century, although it occured with the negative anomalies during these years over the 30°N.

One of the most useful parameters to analyse the circulation of the North Atlantic is the North Atlanctic Oscillation (NAO) index. A high NAO index represents strong westerlies, and low index, a predominant meridional circulation. We have compared our results with two different indices of the zonality over the North

Atlantic. This index is usually defined for the winter season (Hurrel, 1995; Jones, 1993). For Andalusia and Interior winter series, precipitation is significant correlated with NAO index, with a value of -0.7. The inverse relation (high index-low precipitation and viceversa) is also found in other the PC series, which is less clear due mainly to the influence of easterly patterns for the Mediterranean Coast and the influence of the mountains for the Cantabric area.

References

Bárdossy A, Caspary HJ (1990) Detection of climate change in Europe by analysing European atmospheric circulation patterns from 1881 to 1989. Theor. Appl. Climatol., 42:155-167

Beniston M, Rebetez M, Giorgi F, Marinucci MR (1994) An analysis of Regional Climate Change in Switzerland. Theor. Appl. Climatol., 49:135-149

Brázdil R, Samaj F, Valovic S (1992) Variation of Spatial annual precipitation sums in Central Europe in the period 1881-1980. J. Climatol. 5:617-631

Bradley RS (1988) The explosive volcanic eruption signal inthe Northern Hemisphere continental temperature records. Climatic Change 12:221-243

Bradley RS, Kelly PM, Jones PD, Goodess CM, Diaz HF (1985) Climatic Data bank for Northern Hemisphere land areas, 1851-1980. Department of Energy, Washington, D.C.

Bradley RS, Jones PD, eds. (1995) Introduction. Climate since 1500 A.D. Routledge, London.

Capel Molina JJ (1981) Los climas de España. Oikos-Tau, Barcelona.

Esteban-Parra MJ (1995) Contribución al Estudio de laEvolución del Clima de España en el Período Instrumental'. Ph.D. Thesis. University of Granada.

Esteban-Parra MJ, Castro-Díez Y (1996) On the homogeneity of the longest temperature series in Spain: A critical analysis. In: Climate Dynamics and the Global Change Perspective, Jagiellonian University Series, Krakov.

Esteban-Parra MJ, Rodrigo F S, Castro-Díez Y (1995) Temperature Trends and Change Points in the Northern Spanish Plateau during the last 100 years. Inter. J. Climatol. 15:1031-1042

Esteban-Parra MJ, Rodrigo F S, Castro-Díez Y (1998) Spatial and temporal patterns of precipitation in Spain for the period 1880-1992. Inter. J. Climatol. 18:1557-1574

Giorgi F, Mearns LO (1991) Approaches to the simulation of regional climate change. A review. Rev. Geophys., 29, 191-216

Goossens Ch, Berger A (1986) Annual and seasonal climaticvariations over the Northern Hemisphere and Europe during the last century. Annals. Geophys. 4, B, 4:385-400

IPCC (1996) Climate Change 1995: The Science of Climate Change. Contribution of Working Group I to the Second Assessment Reportof the Intergovernmental Panel on Climate Change. Houghton JT, Meira Filho LG, Callender BA, Harris N, Kattenbergand A, Maskell K (eds). Cambridge University Press, Cambridge and New York, 572 pp

Jones PD, Raper SCB, Bradley RS, Diaz HF, Kelly PM, Wigley TML (1986) Northern hemisphere surface air temperaturevariations: 1851-1984. J. Clim. Appl. Meteorol. 25:1213-1230

Jones PD, Wigley TML, Briffa KR (1994) Global and Hemispheric temperature anomalies-land and marine instrumental records. In: Boden DP, Kaiser TA, Sepanski RJ, Stoss FW (eds.) Trends 93. Carbon Dioxide Information Analysis Center. Oak Ridge.

Jones PD, Jønsson T, Wheeler D (1997) Extension of the North Atlantic Oscillation using early instrumental pressure observations from Gibraltar and south-west Iceland. Inter. J. Climatol. 17:1433-1450

Karl TR, Diaz HF, Kukla G (1988) Urbanisation: Its detection and effect in the United States. J. Climate 1:1099-1123.

Karl TR, Tarpley JD, Quayle RG, Diaz HF, Robinson DA, Bradley RS (1989) 'The recent climate record: what itcan and cannot tell us', Rev. Geophys., 27, 3:405-430.

Katz RW, Brown BG (1992) Extreme events in a changing climate: Variability is more important than average. Clim.Change 21:289-302

Kozuschoswki KM (1993) Variations of Hemispheric Zonal Index since1899 and its Relationships with Air Temperature. Int. J. Climatol. 13:853-864

Kozuschoswki KM, Trepinska J, Wibig J (1994) The Air Temperature in Cracow from 1826 to 1990: Persistence, Fluctuations and Urban Effect. Int. J. Climatol. 14:1035-1049

Kulkarni A, von Storch H (1995) Monte Carlo experimentson the effect of serial correlation on the Mann-Kendall test of trend. Meteorol. Zeitschrift 4:82-85

Landsberg HE (1981) Urban Climate. Academic Press, New York.

Maheras P, Koliva-Machera F (1990) Temporal and spatial Characteristics of annual precipitation over the Balkans in the twentieth century. Inter. J. Climatol. 10:495-504

Maheras P, Balafoutis Ch, Vafiadis M (1992) Precipitation in the Central Mediterranean During the last Century. Theor. Appl. Climatol. 45:209-216

Makrogiannis TJ, Sahsamanoglou CS (1990) Time variation of the mean level pressure over the major Mediterranean area. Theor. Appl. Climatol. 41, 149-156

Mitchell JM (1966) Climatic Change, WMO, Tech. Note79, WMO No. 195. TP-100, Geneva, 79 pp.

Olmo FJ, Alados-Arboledas L (1995) Pinatubo eruption on solar radiation at Almeria (38.83°N, 2.41°E). Tellus, 47B:602-606

Palmieri S, Siani AM, D'Agostino A (1991) Climate fluctuations and trends in Italy within the last 100 years. Ann. Geophysicae 9:769-776

Pozo-Vázquez D, Esteban-Parra MJ, Rodrigo FS, Castro-D¡ez Y (1999) A non lineal study of the influenceof the NAO on the temperature variability in Europe. Clim. Dyn., submitted.

Preisendorfer RW (1988) Principal Component Analysis in Meteorology and Oceanography. Elsevier, Amsterdam

Rhoades DA, Salinger MJ (1993) Adjustment of Temperature and Rainfall Records for Site Changes. Int. J.Climatol. 13:899-913

Sashamanogluo HS (1990) Contribution to the Study of Action Centres in the North Atlanctic. Int. J. Climatol., 10:247-261

Sneyers R (1975) Sur l'analyse statistique des seriesd'observations. Tech. Note, 194, WMO, Geneva.

Shuttleworth WJ (1996) The Challenges of Developing a Changing World. EOS, Transactions, American Geophysical Union, Vol 77, number 36:347

Thom HCS (1966) Tech. Note, 81, WMO, Geneva

von Storch H, Zwiers FW (1993) Statistical Analysis in ClimateResearch. Course on Statistical Analysis of Climate Variability. European School of Climatology and Natural Hazards, Elba.

Wallace JM, Zhang Y, Renwick R (1995) Dynamical contribution to hemispheric temperature trends. Science 270:780-783

Wallace JM, Zhang Y, Bajuk L (1996) Interpretation of interdecadal surface air temperature. J. Clim. 9:249-259

Thermal Land-surface Variables From METEOSAT-IR Data

F. Göttsche, F.-S. Olesen

Abstract

There are two series of satellites that provide long and continuous series of data for the Mediterranean Basin: NOAA and METEOSAT. While NOAA/AVHRR has 5 window channels (1 visible, 1 near IR, 1 water vapour (WV; 3.7 μm), and 2 terrestrial IR (8-13μm)), METEOSAT is limited to 3 channels (1 visible, 1 WV, and 1 terrestrial IR). On the other hand, METEOSAT resolves dynamic processes with 48 measurements per day while AVHRR performs 4 measurements per day for a given location. In the framework of climate analyses of surface properties the vegetation cover and the albedo must be derived from AVHRR (spectral capabilities), while the thermal properties must be derived from METEOSAT (temporal resolution). Therefore, only a combined evaluation of many years of AVHRR and METEOSAT can reveal climatic effects.

The large amount of data and the high degree of automation that is required poses a technical challenge, while the development of adequate algorithms and their application are scientific challenges. In order to respond to these challenges best, the Forschungszentrum Karlsruhe and the FU-Berlin agreed to join forces and to share the tasks according to their respective foci of research. The FU-Berlin provides archived satellite data and derives the Normalized Difference Vegetation Index (NDVI) and albedo from AVHRR. The Forschungszentrum Karlsruhe – IMK determines thermal land surface properties from METEOSAT IR data and provides access to an automatic mass storage system. The determination of thermal surface parameters is designed for METEOSAT's temporal and spectral capabilities and consists of the following components:

- The IR measurements are calibrated and satellite instruments are inter-calibrated.
- Cloud covered pixels are detected using dynamic thresholds.
- A neural network is used to determine the atmospheric influence on IR measurements at satellite level. The network is about 5.000 times faster than MODTRAN and uses ECMWF atmospheric data as input. The processing includes the inversion of satellite brightness temperatures to

thermodynamic land surface temperatures (Göttsche and Olesen, 2002).
- The spatial interpolation from a limited number of locations with known atmospheric profiles (ECMWF grid cells) to each pixel (Shepard algorithm; Schroedter et al., 2001).
- The temporal interpolation from 4 times per day, for which the atmospheric situation is known, to all 48 METEOSAT slots (Schädlich et al., 2001).
- A model consisting of a cosine (daytime) and an exponential decay (night-time) is fitted to 10 day or monthly composites of cloud free data in order to derive thermal surface parameters, e.g. minimum temperature, diurnal amplitude, and the time of the maximum temperature (Göttsche and Olesen, 2001).
- The merging of METEOSAT and AVHRR IR data to time series of 2 km spatial and 30 minutes temporal resolution.

All the above components were developed at the IMK. The two most recent components, the atmospheric corrections using neural networks and the model for the derivation of thermal surface parameters, are described in more detail below. In combination with work carried out by the FU-Berlin this will result in one of the most complete evaluations of long term satellite data.

1 Neural Network Versus Modtran

1.1 Principal Considerations

The atmospheric influence on the radiance measured by a satellite can be calculated with a radiative-transfer model, e.g. MODTRAN, provided one has sufficient knowledge about the state of the atmosphere (moisture- and temperature-profiles) and the surface (emissivity). These calculations are very expensive in terms of computing time and are, therefore, not well suited to calculate corrections for large quantities of data, i.e. one year for Europe. On the other hand, faster split-window methods require two or more channels to estimate the atmospheric influence. Until recently these were only available on sun-synchronous satellites. At moderate latitudes these satellites provide a maximum of 4 samples with a variable viewing geometry per day. In order to exploit the long time series (20 years) of METEOSAT data with 30 minutes temporal resolution the possibility of atmospherically correcting single channel IR data using neural networks is investigated. The neural networks are developed and trained using the Stuttgart Neural Network Simulator (SNNS) and the evolutionary algorithm "Evolutionärer Netzwerk Optimierer (ENZO)". The training and validation data sets consist of MODTRAN-3 calculations for a representative selection of atmospheric profiles taken from the

TOVS Initial Guess Retrieval (TIGR) library (globally distributed, quality checked and representative radiosondes). The data were chosen from moderate northern latitudes and the elevation, the scan-angle, and the surface temperature were varied over an appropriate range. The trained network was verified using data sets generated for atmospheric profiles from ECMWF data from 1996 over Europe.

1.2 The Radiative Transfer Model Modtran-3

MODTRAN-3 simulates the transport of radiation in the visible, the infra-red, and the microwave spectral range with a resolution of $1cm^{-1}$. For a given atmospheric situation and with some additional parameters MODTRAN-3 can calculate the radiation density at the satellite sensor, e.g. for calculations in the terrestrial infra-red spectral range also surface emissivity and surface temperature have to be supplied, while the path through the atmosphere is derived from surface location and viewing angle. The measurement in a satellite channel and its corresponding brightness temperature can be determined by applying the sensors response function to the calculated radiation densities. For a known surface emissivity the atmospheric correction is then given with an accuracy of 0.7 K by the difference between the known surface temperature and the calculated brightness temperature (Chinnaswamy, 1999). This forms the basis of the "single channel method" developed by Reutter et al. (1994), which was improved by developing dedicated spatial and temporal interpolation schemes (Schroedter et al., 2001; Schädlich et al., 2001). For mid-latitude summer and winter and with an average surface emissivity of $\varepsilon = 0.975 \pm 0.025$ the temperature error associated with this emissivity is estimated to be ca. ± 1.4 K and the total accuracy of the method is estimated to be ca. ± 2.1 K. Unfortunately, MODTRAN-3 only manages about 90 forward calculations per minute on a Sun Ultra-Sparc. Therefore, the method is not suitable to calculate vast amounts of corrections, e.g. for several months or years of METEOSAT data.

1.3 Feed-forward Neural Networks

Feed-forward networks are the most commonly used type of neural network (NN). Two reasons for their popularity are that they are well suited for pattern matching task and that well established training algorithms for this type of network exist. In its most simple case a feed-forward network consists of two layers, the input and the output layer, but it usually also has at least one hidden layer. In feed-forward networks the information is fed into the input layer from where it is passed via the hidden layer(s) to the output layer. The output of each neuron is specifically weighted for all connections to the neurons of the following layer where it serves as input and is processed further - there is no horizontal spreading of information within the layers. In Fig. 1 the network has one hidden layer. The x_i and o_i are the given inputs and the corresponding outputs of the neurons, respectively. Each neuron sums up (Σ) the weighted outputs of the previous layer to yield its so called 'net input'.

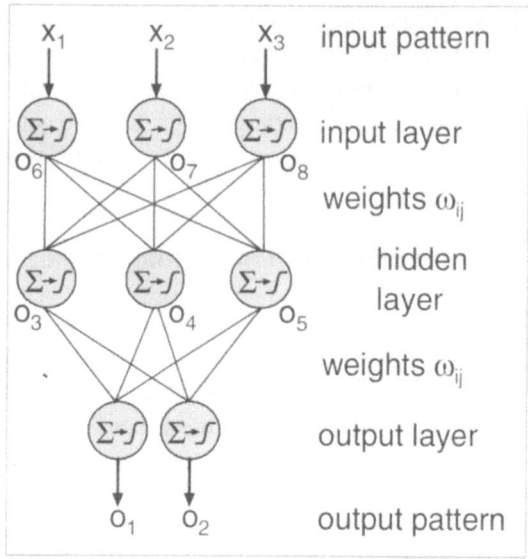

In the more general case the net input, net_{pj}, of a neuron j with respect to the pth input pattern is defined by

$$net_{pj} = \sum_{i=1}^{n} o_{pi} \omega_{ij} \qquad (1)$$

where the o_{pi} are the outputs of the n predecessors of neuron j and ω_{ij} is the weight between neuron i and j.

Fig. 1. Topology of a simple feed-forward NN: Input layer (3 neurons), hidden layer (3 neurons), and output layer (2 neurons)

Local outputs are derived by passing the net inputs through a non-linear activation function, e.g. the S-shaped Sigmoid:

$$f_{act}(net_{pj}) = \frac{1}{1 + e^{-net_{pj}}} \qquad (2)$$

The internal outputs usually have no physical meaning ($O_3 - O_8$). Outputs O_1 and O_2 are the meaningful outputs of the NN; these are compared to the desired (target) outputs.

The total error E of the network is the sum over the quadratic errors for all n input patterns:

$$E = \sum_{p=1}^{n} \sum_{i=1}^{m} (t_{pi} - o_{pi})^2 \qquad (3)$$

where m is the number of output neurons (for the network of Fig. 1: m=2) and the t_{pi} and the o_{pi} are the target values (desired outputs) and the outputs of the network for input pattern number p, respectively.

1.4 Design and Training of Neural Networks

At present the topology of the NN, which performs best for a given task, is most often determined by trial and error. Initially, the networks were developed manually using the Stuttgart Neural Network Simulator (SNNS). The training of neural networks is equivalent to finding the minimum error E given by equation 3. This task can be performed by several algorithms, e.g. Back-propagation, Quick-prop, or Resilient Propagation (Rprop). Here, Rprop with weight-decay (Riedmiller, 1993) was used because it converges reliably and is robust in respect to the choice of its parameters. The best network-architecture was determined during training and validation. This is a time-consuming approach and chances are small that the global optimum is found. Therefore, the evolutionary optimisation scheme "Evolutionärer Netzwerkoptimierer (ENZO)" was utilised to determine a NN which is closer to the global optimum (Braun and Ragg, 1996). More information on neural networks and SNNS can be found in Zell (1994). The simulator, a user manual and information about ENZO can be downloaded from the SNNS homepage at the University of Tübingen: http://www-ra.informatik.uni-tuebingen.de/forschung/snns/.

1.5 Training and Validation Data for Atmospheric Correction

The first step in designing any neural network is to obtain adequate data sets for training and validation. The data has to cover the whole bandwidth of situations which might be encountered during the application phase of the neural network – otherwise it will produce unpredictable results. Here, the atmospheric situations are described by 84 profiles from the TOVS Initial Guess Retrieval (TIGR) library. The profiles were chosen in Eurasia between 45.0 N and 55.0 N latitude. The number of pressure levels of the profiles was reduced to 14 to match ECMWF re-analyses (ERA-15). In order to take into account the influence of atmospheric path-length and the variability of surface temperature, MODTRAN-3 calculations were performed for variations of surface temperature (4), surface elevation (4), and scan angle (3) for all 84 profiles. Then brightness temperatures corresponding to METEOSAT-IR measurements were derived. The generated data was split into training data (68 profiles, 3264 input patterns), which are actually used for the teaching of the network, and validation data (16 profiles, 768 input patterns), which are used to detect over-fitting (Zell, 1994). The structure of the data determines the number of neurones in the input layer and the output layer (Table 1).

In order to avoid problems associated with neuron saturation all data were linearly mapped to the interval (0,1). According to an neural network heuristic the optimum network is the one which has the fewest neurones and the smallest error for the validation data – if more neurons are added the networks ability to generalise, i.e. to produce good results for the validation data, deteriorates, because it starts to memorise the training data rather than to approximate underlying relationships (over-fitting). Starting with one hiden layer and with 5 neurons, Vollmer et al. (2000) determined a suitable network architecture by successively adding neurons or layers. During the training phase the quality of a network can be

judged by observing
- the error for the training data which shows the general ability of the network to produce the correct outputs for the given inputs and
- the error for the validation data which allows to detect over-fitting of the data, which would diminish the ability of the networks to generalise.

After 19500 epochs the training was stopped and the network was saved ("Manual-NN" in Table 2). The error for the validation data is 0.16K, which is small compared to the intrinsic accuracy of MODTRAN-3 (0.7K). The Manual-NN showed

Table 1. Input and output layers as determined by the structure of the data

Input layer (45 neurons)	14 temperatures 14 dew-point temperatures 14 altitude levels 1 land-surface temperature 1 scan-angle 1 surface elevation
Output layer (1 neuron)	1 satellite brightness temperature

Table 2. The NNs determined manually (Vollmer et al., 2000) and using ENZO

Layer	Manual-NN	ENZO-NN
Input	45 neurons	31 neurons
Hidden 1	44 neurons	22 neurons
Hidden 2	18 neurons	4 neurons
Output	1 neuron	1 neuron
	2790 weights	286 weights

no signs of over-fitting. However, the large number of neurons and weights indicated that better solutions might exist. Therefore, evolution was utilised to search for NNs, which are closer to the global optimum. The Manual-NN served as reference for the creation of the initial population of NNs. Due to ENZO's ability to identify redundant input parameters the 14 input neurons representing the altitudes of the atmospheric pressure levels (Table 1) could be removed from the input data (Göttsche and Olesen, 2002). This can be understood when taking into account that the pressure levels are intrinsically coded in the neurons for the atmospheric temperature and humidity values. The final NN determined by ENZO for a desired rms. error of 0.3 K ("ENZO-NN" in Table 2) has 58% fewer hidden neurons and ca. 90% fewer weights than the manually determined NN. The rms. validation error (untrained TIGR data) of ENZO-NN is 0.25 K, which is slightly higher than for the Manual-NN.

1.6 Verification of the Trained Neural Network

Finally, the developed network is verified (tested) with data which were generated analogously to the training and validation data, but using cloud-free atmospheric profiles from ECMWF analyses. These are available globally for more than 10 years and have a spatial resolution of 1.1 degree. The profiles were selected from within the rectangle given by (5°E, 54°N) and (20E, 46°N). The rectangle is contained in the geographic area from which the TIGR profiles used for training were selected. The profiles were taken from 6 ECMWF analyses: 10.03.96 – 12 hours, 12.04.96 – 06 hours, 15.06.96 – 06 hours, 20.07.96 – 18 hours, 18.09.96 – 00 hours, and 26.10.96 – 12 hours. By varying the surface parameters of Table 1, a verification data set of 1200 situations was generated for each analysis. Cloud-free grid cells were identified using cloud-masks derived from METEOSAT data (Schädlich et al., 2001).

For the cloud-free verification data the average error between MODTRAN-3 and the Manual-NN is 0.33 K; the mean temperature errors for the six individual data sets range from 0.16 K to 0.68 K. The corresponding mean errors for all grid cells (includes clouded ones) range from 0.34 K to 1.03 K. ENZO-NN proved to be superior to the Manual-NN: for the cloud-free situations the rms. error ranges from 0.26 K to 0.44 K and for the corresponding rms. errors for all grid cells (no cloud-clearing) the rms. error ranges from 0.31 K to 0.69 K. The rms. verification error (cloud-free ECMWF analyses) of ENZO-NN is 0.31 K, which is slightly lower than for the Manual-NN. This underlines the better generalisation of the impressively smaller ENZO-NN (Göttsche and Olesen, 2002). Furthermore, in spite of the different structure of the ECMWF analyses compared to the TIGR radio-soundings, which were used to train the network, the errors are smaller than the intrinsic maximum error of MODTRAN-3. Taking into account these results, the higher rms. validation error of ENZO-NN compared to the Manual-NN is interpreted as a hint that the latter was over-fitted. A practical advantage of ENZO is the fast development of the NNs (6 days without user interaction). Radiative transfer calculations with ENZO-NN are estimated to be of the order of 10^4 times faster than with MODTRAN-3: this allows atmospheric corrections of long time series of IR satellite data, e.g. the 25 years of METEOSAT single channel IR data.

2 Thermal Surface Parameters

2.1 The Data Basis

For quantitative exploitation METEOSAT IR data have to be calibrated in brightness temperatures. Furthermore, for processing algorithms to produce meaningful results, the data have to be cloud-screened. The calibration information is taken from the EUMETSAT web-pages (EUMETSAT, 1998). The cloud screening is based on the information of the IR and VIS channel during daytime and the IR-channel alone during night time (Schädlich et al., 2001). This information is utilised to deduce monthly temporal and spatial dynamic thresholds. It is assumed that the monthly minimum VIS/maximum IR pixels of each slot are cloud-free values. In order to find a correlation between thermal behaviour and land-surface characteristics, diurnal temperature cycles (DTC's) are extracted from sequences of the pre-processed METEOSAT IR measurements. Data gaps due to cloud-screening and errors are identified and limited to a duration which allows small cloud-fields to pass through an observed scene, but rejects permanently-clouded areas. This process may be regarded as a cloud-clearing correction scheme.

2.2 Modelling of Diurnal Temperature Cycles

Ideally, after calibration and cloud masking only cloud free pixels remain for the determination of LST. For each pixel location more than 10.000 pixels are processed for one vegetation period and about half of them remain after the cloud masking. For such an amount of data tools must be developed to make interpretation feasible. The first step is to model the diurnal variation of LST. The model consists of a harmonic and an exponential term (Fig. 2), describing the effect of the sun and the decrease of the surface temperature at night, respectively. It is fitted automatically to the temperature waves by a Levenberg-Marquardt least-squares scheme (Göttsche and Olesen, 2001). Therefore, model parameters can be determined for data of arbitrary size. Ideally, the diurnal thermal behaviour of every METEOSAT pixel is completely characterised by a set of parameters.

For data of high quality the method works sufficiently well and yields useful parameters. In a case study for the 19.4.96 – a selected day with nearly cloud-free conditions in northern Italy – it was shown that the model is able to describe thermal

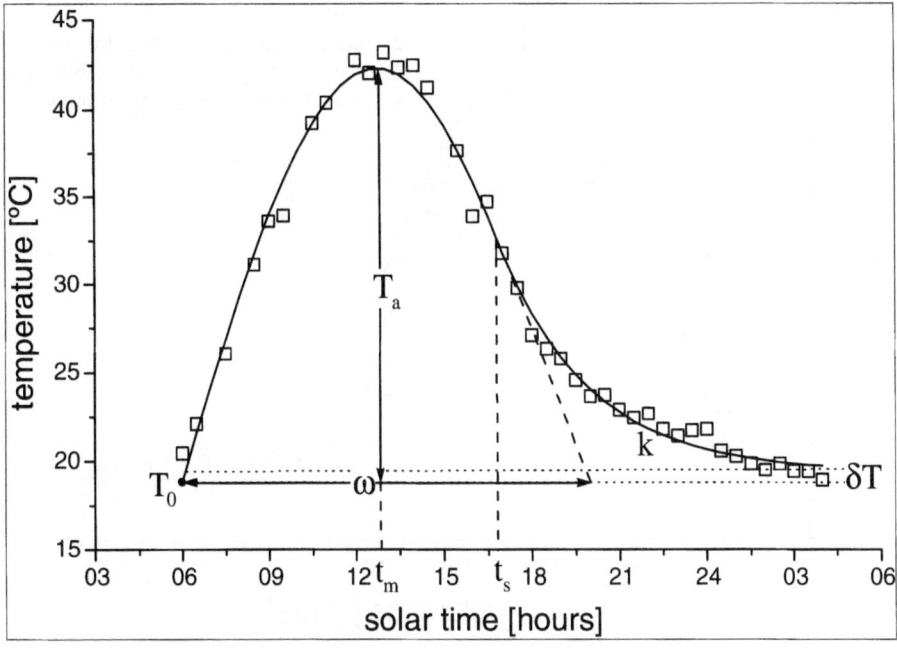

Fig. 2. Least-squares fit of a 7 parameter model to a Diurnal Temperature Cycle (DTC). The boxes represent METEOSAT brightness temperatures for a desert pixel. The smooth, continuous line is the model-fit. For this example the model and the data agree nearly perfectly.

T_0	°C	Residual temperature of previous day
T_a	°C	Diurnal temperature amplitude
ω	hh:mm	Width of cosine function over $\pm \pi/2$
t_m	solar time	Time of maximum temperature
t_s	solar time	Starting time of attenuation function
k	hh:mm	Attenuation constant
δT	°C	$T_0 - T(t\rightarrow\infty)$, where t is the time

characteristics of different surfaces. Fig. 3 shows three fits to DTCs and the corresponding parameters are given in Table 3. The DTCs are modelled well and the fit-errors of T_o, T_a, and δT are of the order of METEOSAT's radiometric resolution of 0.5 Kelvin. The amplitude of the broad-leaved forest is the smallest because of evapotranspiration and shading effects. The scatter after 20 hours, which might be due to a navigation error or undetected clouds (fog), leads to a higher δT for this curve. T_o, the delay of t_m, and k are highest for the town of Milano: this is to be expected because it has the highest thermal inertia of the three examples (urban heat island effect). An advantage of fitting DTCs is the reduced navigation error in the derived parameters: if the navigation error in the satellite data is purely random, then it acts like noise on the DTCs – and noise is reduced by the fitting of the model. This reduction of noise can be well observed with the curve fitted for the broad-

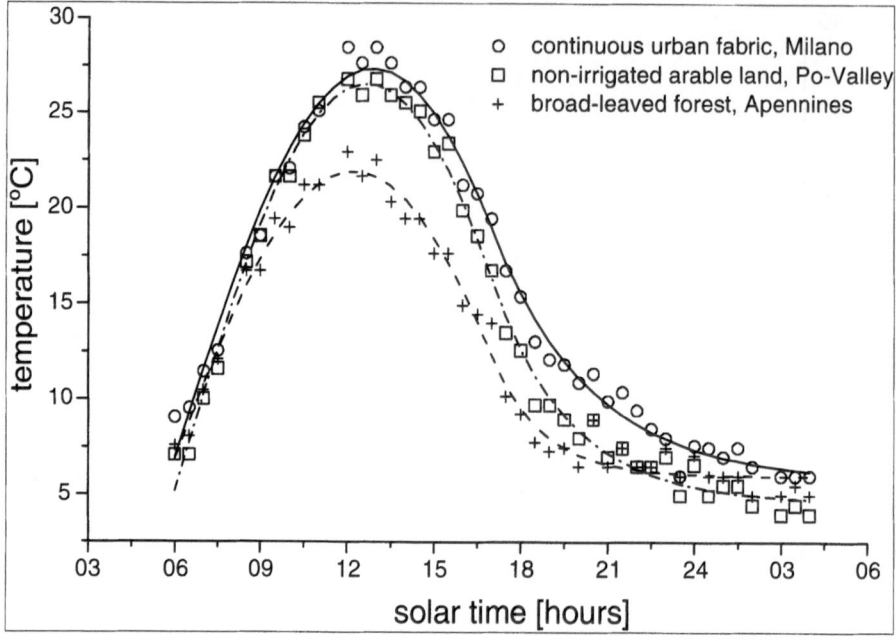

Fig. 3. Examples of modelled diurnal temperature waves for 3 different surfaces as described in Table 3

Table 3. The parameters determined for the 3 temperature cycles shown in Figure 3

Parameter		Continuous urban fabric	Non-irrigated arable land	Broad-leaved forest
T_o	[°C]	6.6 ± 0.6	3.7 ± 0.5	3.9 ± 0.6
T_a	[°C]	20.8 ± 0.7	22.9 ± 0.7	18.0 ± 0.8
w	[hh:mm]	13:56	13:54	13:53
t_m	[solar time]	12:53 ± 00:04	12:39 ± 00:03	12:13 ± 00:04
t_s	[solar time]	17:07 ± 00:18	17:04 ± 00:15	17:02 ± 00:22
k	[hh:mm]	03:21	02:38	01:44
dT	[C]	-0.8 ± 0.7	0.9 ± 0.6	2.0 ± 0.6

leaved forest.

In a further step the method was applied to the vegetation period of 1996 for a data set that covers the northern Sahara as well as vegetated areas in Germany up to the Danish border. In tests on these data it turned out that the fitting of all 7 parameters fails in situations with too few cloud-free slots. Thus, the following

alterations were implemented:
- The width of the cosine ω is calculated from astronomical parameters (daylight hours).
- The model is required to be differentiable at the boundary between the cosine and the attenuation function. With this constraint the attenuation constant k can be calculated.
- Furthermore, a robust estimator of the error function is used to reduce the effect of outliers.

With these improvements the determination of the model parameters was stable for all meaningful conditions. Furthermore, two parameters less have to be fitted. The method is superior to minimum/maximum temperature schemes, because theoretically no information is lost during parametrisation. Small gaps due to clouds as well as outliers due to undetected clouds are smoothed by the modelling.

2.3 Maximum and Median Composites

Diurnal temperature waves of individual days are influenced by the current and the previous synoptic situation, e.g. by the cloud situation and the surface-moisture. On the other hand, more permanent surface properties, e.g. vegetation and soil type, are to be derived from the temperature wave. The task now is to find a method to separate the short term synoptic effects from these more permanent characteristics of the surface. In principle two approaches are possible:
1 Calculation of the model parameters for a number of individual days and subsequent averaging.
2 Generation of IR composites for a number of days and calculation of model parameters for the composites.

Both approaches were applied to the vegetation period of 1996 with a composition period of one month. In months with many clear sky situations both approaches proved to be feasible, but for periods with moderate to high cloudiness only the fitting of IR composites yields reasonable results. Good composites were obtained using the maximum and the median:
1 Maximum composite: the maximum for each pixel location and slot in the composite-interval.
2 Median composite: the median for each pixel location and slot in the composite-interval.

In order to exclude the maximum and minimum from the median the number of cloud-screened values is required to be ≥4; otherwise the median is marked as missing. With this constraint, and because no averaging is performed, it is unlikely that undetected clouds influence the median. Most pixel locations do not fulfil this condition for all METEOSAT slots, but usually the remaining slots are sufficient to model the missing values. In case of too many missing values, the modelled temperature maps become noisy. Most synoptic artefacts are eliminated in the modelled temperatures; in this respect, the results for median composites are superior.

It is now assumed, that the determined model parameters reflect thermal surface properties. The parameters for the maximum composites describe the hottest situation, whereas the ones for the median composites describe the typical (average)situation and characterise the more permanent surface properties better. Furthermore, in maximum composites hot regions "grow" on the expense of colder ones; this leads to some loss of dynamics. Fig. 4 shows modelled temperatures for August 1996 at 12:30 hours UTC for most of Italy.

The model of the DTC was fitted to median and maximum composites of the vegetation period of 1996, which were generated for each month, pixel, and slot. This resulted in two data sets of synthetic DTC's per month, covering the northern Sahara as well as vegetated areas in Germany up to the Danish border. However, the parameters T_0 and T_a proved to be noisier than those of maximum temperature T_0+T_a. This is due to the high temperature gradients in the early morning which limit the accuracy of T_0 and, as a consequence, also of T_a. Therefore, instead of using T_0 and T_a directly, the 'corrected' residual temperature $T_0 +\delta T$ and the 'corrected' temperature amplitude $T_a-\delta T$ are discussed, which are the temperature the model reaches for $t \to \infty$ and the amplitude in relation to this value, respectively. Due to the small gradient of the model for $t \gg t_s$ the uncertainties of these quantities are low. Fig. 5 shows $T_0+\delta T$ and $T_a-\delta T$ for the large part of Europe.

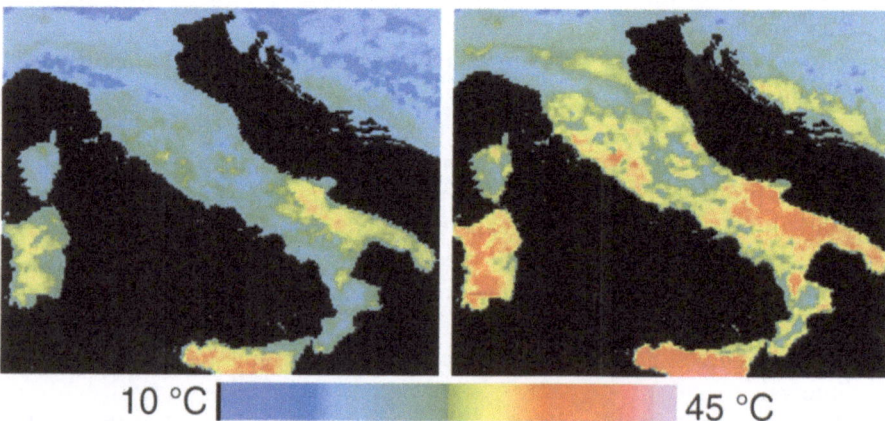

10 °C   45 °C

Fig. 4. Modelled temperatures for 12:30 hours UTC using parameters determined by fitting the model to median (left) and maximum (right) composites for August 1996. No obvious synoptic artefacts can be observed. The temperatures obtained from the maximum composites reflect the hottest situations during the composition interval; while highlighting areas, where high temperatures occurred, they also "flatten" some details, i.e. on Sicily. In contrast, the temperatures obtained from the median composites reflect the typical values during the composition interval and show more details in the colder range.

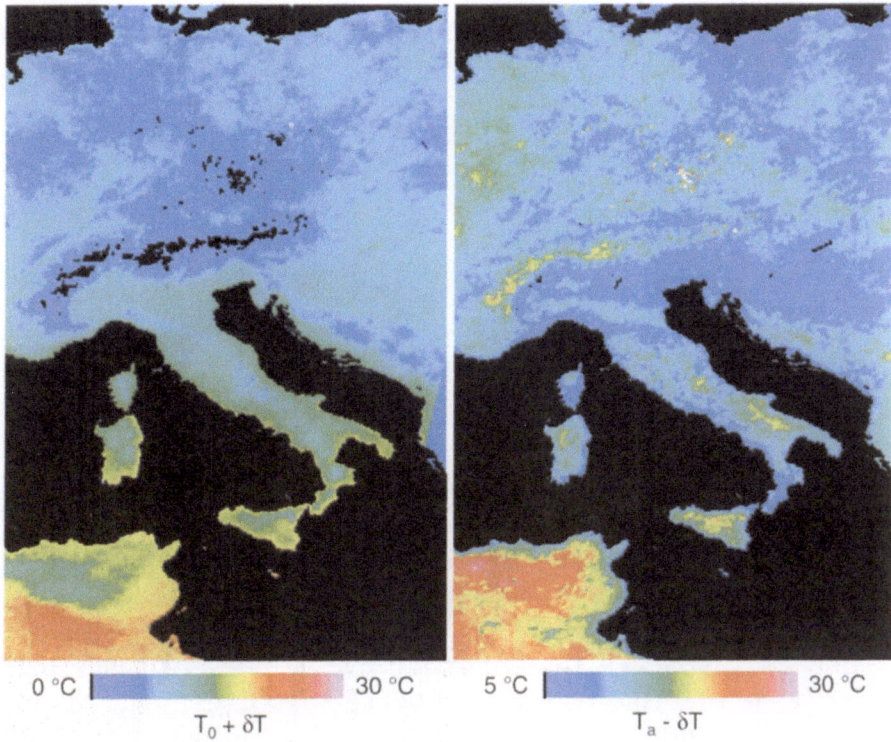

Fig. 5. "Corrected" residual temperature T_0+dT and "corrected" temperature amplitude T_a-δT for the median composites of August 1996 for a large part of Europe/northern Africa. T_0+δT is relatively high in the Rhine valley and for lakes, e.g. lake Constance, but low in mountain areas, e.g. the Alps. T_a-δT is low for densely vegetated areas (see Fig. 6) as well as for water surfaces, e.g. lake Balaton. Some parts of the Rhine are visible. In Tunisia/Algeria high T_a-δT in the mountains, but also relatively low values for salt-lakes and oases can be seen

The parameters determined by the fitting procedure need to be interpreted in combination with other parameters. Here, the surface elevation, a static parameter, and the normalised difference vegetation index (NDVI), a dynamic parameter, are used (Fig. 6). The maximum NDVI was obtained from AVHRR visible and near IR measurements for a composition interval of 10 days (Koslowsky, 1996); here, those NDVI composites are used, that cover the individual months best. It is assumed that the surface-type varies between the extreme cases of a desert and that of dense, moist vegetation:

- For a desert a low NDVI and a high T_a - δT are expected.
- For moist vegetation the NDVI and the T_a - δT should be high and low, respectively.
- Sparsely vegetated moist or shady surfaces have low NDVI, but can also have

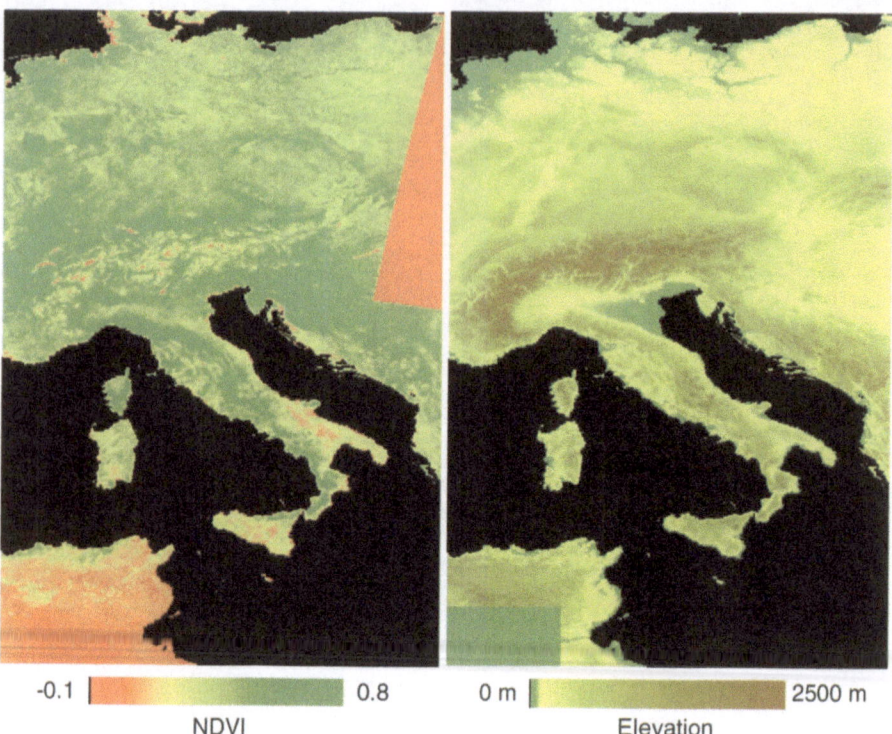

-0.1   0.8   0 m   2500 m

NDVI   Elevation

Fig. 6. Surface parameters used in the interpretation of the thermal parameters. Left: the maximum NDVI for August 1996 based on AVHRR 1 km datasets (data courtesy FU Berlin). Right: surface elevation based on the Globe model with 500m resolution (DMA, 1986). Both parameters were resampled to METEOSAT resolution

relatively low $T_a - \delta T$.

Inverse correlations between NDVI and $T_a - \delta T$ and between surface elevation and $T_o + \delta T$ are clearly visible (Fig. 5 and Fig. 6). A positive correlation between surface elevation and $T_a - \delta T$ can also be seen, e.g. in the Alps and the mountain areas in northern Africa (Tunisia/Algeria); the relationship being partially masked by the effect from NDVI.

3   Conclusions and Outlook

It was shown that land surface temperatures can be derived from METEOSAT IR measurements by using neural networks as a substitution for radiative transfer calculations with MODTRAN-3 to perform atmospheric corrections. The network was trained and validated using atmospheric profiles from the TIGR library and verified using profiles extracted from ECMWF analyses. The mean temperature error of ENZO-NN (Table 2) for the verification data (ECMWF) was 0.31 K as compared to MODTRAN-3. For the validation data (TIGR profiles not used for training) the mean temperature error was 0.25 K. The biggest advantage of the neural network is speed: it is approximately 5000 times faster than MODTRAN-3.

It is planned to use emissivity as a further input of the neural network. The neural network is expected to achieve an accuracy comparable to MODTRAN-3 for an average emissivity of 0.975 with an error of ± 0.025. For the standard profiles 'mid-latitude summer' and 'mid-latitude winter' the temperature error associated with this emissivity is estimated to be 1.4 K. The combined temperature error due to MODTRAN-3 and the above emissivity error is estimated to be 2.1 K (Schädlich et al., 2001). If a more accurate method to determine emissivity can be found (Dash et al., 2001), e.g. by using METEOSAT Second Generation (MSG), the error can be reduced accordingly.

It was shown that diurnal temperature cycles can be modelled in a stable and meaningful way. The modelling yields residual temperature, temperature amplitude, time of the maximum, and attenuation during the night. Low residual temperature anomalies are primarily caused by elevation effects, whereas, high residual temperature anomalies can be observed for water-bodies. The temperature amplitude proved to be the most discriminating parameter for surface characteristics, as it reflects the surface moisture and bio-mass. In order to yield a continuous spatial coverage for change detection, monthly composites of the IR data were generated. Cloud screening of the data combined with subsequent composition and modelling ensures that the thermal parameters are practically free from synoptic artefacts. Good results were achieved using median and maximum value composites, which describe the typical and the hottest situations during the composite interval, respectively. The NDVI and the corrected temperature amplitude $T_a - \delta T$ were compared. For many areas $T_a - \delta T$ is strongly correlated to the NDVI. A deviating behaviour can be a hint for different land use (Göttsche and Olesen, 2001).

The next step will be the calculation of parameters for several years in order to allow change detection. The necessary amount of computing time as well as the amount of data that must be handled are manageable due to the composites. The fitting algorithm proved to be stable and robust in respect to undetected clouds and data gaps and only little user-interaction is required. Therefore, this is a promising method to add IR information to change detection, which is until now based on visible data alone.

The data evaluation of long time series (years) of METEOSAT and AVHRR IR data is feasible and the data as well as the required algorithms are available at the FU-Berlin and at the Forschungszentrum Karlsruhe – IMK. MSG, which is scheduled to become operational at the end of 2002, combines the capabilities of AVHRR and METEOSAT and will allow to expand the existing time series of satellite data. The principles underlying the developed algorithms, in particular the atmospheric correction of IR satellite data with neural networks and the modelling and analysis of the thermal behaviour of single pixels, also apply to MSG. These investigations of the Forschungszentrum Karlsruhe – IMK will be complemented by investigations of vegetation and albedo performed by the FU-Berlin.

References

Chinnaswamy A (1999) Parameterisation of Retrieval of SST from Geostationary Satellites. 3rd Symposium on Integrated Observing Systems, January 11-15, University of Oklahoma, USA.

Braun H, Ragg T (1996) ENZO: Evolution of neural networks (Internal report. Fakultät für Informatik, Universität Karlsruhe. 1996,21). University of Karlsruhe, Institute for Logic, Complexity, and Deduction Systems, Germany.

Dash P, Göttsche F-M, Olesen, F-S, Fischer H (2001) Land surface temperature and emissivity estimation from passive sensor data: theory and practice; current trends. Int. J. Remote Sens. (accepted).

Defence Mapping Agency (DMA) (1986) Defence Mapping Agency Product Specifications for Digital Terrain Elevation Data (DTED). DMA Aerospace Centre, St. Louis

EUMETSAT (1998) http://www.eumetsat.de/en

Göttsche F-M, Olesen F-S (2001) Modelling of diurnal cycles of brightness temperature extracted from METEOSAT data. Remote Sens. Environ. 76(3):337-348

Göttsche F-M, Olesen F-S (2002) Evolution of neural networks for radiative transfer calculations in the terrestrial infrared. Remote Sens. Environ. 80(1):157-164 (accepted)

Koslowsky D (1996) Langjährige validierte und homogenisierte Reihen des Reflexionsgrades und des Vegetationsindexes von Landoberflächen aus täglichen AVHRR-Daten hoher Auflösung. Institut für Meteorologie, Freie Universität Berlin, Wiss.Met.Abh., Neue Folge, Serie A 9, 1, Dissertation

Olesen F-S, Kind O, Reutter H (1995) High resolution time series of IR data from a combination of AVHRR and METEOSAT. Advances in Space Research, Vol. 16:141-146

Riedmiller M (1993) A direct adaptive method for faster back-propagation learning: The RPROP algorithm. Proceedings of the IEEE International Conference on Neural Networks (ICNN), 1, pp. 586-591

Reutter H, Olesen F-S, Fischer H (1994) Distribution of the brightness temperature of land surfaces determined from AVHRR data. Int. J. Remote Sens., 15: 95-104

Schädlich S, Göttsche F-M, Olesen F-S (2001) Influence of land surface parameters and atmosphere on METEOSAT brightness temperatures and generation of land surface temperature maps by temporally and spatially interpolating atmospheric correction. Remote Sens. Environ., 75(1): 39-46

Schroedter M, Olesen F-S, Fischer H (2001) Determination of land surface temperature distributions from single channel IR measurements: An effective spatial interpolation method for the use of TOVS, ECMWF and radiosonde profiles in the atmospheric correction scheme. Int. J. Remote Sens. (accepted)

Vollmer M, Göttsche F-M, Olesen F-S (2000) Correction of the atmospheric influence on IR measurements by satellite with neural networks. The 2000 EUMETSAT Meteorological Satellite Data Users' Conference. Bologna, Italy, pp. 506-511

Zell A (1994) Simulation neuronaler Netze. Oldenbourg-Verlag, München, Germany

Chapter 6
Variability of the Mediterranean Sea

Recent and Expected Future Changes in the Hydrography, Ecology, and Circulation of the Eastern Mediterranean

W. Roether, B. Klein

The classical view of the thermohaline circulation (THZ) of the Eastern Mediterranean has been that relatively less saline near-surface waters intruding from the Western Mediterranean through the Strait of Sicily were converted into more saline (i.e., denser) waters essentially in two regions. These are the northwestern Levantine Sea, where a high-salinity intermediate water mass is formed (Levantine Intermediate Water, LIW) which feeds a subsurface outflow into the Western Mediterranean, and the Adriatic, from which all the deeper waters were replenished (Wüst, 1961). The dominant role of the Adriatic was manifest in a rather uniform salinity of the deep waters. Observations that go back to 1910 indicate persistence of the classical situation since at least then. The Aegean Sea, which was known to be capable of also producing dense waters, was mostly seen as playing but a minor role in the THZ of the sea. But around 1990 the situation changed in that the Aegean suddenly formed large volumes of particularly dense water, which by 1995 had replaced approximately 20 % of the waters below 1500 m and had influenced most of the deep and intermediate waters (Roether et al., 1996, 1998; Klein et al., 1999; Lascaratos et al., 2000). The changes were apparent in deep water salinities being markedly raised compared to the classical situation, with maximum effects in the waters adjoining the Aegean Sea. Whereas previously salinity and temperature decreased continuously below the LIW depth range down to the sea floor, there were now inversions in both properties (in 1995, inversion at about 1500 m depth). The changes are illustrated in Fig. 1, which shows salinity sections along the length of the Eastern Mediterranean for 1987 (classical situation), 1995, and 1999. The sections demonstrate that the Eastern Mediterranean has changed substantially from its classical description, and that it is now in a transient state. Fig. 1 is corroborated by Figs. 2 and 3, which show corresponding sections of the transient tracer CFC-12 (Freon-12) and of oxygen. Both these sections picture the deep intrusion of recently ventilated waters, implying also reduced nutrient concentrations in the deep waters. Particularly the deep waters of the Levantine Sea became more ventilated. Another remarkable fact is that, due to the upward displacement caused by the addition of Aegean overflow, concentrations characteristic of mid-depth waters have penetrated distinctly closer to the surface layer. An important consequence is that the nutricline has moved nearer to the euphotic zone compared to the classical situation (Klein et al., 1999).

The changes were presumably initiated by a change in the near-surface circulation that diverted high-salinity waters from the Levantine Sea into the Aegean Sea and blocked the intrusion of less saline surface waters from the Ionian Sea (Malanotte-Rizzoli et al., 1999). In combination with enhanced cooling during a particularly cold winter, this led to the formation of very dense waters (Lascaratos et al., 1999; Wu et al, 2000). The situation was self-sustaining in that the dense waters overflowing the sills of the Aegean Sea were replaced by more saline near-surface and intermediate waters from the Levantine Sea. A contributing factor has been that the LIW, rather than taking its classical westward route south of Crete, took a path through the southern Aegean Sea. More recently the Aegean outflow has become less dense so that it reached to intermediate depths only, by which it was enabled to intrude the South Adriatic Basin. As the resulting salt import into the Adriatic preconditions dense water production in the classical deep water formation region, the THZ of the sea may in fact move back toward the classical situation (Klein et al., 2001).

The transient effected a net deposition of salt in the deep waters of the Eastern Mediterranean. While part of this salt originated from shallower strata in which salinity was lowered, and a further part can be ascribed to increased net evaporation during the past decade, it appears that a substantial fraction originated from the Western Mediterranean. This means that the entire Mediterranean has been involved in the salt deposition in the Eastern Mediterranean deep waters, pointing to a complex interaction. The westward outflow through the Strait of Sicily has been lowered in salinity, and a similar, albeit smaller salinity decrease may occur in the Mediterranean outflow into the Atlantic, with the faint possibility of affecting the formation of North Atlantic Deep Water. The development of the deep oxygen field (Fig. 3) indicates that deep water oxygen consumption, which was rather high already in the classical situation (Roether and Well, 2001) has become still larger. A possible explanation is that the transient effected a fast downward transfer of dissolved organic carbon from the euphotic zone. The increase is indication that the changes have severly disturbed the entire biogeomechistry of the deeper waters of the sea.

The future will see continued changes in the hydrography and biogeochemistry of the Eastern Mediterranean waters. Considering that the turnover time of the deep waters is on the order of 100 years (Roether and Schlitzer, 1991; Roether and Well, 2001), the Eastern Mediterranean deep waters evidently will remain in a transient state for many decades to come. Should the system move close to its previous, classical situation, this will be accompanied by a higher salt and a lower nutrient supply to the upper waters as the deep waters return toward the surface.

The present and future implication of the ongoing Eastern Mediterranean Transient can be summarized as follows:
1. Textbooks may mislead one: The Eastern Mediterranean has become a different ocean and is now in a transient state.

2. The main effect has been extensive deep water production by the Aegean Sea, but it now looks that the Adriatic may be returning to lead deep water production. The time scale to reach a new quasi-equilibrium should be on the order of 100 years, and it is uncertain whether or not a new, stable equilibrium will ever be reached again.
3. The nutricline became lifted. Reports of ecological effects have so far been sparse, but such effects are to be expected.
4. Oxygen consumption was accelerated, indicating severe disturbance of the biogeochemistry of the sea.
5. Mid-depth waters involved in outflow through Sicilian Passage became denser, less saline and higher in nutrients. This will induce circulation and ecological effects in the Western Mediterranean and possibly beyond.
6. Deep waters have become more saline and lower in nutrients. Upwelling will eventually return these waters into the upper waters, where these changes may produce long-term effects.
7. The evolution of the Eastern Mediterranean Transient must be monitored further, and modelling efforts are required to account for these observations in a quantitative way in order to produce a dynamic explanation of the entire event.

References

Klein B et al. (1999) The large deep water transient in the Eastern Mediterranean, Deep-Sea Res. I, 46:371-414

Klein B et al. (2000) Is the Adriatic return ing to dominate the production of Eastern Mediterranean Deep Water? Geophys. Res. Lett., 27:3377-3380

Lascaratos A et al. (1999) Recent changes in deep water formation and spreading in the Eastern Mediterranean Sea: A review, Progr. Oceanogr., 44:5-36

Malanotte-Rizzoli P et al. (1999) The Eastern Mediterranean in the 80s and in the 90s: the big transition in the intermediate and deep circulations, Dynamics Atm. Oceans, 29:365-395

Roether W, Schlitzer R (1991) Eastern Mediterranean deep water renewal on the basis of chlorofluoromethane and tritium data, Dynamics Atm. Oceans, 15:333-354

Roether W et al. (1996) Recent changes in the Eastern Mediterranean deep waters, Science, 271:333-335

Roether W et al. (1998) Property distributions and transient tracer ages in the Levantine Intermediate Water in the Eastern Mediterranean, J. Marine Systems, 18:71-87

Roether W, Well R (2001) Oxygen consumption in the Eastern Mediterranean, Deep-Sea Res. I, 48:1535-1551

Wu P et al. (2000) Towards an understanding of deep water renewal in the Eastern Mediterranean, J. Phys, Oceanogr., 30:443-458

Wüst G (1961) On the vertical circulation of the Mediterranean Sea, J. Geophys. Res., 66:3261-3271

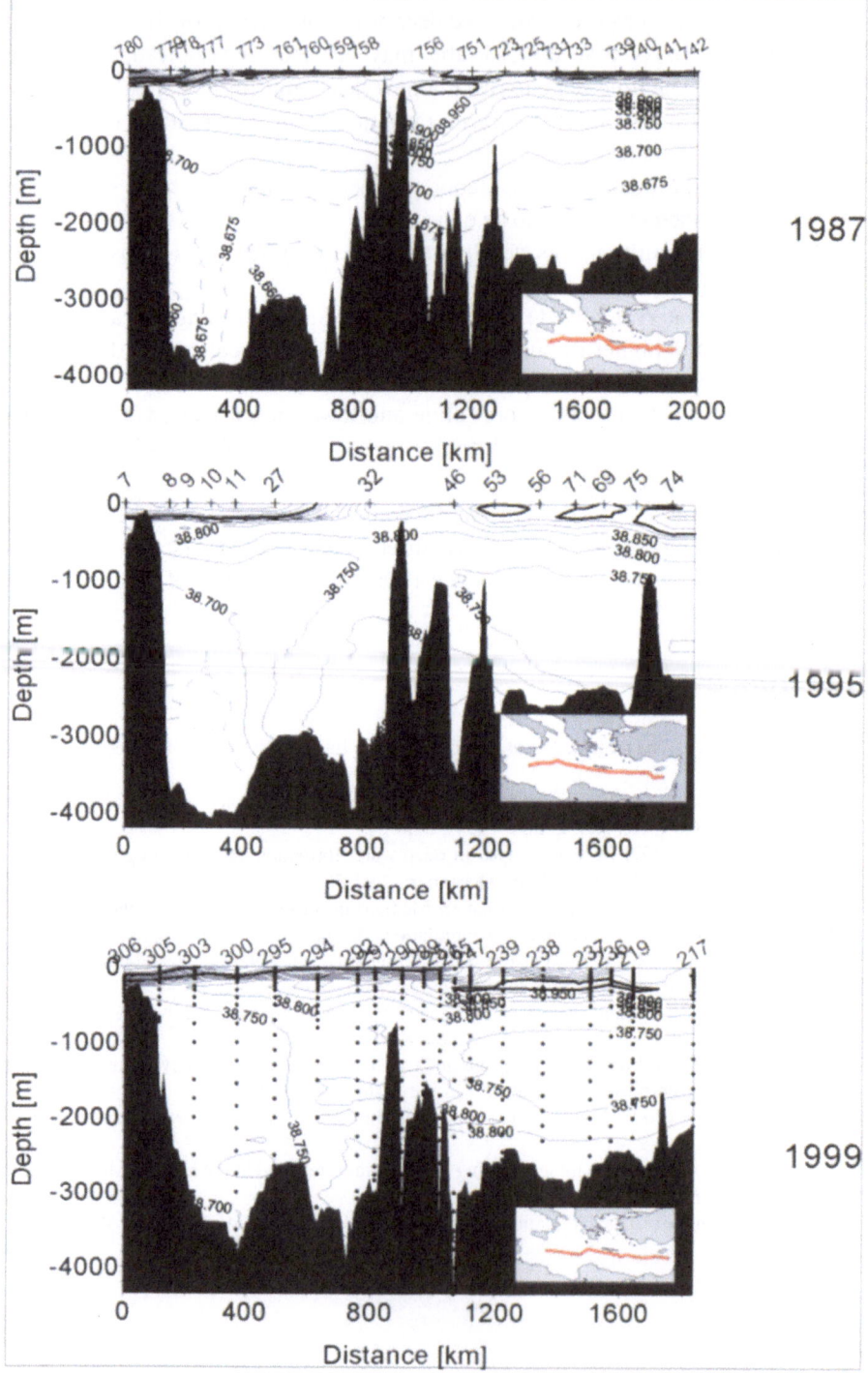

Fig. 1. Salinity sections (psu) along the Eastern Mediterranean Sea in 1987 (METEOR cruise M5/6), 1995 (M31/1) and 1999 (M44/4). Tracks and stations are shown in the insets

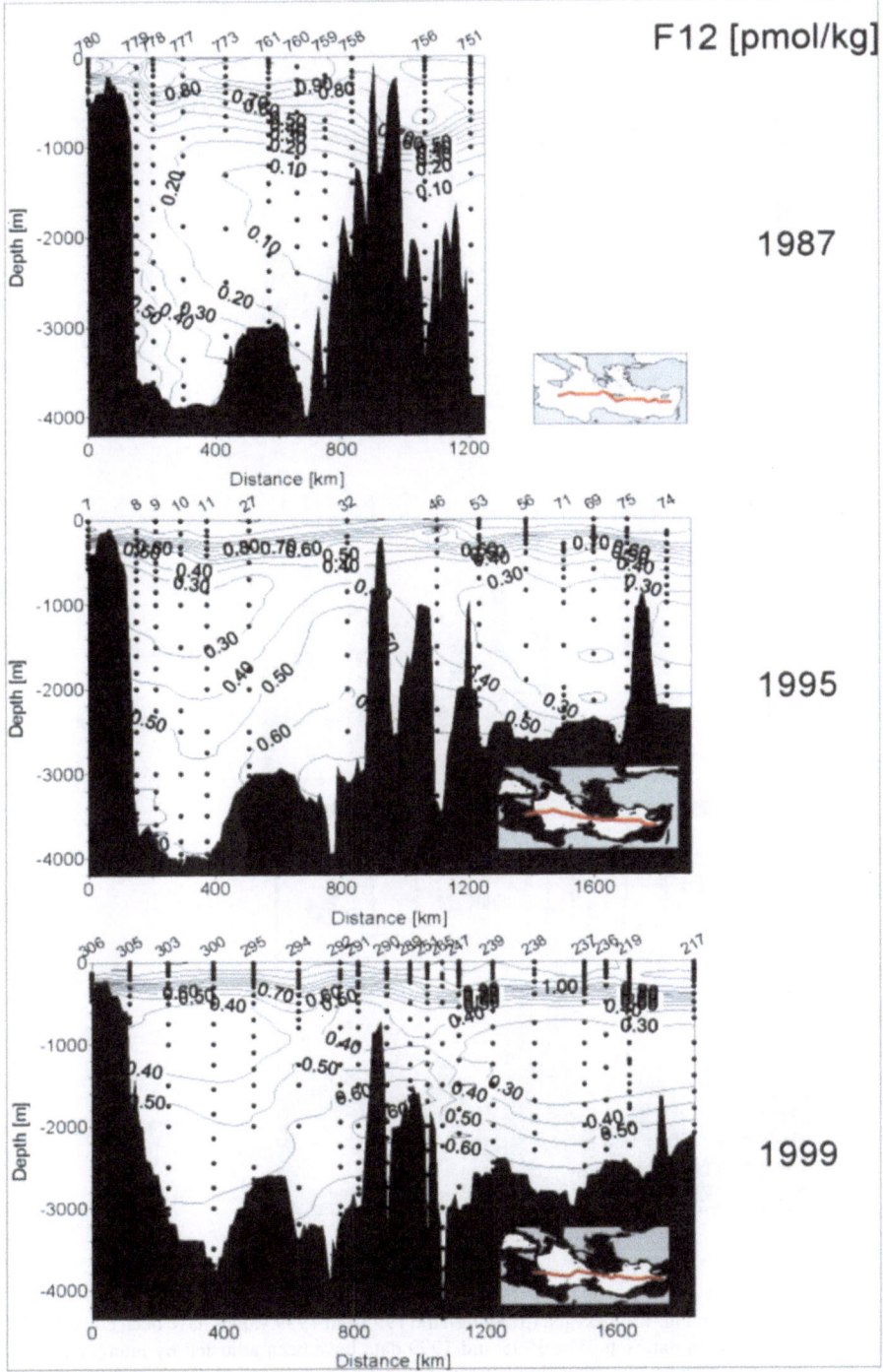

Fig. 2. Sections as in Fig. 1 for the transient tracer CFC-12 (pmol/kg); the 1987 sections did not cover the full zonal range. Near-surface and newly formed deep waters are characterized by higher concentrations

300

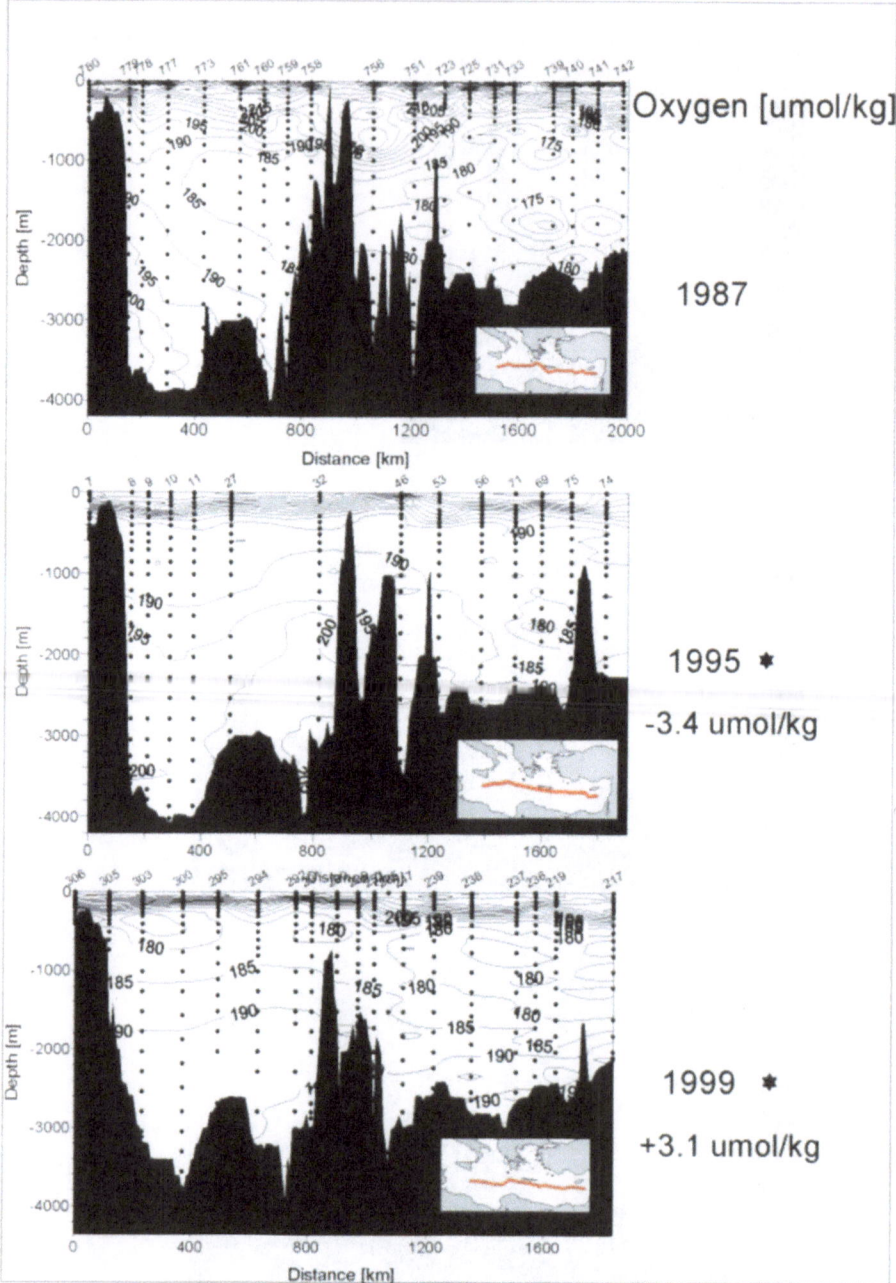

Fig. 3. Sections as in Fig. 1 for oxygen (μmol/kg); the 1995 and 1999 values have been corrected for offsets in the oxygen data sets. The 1995 and 1999 data have been adjusted by intercomparison found by intercomparison with other simultaneous data sets. The corrections are given in the margins

Long Term Data Series on Mediterranean Sea Temperature, Nutrients and Hydrology Changes

J.P. Bethoux

1 Introduction

The Mediterranean is a deep semi enclosed sea, under a continental climate. Its surface water balance is negative (evaporation losses being greater than precipitation) and water dynamics are controlled by winter dense water formation and by two shallows sills at Gibraltar and Sicily Straits. Consequently, the Algero-Provençal Basin (Western Mediterranean) is filled from 500m depth to the bottom, at more than 2800m, by homogeneous deep water. Its physical characteristics (temperature, salinity) are linked to the heat and water exchanges with the atmosphere and the Atlantic Ocean. Heat exchange with the Atlantic being small, the temperature depends mainly from the heat exchanges through its surface, i.e., from the local climate coupled with the northern hemisphere atmospheric circulation. Salinity depends on exchanges with the Atlantic and on evaporation, precipitation and river runoff. Similarly, the geochemical state of the Mediterranean Sea depends on exchanges with the Atlantic and on atmospheric and terrestrial supplies which are linked to natural and mostly anthropic inputs, issuing from more than 200 millions' inhabitants who live on the watershed. Unlike the open ocean, for which the deep water response time to perturbations is of the order of 1000 years, the residence time of the western deep water is about 15 years, as a result of intensive horizontal and vertical circulation. Consequently, the Mediterranean response to climatic or environmental changing trends is perceptible in a few years. Analysis of marine data acquired in the deep waters of the Algero-Provençal basin shows increases of temperature, salinity, nitrate and phosphate since the early sixties. Such evolutions are linked to recent changes in climatic and environmental forcing, as resulting from man's activities. More, in the sediment of the eastern and western basins, there are numerous layers of sapropels, i.e., layers of some centimetre thick containing organic carbon and proving a sedimentation process with anoxic deep layer. This anoxia is a proof of change in the deep layer oxygenation and circulation linked either to change in climate over the eastern basin (the cause of eastern sapropels) or change in the Atlantic circulation off the strait of Gibraltar (the probable cause of western sapropels). These marine events are useful to quantify changes occurred either over the Mediterranean watershed (climate and environment) or in the Atlantic ocean (past iterative changes in the global circulation and paleoclimate) over the last 2 million years.

2 Climatic and Environmental Forcing

2.1 Temperature and Salinity Trends in Deep Waters, and Corresponding Surface Trends

In the Algero-Provençal basin, when the temperature and salinity characteristics were considered as constant from the early century, the accurate measurements made since year 1959 show continuous increases in temperature and salinity. At depth greater than 2000m, over the 1959-2000 period, the annual trends are: 3.4×10^{-3} °C for potential temperature and 1.05×10^{-3} for salinity, i.e., increases of 0.14 °C and 0.043 salinity. More than the last deep layer, these increasing trends concern the whole deep water column, since about 800m depth. As the deep water originates from the winter dense water formation occurring at the sea surface, the trends originate in a multi-decennial evolution in surface temperature and in air-sea exchanges, the origin of trends stands before year 1944 due to a residence time of about 15 years. In the intermediate water (originating from the eastern basin), the trends appears to be about 2 to 3 times those in the deep layer (Sparnoccia et al., 1994; Bethoux and Gentili, 1996), but space and time changes of characteristics complicate the determination of mean trends. In the surface layer, the inter-annual variability of temperature and salinity characteristics are rather high, but in a few places, more or less long time series allow a first determination of surface trends. Over the last decades, temperature trends were found comprised between $9\text{-}15 \times 10^{-3}$ °C yr^{-1} at Villefranche (French Riviera) in the 0-75m layer, and at Medes Islands (Spanish coast, north of Barcelona) the trend was equal to 26×10^{-3} °C y^{-1} in the 0-80m layer. In a 20 box-model, the deep water trends were used to calculate the surface trends in different parts of the Mediterranean Sea (Bethoux and Gentili, 1999). These calculated annual trends for the surface layer stand between 0.9 and 13.8×10^{-3} °C, and 0.2 and 5.4×10^{-3} for the salinity. The lowest values correspond to the Alboran Sea area where the inflowing Atlantic waters smooth the local climatic effect, the highest values concerns the surface layer of the Aegean Sea, i.e., the oldest surface waters due to the Mediterranean circulation. Changes in the dense water formation and circulation in the Aegean Sea in early 1990 (Rother et al., 1996, Della Vedova et al., 1997) result from major temperature and salinity changes in this area and confirm the previous calculations.

From deep water trends in the western basin, it was calculated the corresponding surface changes in heat and water budgets capable to explain the deep trends. Over the whole Mediterranean, it was calculated a change in surface heat budget of 1.5 Wm^{-2} and a change in freshwater budget of 0.1 m. The change in heat budget was linked, via the air and sea surface temperature changes, to an increase in greenhouse effect of 1.7 W m^{-2} over the 1940-1995 period (Bethoux et al., 1998). This estimate from marine data corresponds to the calculation from the changes in the atmospheric content of radiative gas (IPCC, 1995). The change in freshwater budget (about 0.1m) was linked to the increase of anthropic use of freshwater from rivers (e.g., 4 cm over the whole Mediterranean from dams across the Nile and Ebro rivers), to the increase of evaporation (2 cm) and to the decrease of precipitation (3 cm), a decrease observed around the whole sea (Bradley et al., 1987) and estimated to about 10% in the north-western basin over the 1940-1995 period (Bethoux et al., 1998).

In order to increase the accuracy of previous estimates, it would be necessary either to have a better estimate of climate change at the Mediterranean scale (precipitation, air temperature), or to know the true water exchanges across the strait of Gibraltar. For instance, meteorological data over the north western coast give air temperature trends of 23×10^{-3} °C y^{-1} over the 1960-1995 period. With this value, we may calculate a corresponding sea surface temperature trend of 20×10^{-3} °C y^{-1}, and the resulting estimate of the increase in greenhouse effect is 2.2 W m^{-2} instead of the previous 1.7 W m^{-2}, calculated with respective trends of 7.3×10^{-3} °C y^{-1} for surface temperature and 8.4×10^{-3} °C y^{-1} for air temperature (Bethoux and Gentili, 1998).

2.2 Phosphate and Nitrate Increase, Silicate Stagnation

Analysis of available chemical data shows an increase of phosphate concentrations in western deep water since the early sixties, an increase of nitrate since the seventies, and a concomitant constancy of silicate concentration. Increases of nutrient in deep waters mean increases of surface input via the atmospheric and terrestrial inputs. Via the water circulation, the 0.5% per year increase for phosphate and nitrate concentration may be linked to a 3% per year increase of surface inputs (Bethoux et al., 1998), resulting from increasing inhabitants and industrial, agricultural and urban activities around the sea, mainly since the early sixties (UNEP, 1988).

Increases of phosphate and nitrate concentration in deep waters follow similar increases in the rivers, like the Rhône and Po rivers. They are signatures of increasing biological new production in the surface layer, without apparent problem except at some hot spots of coastal eutrophication. Inversely, increase of human activity do not increase the silicate load and even there may be a decrease due to the dams on rivers. Nevertheless, increasing phosphate and nitrate concentration and constant silicate concentration means a change in the molar ratio P:N: Si, i.e., a probable change in the plankton distribution. Due to a rather long residence time of deep waters in the eastern Mediterranean, nutrient ratio measured in the seventies may represent a quasi-steady state prior to the anthropic effect. The Si:P molar ratio was equal to 32 in the eastern basin, when in the western Mediterranean it was equal to 24 in early 70s and 21 in 1994. From an important diatom community, there is probably a shift towards non siliceous community, i.e., flagellates and dinoflagellates. The plankton change will affect the upper level of biological community, the anchovies and sardines, representing about 40% of the fishery, subsist on copepods grazing the diatoms. Inversely, flagellates and dinoflagellates promote gelatinous ecosystem, without fishery interest. Previous studies of ecosystem change due to human activity concerned estuaries of great rivers (Rhine, Mississippi,

Danube rivers) or bay bordered by a great city, but the changes in the nutrient ratios of the western basin are signature that the human impact concerns a whole deep sea.

3 Sediment Signatures of Past Changes in Climate and Circulation in the Mediterranean and in the North Atlantic

3.1 Organic Rich Layers (Sapropels)

Multiple layers of organic rich sediments, the sapropels, were discovered, first in the eastern Mediterranean, in 1947-48 (Kullenberg, 1952), then more recently in the Alboran Sea in 1996 (De Kaenel et al., 1999). Preservation of organic mater in the sediment is explained by anoxia in deep water, acting over several thousand years, with two possible causes: an increase of new production or a change in deep water ventilation. In the present circulation scheme, overconsumption of the oxygen of deep waters by mineralisation of organic mater would require an increase by 4 to 5 times of the present new production, and a similar increase of phosphate input over a long time. We cannot imagine such an increase of a natural input from land weathering. The stoppage of deep water ventilation appears as the probable reason of anoxia in deep waters, with different causes in western and eastern Mediterranean (Bethoux and Pierre, 1999).

3.2 Eastern Mediterranean

In the eastern Mediterranean there are different areas for dense water formation, two of them in more or less coastal area: the Adriatic and Aegean Seas, generally providing the densest deep waters, the last one in the high sea waters of the Levantine basin forming the Levantine intermediate waters. The stoppage of dense water formation in the Adriatic and Aegean Seas, due to an increase of freshwater input (by precipitation and river runoff) may affect the ventilation of eastern deep water, at depth greater than about 600m. More, the water dynamic across the strait of Sicily may be slowed down by a change in the water deficit between the eastern and western basin (e.g., an increase of freshwater input, greater in the eastern than in the western basin) up to limit the deep outflow to the intermediate water situated at a depth lower than the sill (at about 400m depth). Consequently, anoxia may spread over the whole basin, at depth greater than 500 to 800m depth, the reason is the local change in freshwater budget, more or less enhanced by the Nile river spates, referring to changes in the tropical areas. Consequently, the different sapropel layers are signatures of a wetter climate over the Mediterranean and probably over the tropical area.

3.3 Western Mediterranean

A scenario similar to that proposed for the eastern basin is not acceptable to explain sapropel deposits in the Alboran Sea. First, the offshore formation of dense water in the Gulf of Lion is more or less preserved from coastal inputs of freshwater. Secondly, the uplift of deep water and the outflow current of dense water across the strait of Gibraltar depends on the density difference between Mediterranean and Atlantic waters at the sill depth (300m). At present this density difference is high, about 1.6 density units (instead of about 0.2 at the strait of Sicily), allowing the uplift of very deep waters up to the sill depth (Bethoux, 1980). As far as evaporation in the continental Mediterranean is greater than in the Atlantic, a strong decrease of density difference between Mediterranean and Atlantic waters may only provide from a change in the Atlantic circulation. In the Alboran Sea, the last sapropel event occurred between 14.5 and 10.3 Ka (De Kaenel et al., 1999), at the time of maximum freshwater input in the North Atlantic during the last deglacial time, according to the sea level change (Fairbanks (1989). In a previous scenario (Bethoux and Pierre, 1999), it was assumed the decrease or stoppage of dense water formation in the Norwegian Sea, due to the freshwater input, and the increase of the volume of waters from Mediterranean origin in the Atlantic (the behaviour of Mediterranean water being to be used in dense water formation in the Norwegian Sea). A consequence of this Mediterranean increasing layer is a strong decrease of the density difference across the strait of Gibraltar and a possible stagnation of deep waters inside the western Mediterranean. The interest of this scenario and of the discovery of about 70 sapropel layers in the Alboran Sea is to evidence iterative changes in the Atlantic circulation, and consequently in the global climate every 21,000 years, one of the periods of the Earth orbit around the sun. This constitutes a chronology of global climate events over the past 1.5 million years.

References

Béthoux J-P (1980) Mean water fluxes across sections in the Mediterranean Sea, evaluated on the basis of water and salt budget and of observed salinities. Oceanol. Acta, 3:79–88

Béthoux JP, Pierre C (1998) Mediterranean functioning and sapropel formation: respective influences of climate and hydrological changes in the Atlantic and the Mediterranean. Marine Geology, 153:29-39

Béthoux JP, Gentili B, Tailliez D (1998) Warming and freshwater budget change in the Mediterranean since the 1940s, their possible relation to the greenhouse effect. Geophysical Research letters, 25:1023-1026

Bethoux JP, Gentili B (1996) The Mediterranean Sea: coastal and deep-sea signatures of climatic and environmental changes. J. Marine Systems, 7:383-3944

Bethoux JP, Morin P, Chaumery C, Connan O, Gentili B, Ruiz-Pino D (1998) Nutrients in the Mediterranean Sea, mass balance and statistical analysis of concentrations with respect to environmental change. Marine Chemistry, 63: 155-169

Bethoux JP, Gentili B (1999) Functioning of the Mediterranean Sea: past and present changes related to freshwater input and climate changes. J. Marine Systems, 20:33-47

Bethoux JP, Gentili B, Morin P, Nicolas E, Pierre C, Ruiz-Pino D (1999) The Mediterranean Sea: a miniature ocean for climatic and environmental studies and a key for the climatic functionning of the North Atlantic. Progress in Oceanography, 44:131-146

Bradley RS et al. (1987) Precipitation fluctuations over northern hemisphere land areas since the mid-19th century. Science, 237:171-175

Fairbanks RG (1989) A 17,000-year glacio-eustatic sea level record : influence of glacial melting rates on the younger Dryas event and deep-ocean circulation. Nature, 342:637-642

IPCC (1995) 2nd assessment, report, WMO/UNEP.

Jones PD (1994) Recent warming in global temperature series. Geophysical research Letters, 21:1149-1152

Jones PD et al. (1988) Evidence for global warming in the past decade. Nature, 332: 790

Kullenberg B (1952) On the salinity of water contained in marine sediments. Goteborg Kungl. Vetenskaps. Vitt-Samhal. Handlingar, 6:3-37

Roether W, Manca BB, Klein B, Bregant D, Georgopoulos D, Beitzel V, Kovacevic V, Luchetta A (1996) Recent changes in eastern Mediterranean deep waters. Science, 271:333-335

Sparnocchia S, Manzella GMR, La Violette P (1994) The interannual and seasonal variability of the mAW and LIW core properties in the Western Mediterranean Sea. In: La Violette PE (Ed.) Seasonal and Interannual Variability of the Western Mediterranean Sea. AGU, Coastal and Estuarine Studies, 46:177-194

UNEP (1988) Le Plan Bleu, résumé et orientations pour l'action. Technical reports, UNEP Rac/Blue Plan, 94 pp.

Chapter 7
Cooperative Research Into "Global Change" Themes

Desert Aerosol in the Mediterranean

A. di Sarra, M. Cacciani, J. DeLuisi, L. De Silvestri, T. Di Iorio, G. Fiocco, P. Grigioni

Abstract

Tropospheric aerosols may affect climate through different mechanisms. In particular, aerosols intervene in the water cycle, and may influence the hydrologic balance. In the Mediterranean a large role is played by desert dust originating in the Sahara. Mineral dust production depends on soil aridity, i.e., among other factors, on land use, and precipitation/temperature regimes. In the Mediterranean, desert dust is mostly transported northward during spring and summer, driven by low and high pressure systems over north Africa. In addition, depending on the synoptic situation, anthropogenic particles and marine aerosols are found over the Mediterranean. The Mediterranean basin is thus an excellent laboratory to study the complex interactions of different type of particles with the radiative field, the hydrological cycle, and clouds. Some results of measurements carried out at Lampedusa island (35.5°N, 12.6°E) in 1999, showing the presence of desert dust and its effects on the radiative field, are described.

1   Introduction

Atmospheric aerosol produce a significant influence on the Earth radiative budget (e.g. Schwartz and Andreae, 1996), through scattering and absorption of radiation (direct forcing), and by influencing the cloud nucleation processes and the cloud microphysical properties (indirect forcing). Mineral or crustal (dust) particles are among the principal constituents of tropospheric aerosols. It is estimated that a fraction ranging between 30 and 50% of the total mineral aerosols are of anthropogenic origin such as produced in soils which have been disturbed by human activity (Tegen and Fung, 1995; Sokolik and Toon, 1996). Continental aridity is indicated as the main cause of increased dust flux from the deserts (Rea et al., 1985; Pye, 1989), that constitute the most relevant source of these particles. Hyper arid regions however (mean annual precipitation <80 mm) are less efficient dust sources than arid regions (Goudie, 1983). The mobilization of dust in the atmosphere is due to wind erosion in arid regions, and the average size of transported particles depends on wind strength; dust mobilization also depends on the nature of soil and other parameters, and complex mechanisms, as saltation of relatively large dust grains, are involved (Gillette et al., 1974). Anthropic activity, through modification of the land use and/or by inducing changes of climate (primarily through changes of precipitation and temperature regime), may contribute to the production of mineral aerosols. The Sahara desert is one of the main sources of mineral aerosols: the dust particles are captured by the wind at the surface, are raised to considerable altitudes in the troposphere by the strong convective regimes that develop over the desert (Dubief, 1979), and may be transported to large distances. Saharan dust is commonly observed over southern Europe, and, at far distances, in the Carribean, and South America. In rare occasions, Saharan dust reaches northern Europe.

North and central African regions have suffered severe droughts in the recent past (e.g. Middleton, 1985; Prospero and Nees, 1986; Dai et al., 1998), and an increase of the dust export has been correspondingly observed (Middleton, 1985; Prospero and Nees, 1986). The droughts appear to be connected to large scale phenomena occuring in the ocean-atmosphere system, like El Niño and the North Atlantic Oscillation (NAO). A correlated behavior of the dust export from Sahara to the North Atlantic and to the Mediterranean with these phenomena has been also observed (Prospero and Nees, 1986; Moulin et al., 1997). Dust aerosols strongly influence the radiative balance, and affect the solar irradiance reaching sea and land surface. In its turn, a change of the solar irradiance may influence evaporative fluxes and, on a basin-wide scale, its hydrological budget (see e.g. Gilman and Garrett, 1994).

In the Mediterranean dust particles, that are non-hygroscopic and are not expected to interact with clouds, may encounter and mix with different aerosol types. Continental and anthropogenic particles originating from Europe, as well as marine aerosols from the North Atlantic and the Mediterranean itself, are commonly present in the basin. Dust particles coated with sulphate have been recently observed; mixing of the dust with sulphate is believed to occur as a consequence of cloud evaporation processes (Levin et al., 1996). In this way, dust particles become hygroscopic, and may influence the cloud formation and properties. Saharan dust constitutes one of the most relevant inputs of trace elements to the Mediterranean (e.g. Kubilay and Saydam, 1995), and an important source of nutrients for oceanic microorganisms. Desert aerosols may also affect the precipitation acidity (Loÿe-Pilot et al., 1986).

In this paper we discuss some measurements carried out from the Station for Climate Observations of the National Agency for New Technology, Energy, and Environment of Italy (ENEA) in the island of Lampedusa (35.5° N, 12.6° E). Lampedusa is a small, rocky island, relatively isolated in the central-southern Mediterranean, approximately 100 km east of Tunisia, and 200 km North of Libya. The island is 10 km long, has a surface area of about 20 km², and its maximum elevation is 120 m. The ENEA station is located on the north-eastern coast of Lampedusa, on a 45 m high cliff. At the station total ozone, ultraviolet spectral irradiance, aerosol optical depth, greenhouse gas concentration, global solar irradiance and meteorological parameters are routinely measured. A tropospheric lidar has been installed by the University of Rome, providing vertical profiles of aerosol backscattering and depolarization (di Sarra et al., 2001a). Other instruments are under development: a Raman lidar to retrieve water vapour profiles in daytime is being developed jointly by University of Rome and ENEA; a system to launch ozone- and radiosondes will be installed in the near future.

2 Measurements

Fig. 1 shows the evolution of the backscatter ratio R measured by lidar at Lampedusa on June 2, 1999. R is the ratio between the signal scattered back due to atmospheric aerosol as well as molecules and the signal scattered back by molecules only. R is larger than 1 if aerosols are present.

On this day the airmasses reaching Lampedusa originated from central Sahara, and had spent, according to isentropic trajectories, the last 10 days over the desert. The isentropic trajectory for this day ending at Lampedusa at 08 UT shows that the airmass overpassing Lampedusa at 2900 m was close to the surface about 3 days earlier. The airmass at 4000 m or above did not go in the previous 10 days below 1 km altitude, and the airmasses at 5-6 km remained above 4 km altitude throughout the 10 days. The presence of dust particles at 6 km is thus not explained by isentropic motion of the air parcels. This may be partly attributed to the limited resolution (2.5°) of the meteorological analyses used in the calculation of the isentropic trajectories. We expect that strong convection occurring over the desert play a significant role which is not accounted for by the model. At 08 UT the aerosol optical depth, measured with a Multi-Filter Rotating Shadow-band Radiometer (MFRSR) at 415 nm was 0.46, and the Ångström exponent was 0.3, indicating the

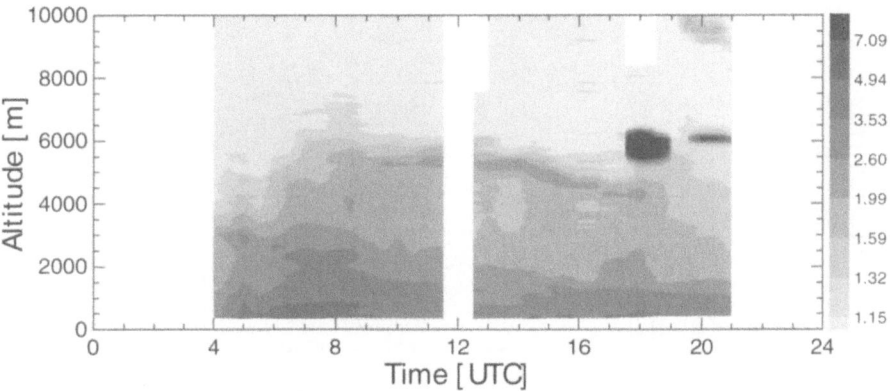

Fig. 1. Evolution of the backscatter ratio measured by lidar at Lampedusa on June 2, 1999, versus altitude and time. The backscatter ratio grey scale is displayed on the right of the graph.

presence of relatively large particles. The presence of desert dust over the Mediterranean produces a large perturbation to the atmospheric structure, and influences its radiative budget. Large effects on the surface ultraviolet irradiance have also been observed (di Sarra et al., 2001b). The development of small clouds within the dust layer is noticeable in Fig. 1 (dark regions at about 6 km between 18 and 21 UT).

For comparison, the evolution of the lidar backscatter ratio of May 27, 1999, is reproduced in Fig. 2. This day is characterized by a relatively low aerosol load (optical depth at 415 nm <0.3 throughout the day, Ångström exponent of 1.2). This behaviour is typical when airmasses over Lampedusa originate from the northern sectors. In these cases the aerosol is generally confined below 3 km, the optical depth at 415 nm is below 0.3, and the Ångström exponent is larger than 1 (di Sarra et al., 2001b). We have compared the global solar irradiance measured with a pyranometer at the same values of the solar zenith angle on May 27 and June 2, 1999. The global solar irradiance is 4-9%, depending on the solar zenith angle and instantaneous values of the aerosol optical depth, higher on May 27 than on June 2.

Large differences also in the spatial distribution of the solar radiation are observed on these two days. The diffuse-to-direct radiation ratio, measured by the MFRSR, at 415 nm and at a solar zenith angle of 40° was 0.51 on May 27, and 0.9 on June 2. Larger differences occur at 868 nm (the diffuse-to-direct ratio at 40° was 0.07 and 0.53 on May 27 and June 2, respectively) The radiation at this wavelength is less dependent than at 415 nm on the scattering by molecules, and the effect of the aerosols is more evident.

At 40° solar zenith angle (in the afternoon) the integral of the irradiance between 400 and 700 nm, which corresponds to the photosynthetically active radiation, is 358 W/m² and 328 W/m² on May 27 and June 2, 1999, respectively, corresponding to a 8% difference. Figure 3 shows the spectra measured on these two days at Lampedusa by means of a Licor spectrometer. The large difference between the two spectra is apparent.

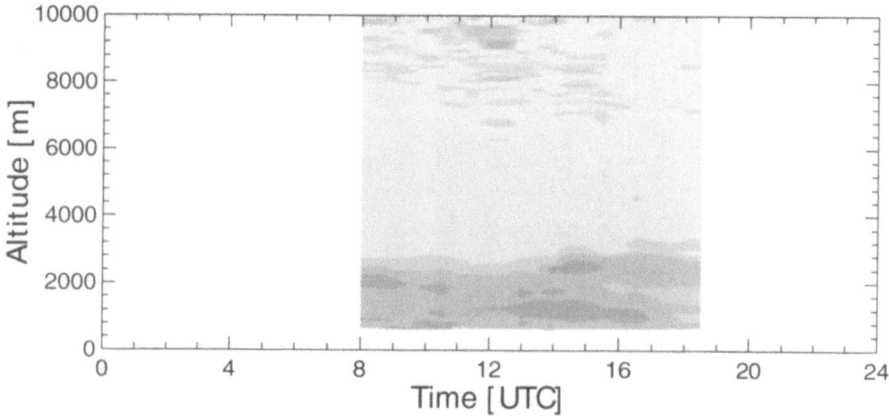

Fig. 2. Same as figure 1, but for May 27, 1999. The grey scale is the same as in Fig. 1

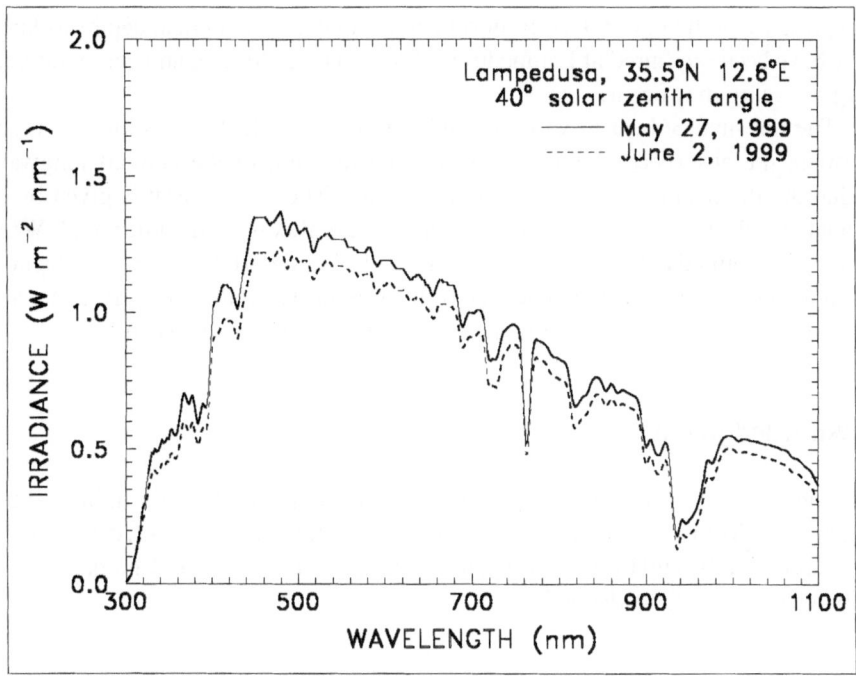

Fig.3. Spectra measured at Lampedusa in cloud-free conditions at a solar zenith angle of 40° on May 27 and June 2, 1999

The daily amount of solar radiation absorbed by the sea surface Q_s is an important term in the estimates of the heat budget. Q_s is generally estimated by means of simplified expressions, such as the formula developed by Reed (1977):

$$Q_s = Q_0 T(1+0.0019h)(1-\rho), \qquad (1)$$

valid for cloud-free conditions. Q_0T (T is a transmission factor) is the clear sky radiation as a function of time of the year and latitude, h is the solar altitude at noon in degrees, and ρ is the albedo. The daily average surface solar irradiance Q_i is given by $Q_s/(1-\rho)$. We have calculated Q_i for May 27, 1999, deriving Q_0T by matching the estimated Q_i with the daily average global solar irradiance calculated from pyranometer measurements. With the derived value of Q_0T we have calculated Q_i for June 2, 1999 and obtained a value of 348.4 W/m², 8% larger than the daily average irradiance determined on the base of the pyranometer data (322 W/m²). Thus, an 8% overestimate of the solar flux is obtained by using the simplified formula, that does not take into account the role of the desert dust. A minor role may be played by atmospheric ozone and water vapour. The daily average total ozone measured by means of the Brewer spectrophotometer at Lampedusa was larger on May 27 (348.7 Dobson unit, DU) than on June 2 (301.7 DU). From the calculated values an estimate of the surface daily average shortwave radiative forcing of about 25 W/m², comparable to that produced by moderate clouds, is derived. The forcing is due to the presence of the desert dust. It must be noted that aerosol optical depths as large as 1 have been measured at Lampedusa (di Sarra et al., 2001b), and comparatively larger forcing may occur.

The net surface heat flux of the Mediterranean is calculated as the difference between the absorbed solar radiation Q_s and the sum of the emitted longwave radiation, the latent and the sensible heat fluxes. The largest term is given by Q_s (about 200 W/m² as long-term average) and the net flux is of the order of 20 W/m² or less. As pointed out by Gilman and Garrett (1994), desert dust largely influences the amount of solar radiation absorbed by the sea surface in the Mediterranean, and may substantially affect the estimates of the heat balance of the basin.

Acknowledgments

The observations at the station of Lampedusa are supported by the Italian Space Agency, the Ministry for Environment, the Ministry for Scientific and Technological Research of Italy, and by the European Union. Contributions by P. Chamard and F. Monteleone are acknowledged.

References

Dai A, Trenberth KE, Karl TR (1998) Global variations in droughts and wet spells: 1990-1995, Geophys. Res. Lett., 25, 3367-3370

di Sarra A, Di Iorio T, Cacciani M, Fiocco G, Fuà D (2001a) Saharan dust profiles measured by lidar from Lampedusa, J. Geophys. Res., **106**, 10,335-10,347

di Sarra A, Cacciani M, Chamard P, Cornwall C, DeLuisi JJ, Disterhoft P, Di Iorio T, Fiocco G, Fuà D, MonteleoneF (2001b) Effects of desert dust and ozone on the ultraviolet irradiance: observations at Lampedusa during PAUR II, J. Geophys. Res., in press

Dubief J (1979) Review of the North African climate with particular emphasis on the production of eolian dust in the Sahel zone and in the Sahara. In: Morales C, Ed. (1979) Saharan Dust, 27-60, Wiley and Sons

Gillette DA, Blifford IH, Fryrear DW (1974) The influence of wind velocity on the size distributions of aerosols generated by the wind erosion of soils. J. Geophys. Res., **79**, 4068-4075

Gilman C, Garrett C (1994) Heat flux parameterizations for the Mediterranean sea: The role of atmospheric aerosols and constraints from the water budget. J. Geophys. Res., **99**, 5119-5134

Goudie AS (1983) Dust storms in space and time. Progr. Phys. Geog., **7**, 502-530

Kubilay N, Saydam AC (1995) Trace elements in atmospheric particulates over the eastern Mediterranean: Concentration, sources, and temporal variability. Atmos. Environ., **29**, 2289-2300

Levin Z, Ganor E, Gladstein V (1996) The effects of desert particles coated with sulfate on rain formation in the Eastern Mediterranean. J. Appl. Meteorol., **35**, 1511-1523

Loÿe-Pilot MD, Martin JM, Morelli J (1986) Influence of Saharan dust on the rain acidity and atmospheric input to the Mediterranean. Nature, **321**, 427-428

Lundholm B (1979) Ecology and dust transport. In: Morales C (Ed.) Saharan Dust, 61-68, Wiley and Sons

Middleton NJ (1985) Effect of drought on dust production in the Sahel. Nature, **316**, 431-434

Moulin C, Lambert CE, Dulac F, Dayan U (1997) Control of atmospheric export of dust from North Africa by the North Atlantic Oscillation. Nature, **387**, 691-694

Prospero JM, Nees RT (1986) Impact of the north African drought and El Niño on mineral dust in the Barbados trade winds. Nature, **320**, 735-738

Pye K (1989) Processes of fine particles formation, dust source regions, and climatic changes. In: Leinen M, Sarnthein M (Eds.) Paleoclimatology and Paleometeorology: Modern and Past Patterns of Global Atmospheric Transport, 3-30, Kluwer Academic Publisher

Rea D, Leinen M, Janacek TR (1985) Geologic approach to the long-term history of atmospheric circulation. Science, 227, 721-725

Reed RK (1977) On estimating insolation over the ocean, J. Phys. Oceanogr., 17, 854-871

Schwartz SE, Andreae MO (1996) Uncertainty in climate change caused by aerosols. Science, **272**, 1121-1122

Sokolik IN, Toon OB (1996) Direct radiative forcing by anthropogenic airborne mineral aerosols. Nature, **381**, 681-683

Tegen I, Fung I (1995) Contribution to the atmospheric mineral aerosol load from land surface modification J. Geophys. Res., **100**, 18,707-18,726

ROSELT/OSS Network in the African arid regions

J.-M. D'Herbes

1 Long Term Ecological Monitoring Observatories Network of the Sahara and Sahel Observatory

ROSELT/OSS is a network constituted by a group of observatories collecting and exploiting field and teledetection data in terms of environment and renewable resources management. It is the first network of this importance in arid circum-Saharian Africa, which organizes and exploites a thematic and statistical follow-up on the environment from local level to national level.

ROSELD/OSS strategy is designed intentionally as an essential contribution to understanding environmental phenomena, and namely desertification in liaison with the problematic of the climate global changes, biodiversity and sustainable development. ROSELT/OSS is simultaneously a tool serving both research and development, through three major concerns:

1) To contribute in improving basic knowledges on the functioning and long-term evolution of ecological and agricultural systems and on the co-viability between ecological systems and socio-economic systems.

Corresponding objectives relate to environmental monitoring and research. They clearly refer to the provisions of articles 16 and 17 of the International Convention on the Control of Desertification:

- To ensure an activity of monitoring the causes an effects of desertification by measuring the state and evolution of climatic factors, of areas concerned on both biological and physical levels (by bringing up to light the functionings and irreversibility thresholds in the ecosystems), of the uses and socio-economic activities in relation with the environment.

- To carry out researches on the mechanisms which lead to desertification by clarifying on one part the interactions between the causes (climatic, anthropic) and the effects of desertification and, on the other, the dynamic relationships between the ecological systems and the social systems, by combining both approaches.

2) To contribute in assuring that these knowledges are used through grouping, data processing and their availability, through the design of indicators and end products at different local, sub-national, national and regional levels. These

products designed as per the state of the environment and its evolution, its relationships with the social and economic dynamics, serve on one side, as tools for the establishment of strategies and sustainable development and protection plans and, on the other, they serve as support to development programs and assistance in decision-making in the field of natural resources management. Within this framework ROSELT/OSS must be a strict scientifical program, technically and financially workable, sociably acceptable and self-financing in the end.

3) To ensure a function in the fields of training experiment and learning of environmental problematics and their inclusion in rural development policies, programs and practices.

ROSELT/OSS strategy enters in a gradually formed and participative approach, preferring by far local and sub-national scales to feed in a coherent manner the needs at national, sub-regional and regional levels. It takes into account the characteristics and the particular nature of OSS intervention zone, whose specific features are of two orders at the same time: eco-climatic order and socio-economic or cultural order.

Thus, ROSELT enters entirely in the spirit of the Rio Declaration and ACTION 21 (Conventions on Biodiversity, Climatic Changes and Control of Desertification), and can constitute a key device in the implementation of the three conventions put forward and approved by CNUED.

2 Structure of the Network

The observatories identified by OSS have been selected according to the following criteria:
   - The different bioclimatic zones characterizing the arid lands;
   - The major particular ecosystems (steppes, savannahs, associated farming systems);
   - The many uses of natural resources in each one of the three sub-regions;
   - The quality of the scientific and technical results.

The members countries have proposed potential candidate territories by supplying the necessary data and by defining their priorities. The identification approach has led to the selection of a first group of 25 observatories and clusters of observatories identified as being part of ROSELT by OSS and responding to pre-defined criteria.

Their total number is certainly insufficient for them to become representatives of thematics addressed on the circum-Saharian zone. Nevertheless, the implementation of such a program can only be achieved in a progressive manner, and a number of preliminary conditions must be met: elaboration and mastery of an harmonized methodology of data collection and processing, training of personnel in sufficient number, defining the needs, compatibility with the national programs and with the ones through cooperation, etc..

In the prospect of a growth of the network towards taking into account new national concerns and towards a greater network representativity (thematic or eco-geographic); these selection criteria could be modulated, classified differently in order of importance, or modified.

Therefore, ROSELT/OSS is founded on an evolutive scenario in three successive phases:

1) An initiation stage or program start-up through a reduced number of observatories selected for implementing and testing protocoles of data collection and processing defined as such following the integration and harmonization of existing methods. These pilot observatories must be useful for demonstration purposes and training of local and national personnels hence they will constitute the beginning of the national networks of observatories.

2) A stage of structuration and consolidation of the network, in the framework of PAN and PASR as scheduled by CCD, which will extend to new observatories (those already identified, to which others will be added, in terms of the diversification of thematics and financings available) aimed at increasing the representativity.

3) A stage of total functioning of the observatories and continued improvement of the network in the long-term, when the network organisation will permit optimal integration and harmonization at different levels of organisation. Elsewhere, the different means of orientating progressively the ROSELT structure towards self-financing will have been explored.

3 Sample Program of Activities of An Observatory

O. Preliminary activities

Preliminary stage of the quadriennal program comprising up-grading of existing data on ROSELT observatories, in a form of synthesis report and the elaboration of a map on land cover (point zero).

A. Long-term environmental monitoring activities

A1: Land cover monitoring and states of soils surface: Periodical up-dating of maps on lands cover and states of soils surface.

A2: Meteorological and climatic parameters monitoring: Local climatic conditions monitoring (rain-falls, sunshine, temperatures, winds, humidity, exceptional atmospherical events, vegetation fire...).

A3: Monitoring of edaphical and resources in soils parameters: Monitoring of the biological, physical and chemical variables of substrates of spontaneous or planted vegetation.

A4: Vegetation and vegetal resources monitoring: Monitoring of the biological diversity and the state of conservation of salvage species, key species, planed taxons (including traditional varieties) and the endemic and/or threatened species (the unlisted).

A5: Fauna monitoring: Monitoring of the biological diversity, of the state of conservation of the salvage and domestic fauna (including local races), and endemic and/or threatened species (the unlisted).

A6: Water resources monitoring: Water availability monitoring (streaming, storage, rains infiltration, useful water for vegetation,...)

A7: Socio-economic monitoring and uses follow-up: Follow-up of the evolution of the activities of the human population having an effect on the observatory and their socio-economic conditions; follow-up of the uses; follow-up of the demography and migrations.

A8: Monitoring of land management and conflicts: Monitoring of the types of rights and conflicts associated with the space and use of the main natural resources and modes of space management.

B. Analysis, interpretation and synthesis of information-study of mechanisms

B1: Impact of climatic evolutions on the environments: Study of interactions between the evolution of climatic data and the state of the physical surroundings (vegetal cover, states of soils surface, hydric assessment of soils,...).

B2: Impact of human activities on the environments: Study of the impact of different uses (biological and land resources utilisation systems) and their evolutions on lands degradation and the evolution of physical surroundings

B3: Study of interactions between the ecological systems (resources) and the social systems (uses): Namely within the framework of diverse economic models and statistical analysis of data.

B4: Setting up and/or experimentation of information systems on environment at local level, and approach methodologies on the mechanisms of desertification

B5: Study of the dynamics of reference ecosystems, farming systems and populations Namely with regard to these disfunctionings to which they are subjected, and the determination of the irreversibility thresholds of these systems.

C. Elaboration of support products in decision-making and development

C0: Environmental data bases and metabases: Storage and harmonizing of the data collected in the field by the observatory teams and by associate partners, meteorological services and satellite data.

C1: Support to the elaboration of the management plan of natural resources

From data production and analysis, definition of the conditions of an optimal efficiency for the use and management of natural resources, function to the biological potential and the dynamics of the ecological systems taken together.

C2: Support to the restoration and management of vegetal and animal species and ecosystems

From data production on the state of conservation of natural resources and their trends towards change, proposal of rational solutions for their restoration in particularly disrupted ecosystems.

C3: Environmental studies accompanying development projects

Environmental informations collection, and technical know-how of the interveneers of the observatories allowing to accompany development projects at the level of their definition and at the levels of the study of environmental impacts.

C4: Desertification indicators production

Desertification biophysical and socio-economic indicators, attached on one hand to the causes of desertification, both natural (climatic factors) and anthropic (man pressure on his surroundings), and on the other hand to the effects of desertification on the biological surroundings (vegetation, fauna, farming systems), physical (soils, waters), and socio-economic (activities, living conditions, etc...).

C5: Biodiversity indicators production

Richness in salvage species and those used by the local populations; assessment between gains and losses of the different constitutive components of biodiversity; list of endangered species and ecosystems (the unlisted); conservation measures and practices.

C6: Indicators production for the analysis of climatic changes

Specific indicators usable for soils contracts analysis and carbon stocks evaluation: states of soils surface, recovery rate of different vegetation stratum and vegetal biomass production.

C7: Reports and thematic maps

Reports productions (support reports to development projects; thematic assessments; methodological reports; prospective reports), and thematic maps at different scale.

Annex 1   Long-term Environmental Monitoring Activities

From the mapping elements of the preliminary stage, reference zones can be selected for the ecological monitoring.

Each dominant "form of vegetation" of the rural landscape, either spontaneous or artificial, each dominant "form of land", will undergo the setting up of a permanent observation device (parcels of some hundreds or some thousands of square metres of surface). The number and the repartition of these devices must be established according to a sampling plan taking into account the heterogeneity of the space and its methods of utilization and management. In all the stations retained, it will be established permanent squares, permanent lines or bands, or any sustainable device that allows a rigorous location of ecological and agro-ecological reference units, to be submitted to repetitive measurings under variable weathers, according to the types of questions that will determine the protocols of measurements and observations to be set up.

Monitoring of lands cover and states of soils surface:
Maps regularly updated with a more up-to-date presentation of:
- Vegetal formations (with identification criteria);
- Forms of vegetation (with their identification criteria);

- The nature and the states of soils surface: percentage of vegetation at 0 level (necks, trunc bases); fine land with or without a fine layer; litter; fine gravels; gravels; stones; free blocs; crusts outcrops; cuirasses outcrops or hard rock base; deflation and accumulation zones;...

Monitoring of meteorological and climatological parameters:
- Rainfalls (daily),
- Sunshine (daily total radiations),
- Maximal and minimal temperatures,
- Winds (speed and direction/day),
- Maximal atmospherical humidity (daily),
- Soil temperature (horizon 0 to 30 cm),
- Exceptional atmospherical events (sandy winds, tornadoes...),
- Vegetation fires (nature, intensity, affected vegetal systems, ...).

Monitoring of edaphic and resources in soils parameters:
- Depth of movable soil profile judged exploitable by vegetal sub-soil systems (at least 90% of sub-soil systems), capable of assuring the safeguard of the useful hydric reserve for the growth of the dominant vegetations at different stratums;
- Global texture and fine texture of movable soil profile horizons;
- Hydric and trophical assessments (fertility) of substrates: useful hydric availability for vegetal growth; rates of organic matter, carbon nitrogen, K20, P205; capacity of exchange and cationical composition;
- Characteristical parameters of the edaphical disfunctionings: percentage of the surface ifluenced by aeolian deflation, aeolian deposits (nebkas, barkanes, massive dunes); levels of saltness; blocking; encrusting;
- Stocks of viable grains of soil superficial horizons (namely for species considered more useful function to pastoral or forest uses, or to ensure the rehabilitation of the ecosystems).

Monitoring of water resources
    NB: In the preliminary stage, a particular attention will have to be brought on to the nature and origin of the resources in waters, and on to the characterization of surface circulations (cf. Streaming networks of rain waters).
    The monitoring addresses the following:
- The hydric regime of rivers and oueds (flow, swellings);
- The thin plate of streamed water, and the rain infiltrated into soil;
- Data collection of the following uses: agricultural (irrigation, waterings, run-off farming); domestical (watering of cattle, family consumption, handicrafts, small industry and commerce.

Monitoring of the vegetation and of the vegetal resources
- Species structure of the forms of vegetation; key species per stratum of vegetation; phenology and biological characteristics of the key-species;

- Species and varieties cultivated: agronomic characteristics; cropping techniques;
- Phytomass[1]:
  - Air phytomass: green phytomass (at production peaks); ligneous plants (alive); dead plants; litter;
  - Phytomass used by men: harvests (seeds, grains, woods, fruits, fibers, roots, leaves, flowers);
  - Phytomass usable by domestic and salvage herbivores;
  - Eventually: sub-soil phytomass.
- Pastoral values: indexes values of species of each vegetation unit according to their importance (recovering; MS weight) in each pastoral system; translation in terms of pastoral production and capacity of animal charge;
- For biodiversity: endemic species, threatened species (unlisted ones) safeguarded species and varieties by traditional agro-systems;
- Threat and status of conservation: current or potential threats (due to climatic and/or anthropic factors); practices or statutory measures of conservation; management plans for protected surfaces.

Monitoring of the fauna (salvage and domestic animal resources)
- Faunistic composition; functional groups; biological characteristics of key-species playing a great role in the functioning of ecosystems;
- Domestic animals species and races; systems of cattle-breeding;
- Animal charges (livestock versus salvage animals); relationships between animal charges and the capacities of charge of the pastoral ecosystems;
- For the biodiversity: endemic species, threatened species (unlisted ones); domestic races to be safeguarded;
- Conservation threats and status: ongoing or potential threats; conservation practices or measures; surfaces protected for their fauna.

Monitoring of the methods of utilization of renewable natural resources
- List of the different current types of utilization of the renewable natural resources: agrosystems; sedantary and transhumantal pastoralism; utilization of wood; piscicultural uses; utilization of the resources in water; medical uses; handicraft; cultural uses;
- Data collection on vegetal resources (see "vegetation": phytomass used by men and their domestic animals);
- Techniques used: inputs participation (fertilizers and pesticides);
- Identification of ongoing or potential threats related to practices.

---

[1] Phytomasses are expressed in most cases in kg of dry matter (MS) per weather unit (vegetation season, year...) and surface (ha, km^2). LAI determination (Leaf Area Index) presents considerable difficulties in the ecosystems of arid zones. This variable will be rarely taken into account in order to satisfy eventually the demand of GTOS (Demonstration Project "Terrestrial Ecosystem Productivity").

Monitoring of socio-economic systems
- Ethnical groups and different users of the renewable natural resources;
- Traditional forms of social and political organisation;
- Types of rights associated to space and the utilization of the main natural resources (land systems and rights of usage);
- Land disputes;
- Human demography; density; pyramid of ages; migrations;
- The active population in relation with the resources of the rural space;
- Activities (time of activity per socio-professional categories and the diverse registered uses of renewable natural resources);
- Methods of management of the rural spaces; production systems;
- Agricultural productions (S.I.): croppings, livestock, forest products;
- Structure of the habitat (scattered, concentrated, sedentary/nomad);
- Human consumption and nutritional methods;
- Satisfaction of energetic needs;
- Commercialized products;
- Micro-economic methods and creation of monetary income for households;
- Health and education infrastructures and structures;
- Institutional or political measures likely to have an influence on social and economic development.

Annex 2   Information   Analysis,   Interpretation and Synthesis-study of Mechanisms

Themes

The data collected are analyzed, interpreted and summarized in order to establish assessments on the state of the environment ("indicators panels") and to attempt and explain the functioning and the evolution of the ecosystems and the agro-systems.
    The following themes will need a particular attention:
- Dynamics of ecosystems, agro-systems and populations (vegetal, animal) of reference (models of functioning, assessment of losses and gains, models of functioning, assessment of losses and gains, potentialities);
- Impacts of climatic changes on the environment;
- Impacts of human activities on the environment;
- Interactions between the ecological systems and the social systems;
- Development and/or experimentation of Information Systems on the Environment and approach methodologies on the mechanisms of desertification.

Elaboration of support products in decision-making and in development

The systematic and repeated data collection allows the analysis and processing of these data, and their translation in terms of ecological describers and/or indicators for each form of vegetation (or lands cover unit, or type of ecosystem, or agro-system), and for all types of environmental impacts (natural or caused by men).

Describers and/or indicators originating from lands cover:
- Evolution of each lands cover unit (superficies covered; rates in percentage of the progressive or regressive evolutions of each unit); models building and validation (of a type "matrix of transition" for example;
- Relationships between the diverse lands cover unit, aiming at demonstrating the major tendencies of the evolution of the uses of space and vegetal resources; for example: relationship between the superficies covered by farming systems (more or less extensive or intensive) and the superficies covered by the integrated pastoral systems or not, in systems of controlled management of pastoral resources and corresponding, if need be, to lands and resources traditional strategies of utilization, or to projects aiming at the restoration of disturbed ecosystems;
- Evolution of the global level of artificial restoration of a given territory undergoing environmental monitoring.

Describers and/or indicators built on the floral composition of vegetal communities:
- Evolution of the indexes of floral richness (number of species per surface unit);
- Evolution of the indexes of floral diversity (several indexes can be put forward, among which only one could be related for example to species already unlisted);
- Relationship between the number of species indicating stability or resilience of the vegetal system, and the number of species indicating the degradation of the system, eventually balanced by the indexes of recovery of these groups of species.

Describers an/or indicators built on the structure of the vegetation (spontaneous or cropped):
- Evolution of the rates of recovery of high ligneous (HL), shrubs (LL) and herbaceous (H) by surface unit and per lands cover;
- Evolution of the total standing phytomasses (dry matter and spurred parts) of high ligneous (HL), of low ligneous (LL) and herbaceous (H) per surface unit and per lands cover;
- Evolution of the rates of recovery at main peaks of vegetation by stratum and by dominant species;
- Evolution of active photosynthetic phytomasses (dry matter production per surface and time unit) at main vegetation peaks, following the stratum and the dominant species, for the main lands cover unit;

- Evolution of the pastoral production and values of the diverse pastoral systems brought to the groups of lands cover units constituting "pasture lands";
- Evolution of plants yields;
- Evolution of the collections made by men of the ligneous resources (wood), and fiber resources for the lands cover concerned.

Describers and/or indicators built on soil-surface characters:
Per unit of surface and for each reference ecological system, find the rates of variation of the following components:
- Evolution of the litter and of the vegetation (level 0);
- Evolution of the physical states of the soil surface;
- Evolution of the movable, mobile components of the surface (percentage of surface influenced by aeolian deflation; percentage of surface influenced by aeolian deposits);
- Evolution of different types of aeolian deposits (voiles, nebkas; barkanes; massive dunes), or sedimentary (following the swelling of oueds, or the streaming on the slopes,...).

Describers and/or indicators of soils hydric features:
- Evolution of the thickness of movable-soil profile fitted in assuring the conservation of rains and streaming waters;
- Evolution of the useful hydric reserve for vegetal growth;
- Modification of the natural or artificial surface reservoirs (temporary ponds, jessours,...).

Describers and/or indicators of features related to soils fertility:
- Evolution of the rates of organic matter of surface horizons;
- Evolution of the rates of C, N, P_2O_5, K_2O...;
- Evolution of the relation C/N;
- Evolution of the indicators of soils clogging.

Describers and/or socio-economic indicators:
- Demographic evolution;
- Evolution of the dominant production systems;
- Human activities evolution (new activities or disappearing or rarefying activities);
- Techniques evolution;
- Evolution of the performances attained;
- Evolution of factors affecting labour organization;
- Evolution of the incomes at the level of living and consumption standards;
- Evolution of the collective and individual representations concerning the natural heritage and their ultimate uses;
- Evolution of land rights application and eventually of the conflicts;
- Evolution of the decision process.

The criteria and parameters indicated above must be inscribed in a correct appreciation of the methods of actions and the impacts of diverse systems of management of the vegetal, natural or artificial resources, with reference to diverse development projects that are applicable to territories considered.

The ROSELT products, in the form of instruments panel, of states of the environment, of raw or statistically processed data, of synthesis reports, thematic maps, pluri-thematic maps, integrated models of sustainable development of indicators and/or describers that must serve as support tools in decision-making for different groups of users:
- Natural resources and space managers;
- Planners at local and national levels;
- Services of technical ministries;
- Researchers;
- Development projects promotors;
- Rural producers, and local socio-economic actors;
- Local collectivities;
- International organizations.

According to the user, environmental problems can be analyzed at different spatial scales that go from local and sub-national scales, to national and international scales. Basic data being often collected on reference eco-systems of test zones, they require to be approved and extrapolated to these different levels within the framework of a strategy of elaboration of the environment indicators and of development of arid zones submitted to the constraints of desertification. Within the monitoring system, the complementarity between ground data collections and data issued from satellite observation will be taken into account for the same token that a sampling device will have to be established correctly on test zones, and that extrapolation rules will have to be fixed. The representability of test zones will have to be perfectly known in order to facilitate the generalization of results to more extended territories. This implies:
- That the whole of the territory concerned must undergo a typology in the form, for example, of a zoning of lands cover (that can be elaborated by satellite means and through field collections), each of the test zones becoming the spatial unit of a stratified sampling device;
- That the ensuing desertification data, describers, on each of the test zones must be processed by statistical methods correctly experimented and validated.

Annex 3   The ROSELT Network Observatories in Africa

Table 1. ROSELT network observatories in Africa

Name of the observatory	Type	Country	Latitude		Longitude	
El Omayed	pilot	Egypte	30 45	N	29 09	E
Matruh		Egypte	31 13	N	27 24	E
Haddej Bou Hedma	pilot	Tunisie	34 28	N	9 30	E
Oued Graguer		Tunisie	36 83	N	9 15	E
Menzel El Habib	pilot	Tunisie	33 58	N	9 16	E
Issougui	pilot	Maroc	31 29	N	-6 24	W
Oued Mird	pilot	Maroc	30 12	N	-5 18	W
Fezouata		Maroc	30 16	N	5 40	W
Tassali N'Ajjer		Algérie	24 35	N	9 25	E
Steppes des Hautes Plaines Nord	pilot	Algérie	34 50	N	0 20	E
Steppes des Hautes Plaines Centre	pilot	Algérie	33 30	N	-0 16	W
Steppes des Hautes Plaines Sud	pilot	Algérie	32 31	N	0 36	E
Melka Werer	pilot	Ethiopie	9 24	N	40 12	E
Awash Park		Ethiopie	8 54	N	39 58	E
Kiboko	pilot	Kenya	-2 17	S	37 32	E
Kibwezi	pilot	Kenya	-2 20	S	38 15	E
Ribeira Seca	pilot	Cap Vert	15 02	N	-23 38	W
Ribeira principal		Cap Vert	15 11	N	-23 41	W
Cercle de Bourem	pilot	Mali	16 58	N	- 0 20	W
Niono, delta Occ.		Mali	14 7	N	-5 43	W
Boucle du Baoulé		Mali	13 43	N	-9 09	E
Nouakchott	pilot	Mauritanie	18 06	N	-15 57	W
Boutilimit		Mauritanie	17 32	N	-15 07	W
Banc d'Arguin		Mauritanie	20 36	N	-16 22	W
Ferlo	pilot	Sénégal	15 55	N	-14 09	W
Thyssé Kaymor		Sénégal	13 57	N	-15 35	W
Torodi	pilot	Niger	14 15	N	2 50	E
Keita	pilot	Niger	14 45	N	5 47	E
Banizoumbou	pilot	Niger	13 35	N	2 40	E

Desertification Information System to support National Action Programmes in the Mediterranean (DIS/MED)

Project document executive summary

M. Candelori

Secretariat to the
United Nations Convention
to Combat Desertification
Bonn, Germany

A. R. Gentile

European
Environment Agency
EEA
Copenhagen, Denmark

L. Genesio

Applied Meteorological
Foundation
FMA
Florence, Italy

1 Introduction

The United Nations Convention to Combat Desertification (UNCCD) was signed in Paris, on 17 June 1994 and entered into force on 26 December 1996. It provides the innovative framework for the sustainable development in arid, semi-arid, dry sub-humid areas, of an appropriate implementation mechanism to combat desertification and the effects of drought. As of 30 November 2001, 177 countries and the European Union have ratified/acceded to the Convention.

The Conference of the Parties (COP), which is the Convention's supreme body, has held five sessions to date, the last of which in Geneva, Switzerland, from 1 to 12 October 2001.

The UNCCD assigns particular relevance to the identification of criteria for the formulation and implementation of the National Action Programmes, as well as for the evaluation of progress accomplished in combating desertification at all levels. The request by the COP to elaborate appropriate indicators stands from the need to check the real effectiveness of the national, sub-regional and regional policies and measures to combat desertification.

Further to a recommendation of the Committee on Science and Technology (CST), the COP, at its second session, invited Governments to initiate testing the application of impact indicators as well as to introduce them in national reporting to the Conference. Similar resolutions have been adopted by the third session of the COP.

Following the first recommendation of the COP, a sub-regional workshop on the Desertification Information Systems for planning needs in the Mediterranean area was held in November 1998 in Marrakech, Morocco, jointly sponsored by the UNCCD Secretariat, the Authorities of Morocco and Italy. It convened representatives of Northern Mediterranean and of Northern Africa countries, as well as international and sub-regional organizations.

The participants to the Marrakech workshop recommended to the Northern Mediterranean and the Northern African countries to explore the possibility of establishing an operational information system for planning purposes, to potentially service all Mediterranean partners, taking into account the existing local capacities and facilities. They also urged countries to establish a close collaboration for the harmonisation of the methodologies of exchange of information related to all aspects of land degradation[1].

The present project document was prepared in co-operation with the European Environment Agency (EEA) and the CeSIA of Italy, both institution having been involved in the preparation of and follow-up to the Marrakech meeting.

2 The Project Area

The Mediterranean basin represents one of the world's most complex systems, as it includes countries bearing many different characteristics in terms of economic structures and productive systems, as well as of ecosystems and culture. The Mediterranean countries have been developing since centuries in a framework of a long history of interaction and integration.

---

[1]The participants to the Marrakech workshop made the following recommendations:
a)   to co-ordinate their existing information systems and to develop and strengthen a permanent system of communication among the different actors involved in combatting desertification and in the reduction of drought effects;
b)   to harmonise the existing database processing and to facilitate reciprocal understanding among the different partners and actors involved in combatting desertification;
c)   to strengthen collaboration and co-operation among the worlds of science, economy and policy, towards effective implementation of the Convention;
d)   to promote and strengthen the transfer of technology between the Mediterranean parties involved, in the fields of analysis and processing of information on desertification;
e)   in conformity with the 22/COP.1 paragraph 2 decision on benchmarks and indicators, to initiate testing the methodology contained in document A/AC.241/Inf.4, as revised in document ICCD/COP(1)/CST/3/Add.1. Participants to the workshop also asked for the evaluation of the possibility and the utility of using such indicators in the national reports to be submitted to the COP;
f)   to identify, process and utilise a minimum set of common impact indicators for the Mediterranean, being they physical, biological and socio-economic. This set of indicators will be representative of the Northern and Southern Mediterranean, describe the key issues. The existing data and the cost/effectiveness of this exercise should be taken into consideration.
g)   to establish partnership agreements to test the above set of impact indicators as well as the methodology proposed to the first Conference of Parties.

Changes in one single country rapidly spread through the whole region, at the environmental as well as at the social and economic levels. The Mediterranean Sea has always been a shared space and one of the most effective - almost virtual - media for the circulation of information at all levels.

The characteristics of this area represent an ideal case for the testing and validation of an operational information system to support planning, where desertification is not so closely linked to the survival of the populations involved, but to a complex economic model.

3 Problems to Be Addressed

The main problems of the Mediterranean soils are irreversible losses due to increasing soil sealing and soil erosion. These processes will continue and probably increase as a result of climate change, land-use changes and other human activities[2].

Although soil degradation is generally recognised as a serious and widespread problem, its quantification, geographical distribution and total impact affected are only roughly known[3].

Notwithstanding the preparation of some National Action Programmes to combat desertification (NAPs), the new initiatives for natural resources management are not the result of a coherent and organised framework of actions. The fight against land degradation in the Mediterranean countries has not yet attained significant and diffused results.

Tackling the problem of land degradation in the Mediterranean is a complex task due to the co-existence of various causes at different levels. In particular, the interlacing of institutional and technical causes entails to address both aspects at the same time.

The main problems to be addressed can be summarized as follows:
- National and sub-regional policies to combat soil degradation are often based on an empirical evaluation and qualitative analysis, rather than on information resulting from data analyses, due to the limited interaction between scientific institutions and policy makers.
- The NAPs of the Mediterranean countries are not based on common and homogeneous information, due to the scarce linkages amongst the national institutions of the different countries.

---

[2] The Mediterranean countries are experiencing severe soil erosion problems, which can reach the ultimate stage and lead to desertification. At present rates of erosion, considerable areas in the Mediterranean and the Alps, currently not at risk, may reach a state of ultimate physical degradation, beyond a point of no return within 50-75 years. Some smaller areas have already reached this stage (Van Lynden, 1995).

[3] The most recent assessment of soil conditions in Europe is an evaluation of the current state of human-induced soil degradation, derived by ISRIC in 1993 from the world map on the status of human-induced soil degradation (Maps of soil degradation in Europe, prepared by ISRIC, are published in EEA, 1998). There is a need for better, and more detailed information.

- Consequently, national and sub-regional policies in the Mediterranean Region are not sufficiently appropriate and consistent.

4 Objective of the Project

The purpose of the project is to improve the capacity of national administrations of Mediterranean countries to effectively program measures and policies to combat desertification and the effects of drought.

This aim will be pursued by reinforcing the communication amongst them, facilitating the exchange of information and establishing a common information system to monitor the physical and socio-economic conditions of areas at risk, assess the extent, severity and the trend of land degradation.

5 Expected Results

The expected results are of institutional and technical nature.

The *institutional expected results* are as follows:
- Information is circulated and exchanged between the relevant institutions at regional level;
- Interactions between scientific institutions and decision-makers are recognized at the national level;
- Available information is suitable for planning purposes at the national level;
- New information technologies are diffused in national services.

The *technical expected results* are as follows:
- ► Standards and procedures for:
    - vulnerability mapping,
    - impact indicators,
    - databases,
- ► are agreed for the Mediterranean area;
- ► Homogeneous and standardized data are available;
- ► Methodologies to produce information suitable for planning and monitoring purposes are fully available, in particular for:
    - crossing data of different types from different sources;
    - managing of analysis at different scales;
    - producing information in useful format.
- ► The common information system on desertification is set up and operational.

To meet these requirements, the information system has to be based on a minimum set of agreed common indicators for the Mediterranean area, which should be:

♦ Already available or that may be immediately derived from the existing information;
♦ At low cost, to allow a frequent updating;
♦ User-friendly and of immediate understanding for the decision-makers;
♦ At a suitable scale, in order to enable planning at a national and regional scale;
♦ Comparable and congenial at the regional level.

Actions should focus on the following three areas:

1. Designing and developing a system for the management of data and the dispatching of information at the regional level. This system will lie in standardised and homogeneous databases;
2. Developing methodologies for the assessment of desertification at the regional, national and local scale. These methodologies will lead also to the definition of a reference framework for the monitoring of trends;
3. Providing decision-makers with a series of operational tools and outputs, which could be directly used for planning.

6 Project Implementation

The project foresees the active participation of a number of different stakeholders, in particular decision-makers and national scientific institutions. It provides for the enhancement of planning capacity of National Coordinating Bodies (NCBs), by facilitating the exchange of information between partners and the transfer of technology, as well as through a common information system for the Mediterranean region.

The project does not provide for any new structure, on the contrary it aims at facilitating the relationships among the existing institutions in order to make their action more serviceable and effective.

To achieve these goals, the following is crucial:

▸ To rely on the existing capacities and structures;
▸ To build consensus on common methodologies and procedures for the processing and circulation of information;
▸ To define a common set of benchmarks and impact indicators.

For these reasons, a number of workshops will be held during the implementation of the project:

▸ A start-up workshop and three yearly workshops for the evaluation of the results achieved (validation workshops);
▸ Nine thematic workshops for the endorsement of technical options (operational workshops);

▸ A final workshop at the completion of the activities for the evaluation of the outputs and the impact of the project. This workshop will aim also at promoting and disseminating the results achieved.

As the project is being implemented through the active participation of a variety of actors, an attentive and supportive co-ordination is required.

The European Environment Agency provides *technical and institutional assistance* that will facilitate the consensus on technical options and overview the co-ordination of the different national institutions.

The Applied Meteorological Foundation (FMA) provides *scientific assistance* which includes elaborating proposal to be reviewed at the operational workshops, on the following issues:

• The technical specification for the system;
• The system design;
• The identification of development needs;
• The technical assistance to the information system;
• The support to the management of the system.

The UNCCD Secretariat provides for the co-ordination of the project activities, the management of technical and financial resources and the follow-up of the project's strategic objectives.

The country partners are requested to will actively participate through their NCBs and the selected relevant scientific institutions. The countries participating in the implementation of the project are:

■ Northern African sub-region: Algeria, Egypt, Morocco and Tunisia;
■ Northern Mediterranean region: Greece, Italy, Portugal, Spain and Turkey.

Further expansion of the information system is expected in term of adhesions of other potential partners, as well as other potential contributors and donors.

The European Union (EU), the Arab Maghreb Union (UMA), and the Sahara and Sahel Observatory (OSS) support implementation and the expansion of the system to other Mediterranean countries.

7 Activities and Project Financing

The activities to be undertaken include:
1. Technical and institutional assistance
    1.1 Institutional support
    1.2 Holding of validation workshops
    1.3 Holding of operational workshops
2. Scientific assistance
    2.1 Technical support
    2.2 System preliminary design
    2.3 System development
3. Equipment
    3.1 Supply of hardware and software

4. System running

4.1 Supplies and services

The financial requirements to implement the project are met on the basis of the following assumptions:

- The Northern Mediterranean countries will cover their own expenses related to participation to workshops and for national activities;
- The European Environment Agency will cover the costs of the start-up workshop and made available its expertise, telematic infrastructures, and relevant existing data;
- The Italian Co-operation will fund the implementation of the project in Northern African countries;
- The Northern African countries will contribute the salary of the personnel involved in the project.

Additional funds from other partners or from those already engaged in the implementation of the project would ensure increased flexibility and effectiveness of the project, as well as the expansion of the information system to other interested countries of the sub-region.

Research in Global Change in the Mediterranean: A Regional Network

I. Raev

1 Aims of the Regional Network

From the Bulgarian point of view it is emphasized that regions around the Black Sea and beyond belong to the zone of Mediterranean climate. A regional Mediterranean research therefore should not be limited to the coasts of the Mediterranean Sea. The network should concentrate upon the following tasks:

(i)   The collection of climatic information by stations along the Mediterranean basin, as well as in the area of a typical Mediterranean climate together with the border zones of a continental climate respectively an arid climate. Two types of climate data series seem to be useful:
  • longest possible time series of data;
  • data on a moderate climate (1970-2000).

(ii)   Mediterranean Sea circulation status and tendencies reflecting terrestrial ecosystems.

(iii)   To search for the answer to the question of land and marine response and biodiversity to global climate change. Assessment of vulnerable zones.

(iv)   Exploration of the potential for adaptation of ecosystems;

(v)   Study of causes and impacts of drought periods in modern climate as a model of future climatic changes.

(vi)   Anthropogenic effects on key ecosystems in the land -ocean -atmosphere system.

(vii)   Atmospheric dynamics - norm of synoptic processes in the Mediterranean basin, analogous periods in modern climate (1970-2000) and trends.

(viii) Atmospheric chemistry and aerosols, chemistry of precipitation, chemistry of lysimetric waters and chemistry of river water in watersheds.

(ix)   Dendrochronological, paleoclimatic and geophysical records.

2 Requirements for An Anchor Station Network

The following steps would be necessary to activate a Mediterranean Anchor Station network:

▸ Inclusion of existing climatic stations, representative for typical terrestrial ecosystems in each country;

▸ Inclusion of stations for comparative data from Mediterranean regions continental regions, ocean regions or arid climatic regions, especially in boarder zones;

▸ Development of a common methodology in all stations for obtaining of comparable data;

▸ Introduction of modern facilities and equipment for coordinated delivery of data and comparative information at a small number but important anchor stations for each country;

▸ Technological support in order to receive uniform directives in all countries for the use with the equipment of the stations;

▸ Inclusion of representative watershed basins for the study of the integral impact, as well as climatic and the anthropogenic influence.

3 The role of the Bulgarian Forest Research Institute

The Forest Research Institute (FRI) at the Bulgarian Academy of Sciences (BAS), Sofia, has a longstanding tradition in the study of representative ecosystems and watersheds in Bulgaria since 1961. The information from the ecological stations established for the purpose could be useful, both for revealing climatic changes and their impact on the water balance of the country; the vulnerability of the basic forests, for the study of vegetation and wildlife under the impact of climatic stress. FRI-BAS would therefor play an important role among the participating institutes in a multinational Regional Network.

The existing Bulgarian network *inter alia* includes representative forests of the following stands: *Pinus sylvestris L., Picea abies/L.,Karsten, Abies alba.Mill, Pinus mugo Mill, Quercus cerris L.,Fagus sylvatica L.*.

The necessary infrastructure for the carrying out the investigations, the necessary buildings, technical facility and research staff is available, however contemporary equipment will have to be supplied.

It is necessary to adopt realistic projects for building up the Anchor stations in each of the countries. In Bulgaria at this state at least the following two stations could be implemented at the first stage: Vassil Serafimov and Govedartsi.

4 Previous "Global Change" Activities in Bulgaria

The following projects so far have been performed in the framework of "Global Change" Research in Bulgaria:

The Bulgarian Country Study to Address Climate Change (1994-1996)
This Bulgarian project is part of the World Country Study to Address Climate Change and is in line with the Framework Convention on Climate Change (1992).
Sponsor: United States Department of Energy.
Bulgarian participation: National Institute on Meteorology and Hydrology at the Bulgarian Academy of Sciences (BAS), Forest Research Institute (FRI-BAS), Institutes for Nuclear Energy (INE-BAS), Institute of Economics - BAS etc.
Result 1: The First National Communication on Climate Change, Republic of Bulgaria, 1996, Ministry of Environment, PenSoft, pp.88, Sofia.
Result 2: The Second National Communication on Climate Change, Republic of Bulgaria, 1998, Ministry of Environment and Waters, pp.170, Sofia.

National Action Plan on Climate Change (1998-2000)
Sponsor: The Ministry of the Environment and Waters
Bulgarian participation: National Institute on Meteorology and Hydrology at the Bulgarian Academy of Sciences (BAS), Forest Research Institute (FRI-BAS), Institutes for Nuclear Energy (INE-BAS), Institute of Economics - BAS etc.
Result 1: Ministry of the Environment and Waters, 2000. National Action Plan on Climate Change. Part I, pp. 50.
Result 2: Ministry of the Environment and Waters, 2000. National Action Plan on Climate Change. Part II, pp. 75.
Adopted by the Council of Ministers of Bulgaria (Decision for Realization No. 393/6th July 2000).

Bulgarian - American Conference "Global Change in Bulgaria", Blagoevgrad, 17-19 June 1997
Sponsor: CIRA- Pennsylvania State University, USA
Particpants: 7 institutes of BAS
Result: Knight CG et al. (1999) Global Change in Bulgaria. BAS, PenState, pp. 595.

The Drought Period 1982 -1994 in Bulgaria - an Analogy for Future Global Changes. Natural, Economic and Social Consequences from Drought (1998-2001)
Sponsor: CIRA - Pennsylvania State University, USA
Participants: 29 researchers from 11 Institutes from BAS, Bulgaria and 6 researchers from PenState University.
Status: Presently the book of proceedings addressed to the scientific communities and decision makers is sent for reviews in Bulgaria and USA.

Study of the representative forests from Picea abies/L./Karsten in the Rila Mountains: characteristics of the environment, structure of the ecosystems and global change (1999-2002)
Sponsor: Ministry of Public Education and Science
Participants: 22 researchers from 6 institutes in BAS

A Vision From the Rudjer Bošković Institute Regarding Cooperation with the Mediterranean Research and Application Network (MERAN)

T. Legović and M. Ahel

1 Introduction

The Rudjer Boskovic Institute is the largest and most productive scientific institute in Croatia, mainly in the areas of physics, chemistry, biology, ecology and oceanography. The institute also hosts two departments for marine and environmental research located in Zagreb and Rovinj, involving about 160 scientists, technicians and administrative staff. The Institute is well connected by intranet and internet, and hosts an array of sophisticated instruments to aid scientists in their work.

In its 50-year history, the Institute regularly formed and often led consortia of national institutes, especially in the field of marine and environmental research. The Republic of Croatia has one more institute with divisions in Split and Dubrovnik, and four departments at universities where marine science is being taught.Scientists at the Rudjer Boskovic Institute are regularly invited an consulted in strategic environmental issues of the national interest and readily respond to requirements set by the government and public. In addition, a number of applied studies are produced annually that tackle developmental issues ordered by national and international agencies.

Croatia has a jurisdiction over the most of the east Adriatic coast and coastal sea. Toward the inland Croatia extends over the Dinaride Mountains and into the Panonia continental flatland, covering 3 differet climate types. The total length of the Croatian coastline is 5790 km of which 1778 km belong to the continental coast and 4012 km to the insular part. In the coastal region of Croatia are situated 4 National Parks hosting 44 protected varieties of plants and 381 animal species. There is a growing evidence that change in climate as well as in land-use will impact our forests, grasslands, non-coastal wetlands, freshwater ecosystems, littoral pelagic zones, fisheries, hydrology and water management. However, unless we can put the observed changes into a broader context of the global change in the Mediterranean region, we cannot predict with certainty to what extents will they occur and what their dynamics is going to be. Therefore, it is highly desirable to establish links between the research efforts in the Adriatic region with relevant international programs including the Global Ocean Observing System (GOOS), the

Global Terrestrial Observing System (GTOS), the International Long-Term Ecological Research (I-LTER), and Land-Ocean Interactions in the Coastal Zone (LOICZ). Along these lines, the Institute is interested to participate in the MERAN network, offer its expertise, learn from the others and joins forces in researching the causes and consequences of global change in the Mediterranean region.

2 Vision

The combined effects of global climate change and human alterations of the environment will be especially pronounced in the coastal, shelf and estuarine zones were population density is increasing most rapidly. It is here that the conflicts between economic development, sustaining living resources, protecting and restoring ecosystem health, mitigating natural disasters, and protecting public health and safety will become most pronounced over the next several decades. As in many coastal environments throughout the world, the watersheds of the Adriatic Sea are regions of rapid population growth and changing land-use patterns that have led to increases in nutrient loading and changes in freshwater flow patterns to coastal waters.

The Adriatic region is characterized by intensive land-based and sea-based activities, including urban growth and development, agriculture, commercial and recreational fisheries, tourism, and multinational commerce. Changes in these activities are widely believed to have elicited significant degradation of water quality, manifested as mucilage events, oxygen depletion of bottom water, harmful algal blooms, outbreaks of gelatinous zooplankton, invasions of non-indigenous species, loss of habitat and instability of fisheries. Individually these phenomena may not be cause of major concern. But taken as a whole, they may be indicative of a pattern of environmental stress that threatens the health of coastal ecosystems of the Adriatic as well as Mediterranean at large.

It is already apparent that changes in the coastal ecosystems are making coastal zones more susceptible to environmental hazards (e.g. flooding, droughts, harmful algal blooms, mucilage events, hypoxia). These hazards, which may occur on local to regional scales, must be observed in te context of larger scale changes in circulation, climate, and land-use practices in order to resolve natural perturbations from man-induced changes. In addition, a predictive understanding of the causes and consequences of environmental variability and of the man- or climate-induced changes has to be achieved in order to develop effective and scientifically sound rationale for different adaptation strategies.

Historically, efforts to understand, manage and protect the ecosystems and living resources of the Adriatic Sea have been conducted mainly on a case-by-case basis by individual nations and institutions for specific purposes with insufficient coordination and collaboration. Today we see that the causes and consequences of environmental change can only be understood and mitigated through a coordinated international effort that considers the Adriatic system as a whole and regards it as an itegral part of the Mediterranean basin.

Achieving these goals also requires an integrated system of sustained environmental observations that provides the data and information to broad range of user groups, including government authorities, responsible for environmental protection, management of natural resources, public health and safety, industry, the scientific community, the news media and the public at large.

3 General Problems of the Mediterranean Area in the Adriatic Domain

There are two phenomena, which are of key importance when assessing the effects of global change in the Mediteranean Sea, that should be looked at in the Adriatic basin: circulation patterns, which are important for water and energy balance considerations, and eutrophication-related phenomena, which have a strong impact on carbon balance.

Since the Adriatic Sea is well-known to be the major source of the deep waters of the Mediterranean Sea, the processes of dense water formation in the Adriatic influence profoundly the overall circulation patterns and trigger a chain of biological responses. In turn, the intrusions of Levantine intermediate waters into the Adriatic represent an important component of the hydrodynamics of the basin with strong effect on the Adriatic ecosystem as a whole. Obviously, determining the variability in the exchange of waters between te Adriatic Sea and the Mediterranean Sea and resulting changes in circulation are mandatory when addressing the issue of global change. Given the importance of circulation, waves and turbulent mixing in structuring marine ecosystems, monitoring physical processes and forecasting changes in the physical environment is of a fundamental importance.

The Adriatic Sea, notably the northern Adriatic, is the area exposed to the highest eutrophication pressure in the Mediterranean Sea. The Adriatic Sea, which represents only 1% of the total volume of the Mediterranean Sea, receives up to 25 % of the total freshwater inputs. Therefore, eutrophication represents one of the most obvious signs of the deteriorating quality of the Adriatic ecosystem.

Major indicators of change in the Adriatic Sea are related to eutrophication:
- oxygen depletion of bottom waters
- harmful algal blooms (HABs)
- mucilage events and
- sustainability of fisheries

Although there are still many uncertainties, it seems that the scientific evidence from the Adriatic Sea and elsewhere supports the hypothesis that all these indicators can be related to each other and to increases in nutrient loading from human sources (i.e. man-induced eutrophication).

In addition:

There will be problems of rising sea level for flat areas on islands and some parts of the coastal area, which can be even enhanced if associated with some specific meteorological situations that will also be related to global climate change.

Toward the inland, Croatia extends over the Dinaride Mountains and into the Panonia continental plains, covering three different climate types. A change in temperature and in yearly precipitation will induce changes in agriculture but even more in biological reserves where the highest concern is about persistence of endemic and endangered species. It is hypothesized that global changes will impact differently these parts of Croatia, while existing climatic gradients provide a useful scale to trace subtle changes.

4 The Potential Role of the Institute As A Part of the MERAN Network

- It may co-ordinate national efforts of Croatia towards global change research and monitoring with the special aims:
  - to organize an Anchor station of the MERAN network
  - to establish an advanced data/information management system by integrating GIS and relational database for the Adriatic region
- It will try to get access to all national data relevant for the ground truthing of selected indices of the global change. In the Adriatic Sea, there are two long-term oceanographic data bases, representing time series of >30 years, which is of crucial importance for the goal of detecting trends in the variability caused by global change. One data series was obtained along the northen Adriatic transect (Po River- Rovinj), while the second one covers the transect in the middle part of the Adriatic (Split - Gargano). Both transects are very suitable candidates to become MERAN Anchor Stations and could be used for ground-truthing of the remote sensing data base, which is available for the more recent periods.
- It envisions recommending to the government of Croatia updating its research capabilities and installation of instruments based upon findings and agreements at the Workshop in Casablanca. In this context it is important to note that all existing data bases derive from observations obtained during regular cruises using research vessels, while oceanographic buoys for continuous telemetric measurements *in situ* are still not available.
- It intends to disseminate data and results of its research to its partners in a timely fashion.

Data Accessibility Organised by CTM ERS/RAC in the Mediterranean Area

M. Viel

1 CTM ERS/RAC Objectives in the Map Framework

The Mediterranean Action Plan (MAP) for the protection of the Marine Environmental and the sustainable development of the coastal areas of the Mediterranean is legally supported by the Barcelona Convention (1975) and its protocols, adopted by the 20 Mediterranean Countries bordering states and the European Union.

Since 1993, the CTM plays the role of Regional Activity Centre for Environment Remote Sensing (ERS/RAC) of the Mediterranean Action Plan (MAP)/United Nations Environment Programme (UNEP) that, upon mandate of the Contracting Parties to the Barcelona Convention, is committed to assist, and cooperate with, 20 Mediterranean bordering countries and the EC in producing, using and disseminating environmental information derived from remotely sensed data.

Among the other, CTM ERS/RAC activities are aimed at:

- increasing knowledge of environmental state and changes in the Mediterranean by resorting to remote sensing and its integration with other sources of information;
- raising awareness and ensure access to that knowledge for a proper management of Mediterranean environmental issues, also taking advantage from EU projects experienced achievements.

To this purpose, CTM ERS/RAC has launched several initiatives:

(i) the RAIS project in 1994, which was oriented towards gathering updated and complete information about Remote Sensing Centres in the Mediterranean and their activities;

(ii) a Review in 1998, sponsored by ESA (European Space Agency), of Mediterranean environmental management actions using space techniques;

(iii) the setting-up of the STEPINMED database, published on the Internet in January 1999, on programmes and sub-programmes addressed to the Mediterranean Environment, projects in the Mediterranean using remote sensing, positioning and telecommunications as well as organisations involved in the different surveyed projects (http: //www.ctmnet.it/stepinmed).

(iv) the arrangement of specific Forums in Mediterranean countries on support of remote sensing techniques to planning and decision making processes for sustainable development;

(v) the issuing of publications analysing the use of remote sensing in thematic sectors.

(vi) Inventories on remotely-sensed projects and relevant organisations based in specific Mediterranean countries and analysis of such information to support the measurements of some sustainable development.

In 2000, ERS/RAC conceived the MERSI.Web ("Mediterranean Environment Remotely-Sensed Information Web") initiative which is aimed at involving and linking Mediterranean Countries, through a network of Mediterranean specialised centres dealing with satellite remote sensing and its environmental applications, for providing a frame, as complete as possible, on available information on the state and changes of marine and coastal environment.

2 The MERSI.Web Initiative

2.1 Motivation for A Mediterranean Environment Remotely-sensed Information Web

Today, the need of a better exploitation of existing data, information, projects, programmes on environmental issues is strongly felt and expressed by decision-makers and researchers on the one hand, and by the potential international and national financing bodies on the other. This demand stems from the necessity to rationalise resources, research and funds and to avoid as much as possible duplications.

The speeded-up improvement of technology in the field of data management, data exchange, and communication, makes it possible today to regularly collect existing information from many sources, to implement effective databases for its management, and to provide to the public standardised, organised and homogeneous sets of information - tailored to specific uses - relying in particular on the Internet.

In order to allow the usefulness and maintenance of such databases, the setting-up of them needs to be designed and managed by a specific group of "actors" (network core) who represent the "information providers" and are well aware of requirements of "potential users" of such information.

2.2 General Objectives

The proposed initiative aims at involving and linking Mediterranean Countries, through a network of Mediterranean specialised Centres dealing with satellite remote sensing and its environmental applications, in building-up a "Mediterranean Environment Remotely-Sensed Information Web" (MERSI.Web).

Through MERSI.Web, suitable description of existing datasets and relevant environmental information resulting from the application of remote sensing and its integration with conventional methods and techniques in the Mediterranean, will be:
3.   organised in a standard format
4.   stored in an easily accessible database
5.   made available, to the general or targeted public, through the Internet

The database will be populated, controlled and continuously maintained by the network of Centres.

2.3 Stored Information

The information to be stored, and organised relying on meta-data[1] structure will answer to the following questions:
♦   WHAT has been done in the field of environmental monitoring in the Mediterranean with the support of remote sensing;
♦   WHERE the activities have been (or are) carried out;
♦   WHO have been (or are) the developers;
♦   WHO have been (or are) the users;
♦   HOW this information is (or could be made) available to other researchers and users.

In particular, it will include description of:
•   geographical location;
•   application field;
•   used satellite, image characteristics (date, geographical coordinates, etc.), raw data;
•   used ancillary data;
•   methods and standards of reference;
•   value-added products derived from remote sensing data (thematic observations, classifications, multi-temporal analyses);
•   products derived from the integration between remotely-sensed data and other data referring also to products processed using Geographic Information System;
•   availability of processed data and their characteristics;
•   information providers and information users with appropriate links;
•   framework (national and international projects, programmes, conventions, etc.)
•   links to other specific websites;
•   etc.

The information to be selected and the meta-data structure to be used have to be established by the actors of the network core.

---

[1] Meta-data is ancillary information characterising and describing products and data sets

2.4 Structure

To ensure efficiency and duration of the MERSI.Web system, its structure will resort to:

- a centre in each Mediterranean country which will act as a national reference centre. It will be responsible for the management (search, evaluation, storage, up-dating) of the descriptive information at national level. It will ensure links with other organisations working in this field in its country and with main national users;
- a centre performing a Mediterranean role which will manage information at regional scale (also derived from information at national scale, as well as relevant to multi-national projects, etc.) and connections with regional organisations and regional users. It will ensure links among the national centres and will be responsible for the initiative co-ordination. It will foster the popularisation of remote sensing applications through the drawing-up of special pages and presentation of successful stories on the web.

The system design and implementation will be made through close co-operation among the regional and national centres.

The creation of such a network relying on centres skilled in satellite remote sensing techniques and applications will ensure, during the implementation phase, the homogenisation and quality of the system and the maintenance and the "life" of MERSI.Web during the operational phase.

Access of the Public to the information stored at regional and national levels will be made possible through the Internet. Innovative structures will be used for updating, searching and presenting the database on the Internet and ensure links to the different network nodes.

This process will allow the diffusion and the use of the information set and will guarantee the necessary feedback to improve the system and to increase the information.

3 The Regional Forum in Morocco

A regional Forum has been organised in October 2000 in Morocco, together with the Centre Royal de Télédétection Spatiale (CRTS) of Morocco (ERS/RAC Focal Point[2] to gather representatives of specialised centres based in Mediterranean countries, dealing with remote sensing and its environmental applications, and to present to, and to discuss with them a plan for the joint setting-up of a network to build up a "Mediterranean Environment Remotely-Sensed Information Web" (MERSI.Web).

The following 14 national Centres/Organisations, remote sensing centres or institutions working in the environmental field and resorting to remotely-sensed

---

[2] Focal Point: Person/Institution nominated by the country to act as national reference for ERS/RAC in the MAP framework.

information for environmental monitoring and planning, have been appointed by ERS/RAC National Focal Points:

- CNTS (Centre National des Techniques Spatiales), *Algeria*
- MAP office, *Bosnia-Herzegovina*
- Ministry of Agricultural, Natural Resources and Environment, *Cyprus*
- NARSS (National Authority for Remote Sensing), *Egypt*
- MEDIAS-FRANCE - CNES, *France*
- Ministry of the Environment - Planning Division, *Israel*
- NCRS (National Center for Remote Sensing Lebanon:), *Lebanon*
- LCRSS (Libyan Centre for Remote Sensing and Space Science), *Lybia*
- IcoD, Foundation for International Studies, *Malta*
- CRTS (Centre Royal de Télédétection Spatiale), *Morocco*
- IGN (Instituto Geografico Nacional) - Remote Sensing Unit, *Spain*
- GORS (General Organization of Remote Sensing), *Syria*
- CNT (Centre National de Télédétection], *Tunisia*
- Ministry of Environment - Foreign Relations Department, *Turkey*

International Organisations (CEDARE, C.R.T.E.A.N, ESA, EURISY, GRID, OACTS, OSS) working in the Mediterranean area have been also invited to participate in it and to make comments and suggestions on the MERSI.Web concept. Representatives of Moroccan National Institutions (Ministère de l'Aménagement du Territoire, de l'Environnment, de l'Urbanisme et de l'habitat; Ministère chargé des Eaux et des Forêts; Institut National de recherche halieutique; Ministère de l'Agriculture, du Développement Rural et des Eaux et Forêts) dealing with environmental management have been invited to provide their point of view as potential users of such an information network.

The Forum on MERSI.Web allowed to achieve some important results:

- an overall consensus from all the participants was reached on the MERSI.Web concept, and the Forum boosted the starting-up of its implementation, laying the foundations for the establishment of an operational Mediterranean network. A statement of interest witnessing this consensus was agreed upon by the participants.
- the discussion with Regional and International Organizations was fundamental, since it highlighted that MERSI.Web in its first steps should aim at very targeted and not too ambitious objectives, and in particular that the network should be initially focused on setting up remotely sensed meta-data information at national and regional level, according to international standards and user friendly-interface;
- most of the participants underlined the importance to progress in the dialogue among national and regional participants, and that to this purpose a steering committee composed of their representatives should be established, in order to drive MERSI.Web towards a concrete support to decision and policy making in the field of the environment and sustainable development in the Mediterranean;
- the needs of seeking financial and operational support for implementing MERSI.Web.

Towards a Mediterranean GEWEX

H.-J. Bolle, S. I. Rasool, P. Try

1 The GEWEX Strategy

GEWEX stands for "Global Energy and Water Cycle Experiment"(WCRP, 1990).
It is a sub-programme of the World Climate Research Programme (WCRP). Its goal
is to *determine the hydrological cycle by global measurements*. To reach this goal
studies have been initiated under the GEWEX Continental-scale International
Project (GCIP) (WCRP, 1992) with the objectives as illustrated in Fig. 1:
- Advanced modelling of the hydrological cycle and its effects;
- Prediction of the response of the hydrological cycle to environmental change;
- Improvement of observing techniques and data assimilation systems.
Global cycles generally can be investigated by global models or global measuring

WCRP //////   **Global Energy and Water Cycle Experiment**

OBJECTIVES

· Determine the hydrological cycle and energy
fluxes by means of global measurements of
observable atmospheric and surface properties.

· Model the global hydrological cycle and its
impact on the atmosphere, oceans, and on the
land surface.

· Develop the ability to predict the variations of global and regional hydrological processes
and water resources, and their response to environmental change.

· Foster the development of observing techniques, data management, and assimiliation
systems suitable for operational application to long-range weather forecasts, hydrology, and
climate predictions.

Fig. 1. The major objectives of the WCRP - GEWEX programme

networks. Energy and water fluxes cannot be measured directly, they have to be computed using other measurable quantities which determine these fluxes. The global models used for this task are those which simulate the General Circulation (GCM's) of the atmosphere and the world ocean. They exist also in coupled versions and as climate models which are GCM's that can be integrated over long time periods. Global measuring networks to be considered here must be operational ones. Only this guaranties continuous operation and replacement of instruments in case of failure. Two of these systems exist, the meteorological network at the ground including aerological stations and the operational meteorological satellite system. The classical investigations of energy and water fluxes one can think of are those of Oort (1971), Vonder Haar and Suomi (1971), Oort and Vonder Haar (1976), Ellis and Vonder Haar (1978), and Peixoto and Oort (1982), and the more recently the NOAA Pathfinder Project.

Both the modelling as well as the measuring approach have deficits in assessing climate variability and trends. In the models some of the important processes are only marginally represented - in other words parameterized - especially if one comes down to the smaller scales. The operational ground based and aerological measurements are not as dense as one may wish and are afflicted with instrumental errors. Only standard parameters are measured here. The observations made from space is the only data source continuous in space and time but the remotely measured signals have to be converted into the required information. The two methods applied to determine the global exchange of water and energy consequently have to be validated and to be improved stepwise. On a global scale this is an impossible task. The potential to install a dense network equipped with sophisticated instrumentation just does not exist. Therefore these tasks have to be done at smaller scales in representative or critical areas where the most essential problems can be treated with great accuracy.

GEWEX experiments have been structured in a way that the area under investigation is large enough to compare the results with those obtained by global climate models. Since one grid cell is by far not enough for such a comparison the area must be of the order of 10 grid sizes but regional differences should as well be assessed at much smaller regional scales. Because hydrological processes at the land surfaces have response times of more than a year such an experiment should ideally expand over a number of years. These requirements are already incompatible with the available resources. Therefore a strategy to further reduce the requirements was developed. Only those comparative measurements are extended over two, three years, which are necessary to assess the long term processes like the cycles of soil moisture and aquifer replenishment. On the other hand the investigations needed to improve the parametrization of regional and local processes are only carried out in specific areas nested into the larger experimental area. Satellite data are used to bridge the results obtained in these selected areas over to the whole area and to provide continuous information at least about a few state parameters. The experiments need thorough information on land-use, land-cover, soils, geo-hydrology and vegetation which integrates many disciplines into this effort.

In order to establish a GEWEX one will draw in a number of existing programmes and projects such as the MED-HYCOS network which can be looked at as a backbone of the hydrological branch of a GEWEX in the Mediterranean area which can be called "MEDEX" (Fig. 2).

GEWEX continues to support of potential developmental efforts for the land-ocean-atmosphere coupling studies involving the Mediterranean region. With respect to the Mediterranean area its complex topography and its geographical position are very attractive from the global point of view for a GEWEX study. GEWEX is not able to take on all related projects, it is very interested in any that will help to advance the science, especially in regions with unique characteristics such as the Mediterranean. It exists the MED-HYCOS network which can be looked at as a backbone of the hydrological branch of a "MEDEX".

Moreover at least three global models are available in Europe for numerical studies and these are supported by a number of regional models which can be nested into the global climate models. A wealth of information has been gathered during recent years within regional projects. Twelve years continuous day by day satellite data series exist. The instrumental potential is very high and specific areas exist which could be selected for process studies, we may call them "Anchor Stations".

In brief, for the scientific significance and the regional logistic point of view the

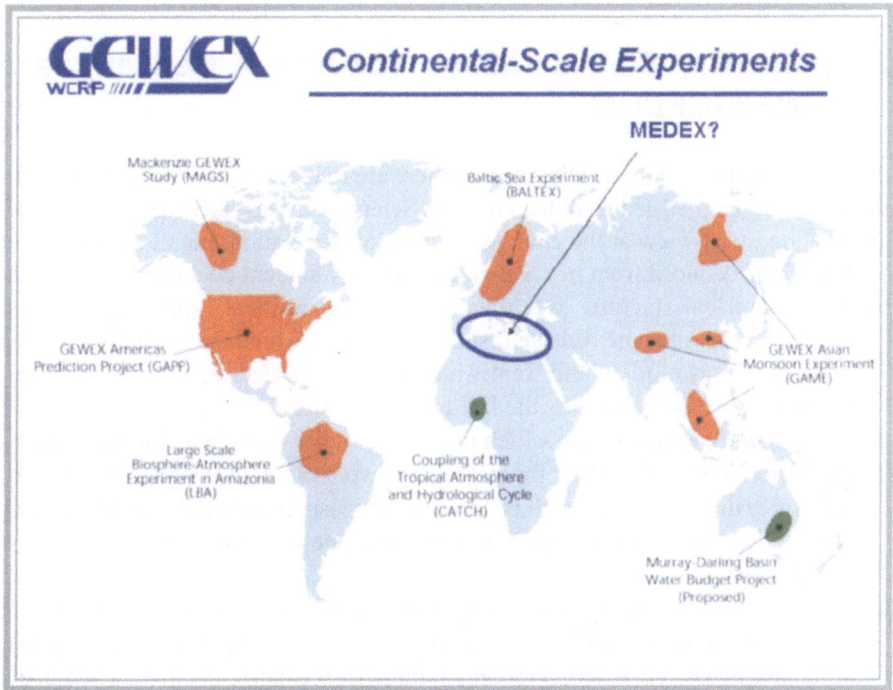

Fig. 2. Presently active Global Energy and Water Cycle Experiments. MEDEX would contribute to the GEWEX family an experiment in a semi dry environment with complex topography and an enclosed warm sea . It would be the first experiment in which potentially more than 20 countries would be involved

available background information in the Mediterranean area seems to be ready for a "MEDEX". But as in the case of other GEWEX Continental-Scale experiments (CSEs) - BALTEX, GAME, GCIP/GAPP, MAGS, and LBA (Fig. 2) some other requirements have to be fulfilled as well, to become full CSEs under GEWEX.

These criteria are:

1. Cooperation of a Modern NWP Center
2. Commitment of Resources/Personnel
3. Regional Cooperative Mechanism for Data Collection
4. Collaborative Activity with a Water Resource Agency
5. Allow Free and open Exchange of Data.

Only if there is a unique potential to provide some transferability study support in a unique climatic zone projects may be included which are not be able to achieve full CSE status like CATCH in West Africa.

It is recognized that it often takes many years of planning to develop support and infrastructure for experiments. Therefore, GEWEX is supportive of the concept of a potential Mediterranean project, keeping in mind the criteria above as guidance for what one would wish to see in a project. Basically it is focussed more on the broader-scale, multinational activities that by their nature must involve major NWP participation to couple with the focussed process studies of the coupled land-ocean-atmosphere interactions.

2 Preliminary Considerations Towards A Mediterranean GEWEX (MEDEX)

As shortly mentioned in the introduction there are four views under which an Earth system and thus the Mediterranean System Science can be visualized:

a) The long term view at the changes of climate, vegetation, man's activities and behaviour deduced from *proxy data* respectively ancient documents.

b) The empirical picture we obtain from operational and experimental *measurements* of some (selected) state parameters which are spatially sampled more or less at random and from which one tries to deduce area averages and temporal developments over the last few centennials.

c) The *remotely sensed* image which provides us - because the technique is new - at shorter time scales than the preceding two tools with completely different quantities than from the Earth bound *in situ* measurements but with spatially as well as temporarily continuous information one could only dream of a few decades ago.

d) The *model* world which tries to copy and to understand nature by a structure of mathematical relationships and to investigate the response of the system simulated this way to external forcing by numerical experimentation. Though the modelling world tries to reproduce the measured state parameters as good as it can and even though it calls its model outputs "data", it must be emphasized that these are artificial, mathematically produced, area averaged or interpolated values which have not much in common with the physically measured "spot" values.

Each of these "visions" provides non-identical images of the world at different scales in space and time. Neither of them provides a complete picture of nature but all of them taken together may result in a better approximation and understanding of the reality than any single one. Scientific communities still tend to separate into these four domains, the proxy data world, the world based upon *in situ* measurements, the remotely sensed world, and the model world, to do their specific investigations and to draw the conclusions of their work without much worrying about other disciplines. This particularly is questionable in the case of the Mediterranean system because of its complexity and sensitivity. A task of high priority within the framework of an *European Research Area* should be to make these four views at the Mediterranean subsystem congruent.

Scientific reasons why an effort should be made to integrate these four views in the Mediterranean area are discussed below. In addition there are other strategic objectives:

- The research infrastructure will be strengthened in a boundary area of the EU.
- The joint multi-disciplinary research into the identified scientific priorities will lead to networks of excellence.
- Jointly with the activities ongoing in the Baltic area it will unify Global Change, climate, and biodiversity research over most of Europe and such build an enormous potential of knowledge in all components of environmental research.
- The research will include partners from newly associated States and reach out into the eastern European region by including the area around the Black Sea.
- The project will finally include countries around the Mediterranean Sea and thus be a touchstone for the confirmation of the international role of Community research.
- Due to the intense use of observations made from space there will be a great challenge to push the methodology to interpret these data beyond the present still existing limitations thus making a strong European contribution to the global observation system (GMES).

The integration of Mediterranean research is not a straight forward task because the problems cannot be solved by just adding the expertise and results of the four "views" or research tools. Actual research topics need new integrative tools. The problem which is addressed in this paper is how existing co-operations and discipline oriented work can be guided into such an integrative "Global Change" approach.

3 Motivation for Research Into the Mediterranean Bio-physical-chemical System

At the level of the research fields *inter alia* the following "Global Change" themes appear:

A   What will be the impact on the Mediterranean area of global warming due to increasing concentrations of radiatively active gases in the atmosphere? Specifically:
   • Does the Mediterranean water cycle change and what would be the consequences with respect to the future water resources available per capita in the various regions and for the thread of desertification?
   • Does global climate change manifest itself in the Mediterranean area by an increase of extreme events with their threats for human life and welfare?
B   What impact have man-made changes at the land surfaces on regional climate and the availability of natural resources? Specifically:
   • Do changes of albedo due to land-use changes matter?
   • How does the need for irrigation affect water availability?
   • Does the generation of dust and pollution change and which effect does this have in the climate system?
C   Is the interaction between the global system and the Mediterranean subsystem changing with changing climate and changing land-surface properties respectively does the feedback to the global system change? Specifically:
   • What is the contribution of the Mediterranean Basin to the global carbon budget and thus to the global greenhouse effect?
   • What is the role of the Mediterranean Basin in the global water cycle (its contribution to the oceanic conveyor belt and the net water vapour flux above the basin).

   The processes to be considered at the next lower level are listed in Table 1 where their relevance to the three thematic blocks (A, B, C) is stated.

4   A Mediterranean Bio-physical-chemical Research Agenda

It would lead too far to go into details of research task but some thoughts how this very complex research agenda may be structured into manageable research components which in themselves are logical and based on joint goals of and close communication lines between partners may help to judge the feasibility of an integrative research approach which should incorporate partners of the whole Mediterranean area. In this approach it is attempted to amalgamate the traditional cooperation as practised within the "four views" with the need of an integrative approach. The following research complexes seem to follow this definition. Their tasks, relationships to the other research communities, and their roles in the overall research approach can shortly be described below.

The research topics listed under the following items 1 to 5 though already highly interdisciplinary nevertheless do not cover the whole spectrum of Mediterranean research problems. Fig. 3 visualizes their relationship in the context of the exploration how the Mediterranean may develop in the future. The figure contains in addition items which treated by other scientific communities such as the

Tab. 1.   Mediterranean state variables and processes potentially affected by changes of global
climate and man's activities on land. A: Impact of global warming on the
Mediterranean. B: Man's impact on regional climate. C: Large scale interactions

State variable/Process of concern	Related to			Comments for clarification of their role
	A	B	C	
Surface temperature and emissivity	X	X		Effective surface radiation temperature, surface emissivity as well as air temperature at 2m height are important for magnitude of heat fluxes
Surface net radiation	X	X		Provides the energy for fluxes between surface and atmosphere and depends on solar output, atmospheric aerosols, water vapour and other trace gases, clouds, surface albedo, surface temperature
Surface energy budget	X	X		Includes the problem of partitioning of the net radiation flux into heat fluxes and the role which soil type, soil moisture and vegetation plays in the water cycle
Artificial changes of the hydrological regime		X		Dam construction, channels, irrigation, water ponds
Soil moisture	X	X		Depends on soil structure, precipitation, water consumption, net radiation, depth of aquifer
Desertification	X	X		Land-surface degradation in its widest sense which feeds back to the atmosphere by changes of albedo, temperature and heat fluxes
Extreme events	X			Floods, dry periods, cold spells, storms
Sea level rise	X			Causes problems in coastal areas
Sea Surface temperature (SST)	X			Factor that determines evaporation from and convection atop the sea surface
Sea currents	X			Depend on atmospheric forcing, temperature, and salinity and influences SST
Land -use & land cover change		X		Change of biosystems with different properties like evaporation and albedo are effective in the climate system
TOA net radiation	X		X	The TOA net radiation is a measure of the strength of the Mediterranean sea as a energy sink because of high absorption of solar radiation in the sea

Table 1. Mediterranean state variables and processes potentially affected by changes of global climate and man's activities on land. A: Impact of global warming on the Mediterranean. B: Man's impact on regional climate. C: Large scale interactions
Continued

The Mediterranean Oscillation and the General Circulation			X	Interdependencies with the General Circulation System, North Atlantic Oscillation and Indian Monsoon; transport of water vapour across the Mediterranean area; evaporation depends on wind velocity
Fluxes of trace constituents; carbon budget			X	Mediterranean as sink or source of carbon
Air pollution, air chemistry		X		Have important effects not only on human health but also on the surface net radiation
Water exchange with North Atlantic and Black Sea			X	Component of the overall water budget of the Mediterranean
Change of biodiversity	X	X		In this context regional changes of dominating agricultural species are of interest as they affect evaporation

economical development in a globalizing environment and its social consequences, biodiversity respectively changes of the population of species, and demographic issues. It seems to be utopia to integrate al topics, processes and forcing factors from climate to socio-economy and globalization at once in one "Global Change" approach. Consequently it seems to be advisable to subdivide the problem into overviewable and presently manageable research fields closely related to GEWEX. The results of this sectoral research then has to be integrated at a higher level.

In the centre of Fig. 3 the research aspects internal to the Mediterranean basin are shown in green boxes and its external relationships in the blue box at the right. The four views at the system which are equivalent with the applied research tool are shown in the red boxes. The relationship between the five research areas and the four tools can be described as follows:

1 The work done by the community working on *proxy and* instrumentally primarily operationally *measured* Mediterranean climate and hydrological *data* is the basis for the understanding of the Mediterranean system and for the improvement of the tools necessary to predict the future.

The analysis of proxy data is the only possibility to establish a long term data base of paleo-climate changes. These data are needed to document past climate changes as well as climate variability and to study their causes. The performance of climate models can be tested against these proxy data and the interaction with

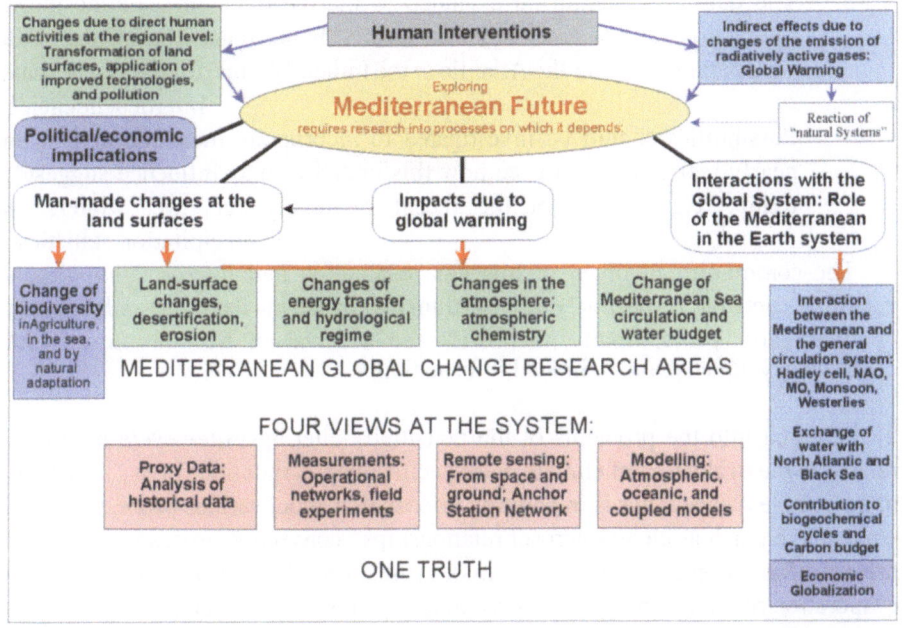

Fig. 3. Research agenda following the intention to keep the Mediterranean socio-economical system in equilibrium under changing global boundary conditions

climate of large scale land surface changes, which occurred in the past, can be studied.

The analysis of data from the operational network provides more accurate and detailed information about more recent fluctuations of the climate. Nevertheless more efforts have to be invested on the following items:

1. Intensification of the cooperation between operational services and research groups dealing with data evaluation and interpretation.
2. More stations should be entrained to enable a more detailed study of spatial differences (such as the influence of the sea in coastal regions or of mountains which is done in some countries already). Efforts should be undertaken to include in addition to data of the operational national networks also data measured by other agencies and industry.
3. Accuracy criteria and validation procedures have further to be developed and strictly applied:
   - Aggregation techniques to build up area averages compatible with the grid widths of models may have to be refined.
   - The various applied evaluation schemes and statistics should be intercompared and causes for differing conclusions be studied and published.
   - Work should be concentrated on trends of amplitudes and frequencies of extreme events.

-   The data base should be extended to marine data on sea-atmosphere interactions and to areas adjacent to the Mediterranean basin (e.g. with state parameters over the Atlantic or the Indian Oceans) to study large scale interdependencies.
-   Possibilities should be investigated to incorporate most recent data in nearly real time and to see how this updating may influence suggested trends (e.g. impact of the recent regionally very wet winters on precipitation statistics).

The community involved in this data evaluation and interpretation should establish a network respectively a consortium to broaden the data base and the scope of investigations and, finally, to provide the wider scientific community with continuously updated and validated information.

2   Research into the problems of the potentially altering *water cycle* is closely related to the evaluation of climate data specifically precipitation (item 1). But it is more complex since it needs in addition information about various processes in the atmosphere (such as cloud - aerosol relationships, convective processes and more precise information about extreme events), at the surface (energy budget, ecosystem-evaporation relationships, area evaporation) and in the sea (exchange with Atlantic ocean and black Sea, thermohaline circulations, gyres). Both carbon dioxide fluxes and aerosol concentrations are for different reasons strongly related to the water cycle.

The marine community has made great progress in understanding and modelling the water flow in the Mediterranean Sea and is working towards linking these phenomena with atmospheric processes. Large research projects like the Mediterranean Forecasting System Science have been initiated as a contribution to the Global Ocean Observing System (GOOS) which is exemplary in linking regional processes with the global system.

Many data have been assessed by individual researchers, national water authorities, the Plan Bleu, the MED-HYCOS[3] and "Global Water Partnership" initiative, and during field experiments in which aircraft measurements as well as

---

[3]see appendix

remote sensing data from the surface and space have been involved. At some places precipitation radar is being operated. This information, if synthesised, would render new insights into the Mediterranean water cycle. Many questions nevertheless remain open to solve the problem of the future water resources. The role of the sea in the context with extreme events has more carefully to be studied (e.g. is it correct to say, more frequent and intensive extreme precipitation events are caused by the increasing SST?). Do models represent atmospheric water vapour concentrations and regional precipitation accurately enough (what at all is their accuracy?). Little is known about the regional development of the soil water content and the ground water level. Are the available data about water budgets consistent with each other? Is the Mediterranean water cycle accelerating?

Beside these basic problems this sector also includes a number of application directed questions such as:

- Are the fresh water resources more endangered by precipitation changes or by exploitation? This relates to the problem of the retention capacity of soils, aquifers and artificial storage for winter/spring rains which are needed in summer.
- Is the sea-atmosphere interaction changing? What is the function of the sea with its temperature stabilising function and its high potential evaporation rates in summer which may be reduced by an increasing capacity of the atmosphere to take up water vapour with increasing temperature? Model simulations indicate even regional negative water vapour fluxes at high air temperatures and the question is, whether these results can be validated by measurements.
- To what degree is it possible to enhance the storage capacity for water needed in drought situations without creating other problems?
- If a system undisturbed by man is compared with one that uses water for increasing the economy (irrigation, tourism, industry), what are the changes in the water flows and how does the consumption of water matter in the budget?

Much independent information has to be brought together to deal with these problems and with the improvement of predictive tools. A first inventory should be undertaken how the mosaic of information pieces can be amalgamated and which gaps exist that must be closed by additional measurements for which Anchor Stations (see item 5) could play an important role. This may then lead to a GEWEX-type experiment or a decade of research into the Mediterranean water cycle (Fig.3).

The water cycle problem outreaches to atmospheric chemistry, pollution and geo-biochemical cycles. Since a number of chemical processes are intimately connected with hydrological processes a close cooperation between these two communities is imperative. Certain processes have jointly to be studied in depth (such as the relationship between carbon and water fluxes from the biosphere, soil transformation due to drought periods, fires and application of new agricultural techniques) and data connecting ground based and satellite measurements have to be produced. Obvious is also the link to the marine community.

Notwithstanding that a narrower network of hydrometeorological stations should be and may be established, an integration over the whole basin to assess the water cycle over the Mediterranean is only possible with the aid of calibrated and tested models.

3   An initiative on *atmospheric chemistry, pollution and geo-biochemical cycles* in the Mediterranean must necessarily include the exchange between the sea surface and the atmosphere which would concern both the atmosphere-climate communities as well as the marine sciences. The combination of biomass burning, industrial emissions, natural sulfur and carbon dioxide sources, marine emissions, UV and PAR radiation, BVOC emissions, wind erosion and local transport phenomena (e.g. by land-sea circulations) provide an unique atmospheric trace constituent environment the impact of which on health and the biosphere is not yet studied in a thorough manner. The Mediterranean area provides an ideal natural laboratory to study involved transformation processes.

Process studies are needed to deepen the knowledge of specific processes and to reduce the uncertainties on basic phenomena. Experiments to study such processes include sophisticated sets of observations (from the ground, aircraft and satellites), and a modelling activity which provides support for the experiment and for the correct interpretation of the measurements. Experimental process studies may be imbedded within long term monitoring activities of key parameters, or can be seen as a component of a large field experiment such as GEWEX

The problems which connect this community with the other atmosphere - climate groups are manifold:

- Atmospheric aerosols interact with the water cycle due to the dependence of cloud formation on the aerosols.
- The aerosol content (including Saharan dust) influences radiative forcing and thus climate.
- The exchange of carbon compounds between biosphere and atmosphere contribute to the greenhouse effect and while increasing levels of trace constituents affect the biosphere.
- The modelling community is challenged to include chemical reactions and trace constituent transports in models.
- The quality of corrections of remote sensing data depends on the knowledge of properties and concentrations of trace constituents.
- Process studies experiments should be carried out jointly with the community working on the water cycle.

There have been and is a number of uncorrelated activities on atmospheric chemistry in the Mediterranean area (such as the early studies of the JRC ISPRA on pollution in the Po valley, the CEAM activities on meso-meteorological cycles of air pollution in the Iberian Peninsula, the intensive but short term measurements of BVOC emissions in Italy, the ongoing semi-permanent European and Mediterranean Flux Nets, the present activities of ENEA in cooperation with the University of Rome at Lampedusa, the Israeli space experiment MEIDEX, a joint activity at Crete to study chemical composition as well as radiative properties of the polluted Mediterranean atmosphere in August 2001) but to obtain an integrative view on spatial and temporal synergies as well as regional idiosyncrasies the various activities have to be coordinated and more has to be done, specifically long term measurements are needed to obtain information on the variability of the investigated processes.

The approach described here concerns the "measured world" but it has strong links to the modelling domain which needs these data to construct atmospheric chemistry models.

4   This leads to the question of the advantage of a dedicated *network of Mediterranean modelling laboratories*. A hierarchy of models is used to simulate various aspects of climate change starting from local SVAT models of high temporal resolution to low resolution global climate models. It is understood that in general modelling centres cooperate well in various programs. The goal for a specific Mediterranean simulation is to

- better simulate regional climate idiosyncrasies such as different developments of the temperature in adjacent areas,
- simulate occurrence, intensity, and frequence of extreme events and to understand their causes,
- simulate the impact of land cover changes on regional climate,
- simulate and understand the dependency of the Mediterranean climate on large scale phenomena such as NAO and the Indian Monsoon (which in turn are relared to tropical systems),
- understand the causes of the MO,
- model the Mediterranean water cycle.

For these purposes the major goal of the Mediterranean modelling community is the development and improvement of coupled sea - atmosphere models. This is one goal of the before mentioned MFS project.

The wider scientific community would be interested to know how accurate the models in use simulate present climate state variables at the regional scale and could answer the questions formulated above. More contacts should be made with those groups that could contribute data for validation and consequently could help to detect discrepancies with observed quantities. Furthermore the cooperation of the water cycle group with modellers must be very close. From the research into atmospheric chemistry new challenges are expected for the improvement of models.

5   The satellite *remote sensing community* could contribute the most integrative data set for the whole Mediterranean basin though this data set consists of a different kind of information than provided by all other communities. The first task would be to standardize calibration, correction, and evaluation methods for the specific Mediterranean conditions. It must be aimed at the longest possible consistent data series. Secondly the relationships to measurements made at the ground must be validated and it must be proved by intercomparison of data sets whether the analysis of satellite data matches the local observations (this is not obvious in the case of soil moisture and surface energy budget nor even in the "simple" cases of surface temperature and albedo). Therefor "Anchor Stations" are inevitable for this validation task. Thirdly the time series of information inferred from the satellite data

must be compared with the initialization (e.g. surface albedo) and the output of models (e.g. cloudiness, surface temperature, atmospheric vertical structure, water vapour distribution). On the other hand remote sensing from space depends on information on atmospheric structure for corrections. Surface bound vertical profilers as well as aircraft missions must be integrated in the validation process of satellite data.

Furthermore it is very desirable to expand the few available ground based remote sensing stations into a denser network to measure state variables which cannot with the required accuracy be inferred from measurements made from space such as areal precipitation, atmospheric spectral optical depth, aerosol composition, cloud parameters, water vapour, temperature profiles, trace constituents in the lower atmosphere. This network should include rain radar stations, profiles of different kind and remote sensing measurements from aircraft. The information obtained from such measurements is important on its own but would in addition support the correct evaluation of satellite data.

Though remote sensing is "only" considered as a tool needed in the other research areas, it is recommended that the remote sensing community because of its different approaches tightens cooperation to exchange data, methods, and experiences. From this basis the communication with other communities would be much more solid than a cooperation between only, say, one remote sensing group with limited resources and marginal spatial knowledge and one ground station or one modelling group. The CTM ERS/RAC probably would be a good starting point for a communication among remote sensing groups around the Mediterranean sea but more emphasis has to be laid on the comparison of algorithms, the merging of high resolution and medium resolution data, the dissemination of remote sensing data products and their validation with the aid of the network of Anchor Stations, at which the necessary specific measurements are carried out. These stations still have to be established and the Mediterranean remote sensing community presently is far from a true exchange, merging, comparing, and validation of measurements made from space. It would be specifically useful for a wide users community, if a data bank of remote sensing research results could be established and maintained by one of the cooperating institutions.

Strong links should be established with the atmospheric chemistry, pollution and geo-biochemical cycles community which may contribute to the network of ground based remote sensing stations as future spacecraft will carry instruments to measure atmospheric constituents. Furthermore the link to the water cycle community is mandatory without question. But the correct interpretation of remote sensing measurements made from space in addition depends strongly on collateral as well as corroborative measurements made at Anchor Stations by the land oriented community. These are hydrological measurements such as soil moisture determinations, albedo measurements, and flux measurements.

In summary, a grouping of five interacting "rings" of 'overviewable' size circling around specific research tools would be beneficial to lead the field of Mediterranean atmospheric and climate research into an "European-North African-Near East research Area". An important sixth component is essential but will not further be discussed here: The joint outreach to the wider scientific and user's communities. The results of the communities addressed here certainly are of considerable importance for hydrological development strategies and the problem of desertification in the Mediterranean area furthered by the United Nations Convention to Combat Desertification. A logical consequence therefor would be that this part of the information is directly fed into the EEA/FMA project DIS/MED (Desertification Information System to support National Action Programmes in the Mediterranean).

5 The Potential of Observations Made From Space

Remote sensing methods are limited to a few measurable quantities which are as stated before of a completely different nature than the information which models and *in situ* measurements produce. They deal with reflection, scattering, absorption and emission of electromagnetic waves, while *in situ* state variables such as pressure, temperature, constituent concentrations, material and energy fluxes are measured. From the radiance, polarization and signal return time measurements made in space in regular intervals nevertheless a wealth of information about the environment is inferred. In Table 1 quantities are listed which are necessary to determine changes and study processes immanent to the system. These are or can be affected by global changes and are used in mathematical descriptions as for empirical diagnostics of the system. To the assessment of a number of them remote sensing can contribute substantially. By analysis of long term data series these quantities also can be used as describers or "indicators" for changes. But if one wants to know what precisely changed and why, more detailed information is required. In the past few years methods have been developed to assess this information from remote sensing with a minimum of collateral data.

As an example let us look at the easily to construct Normalized Difference Vegetation Index (NDVI) that indicates changes of the "greenness" of the landscape, such as the mean annual vegetation indices. To the variability from year to year a number of causes can contribute: Change of precipitation, change of groundwater table, variable length of the vegetation period, vegetation stress due to high temperatures, changes of the density of the vegetation cover, variability of biomass production, a change of crops, irrigation, or fires. To entangle these causes and to obtain a clearer view at the processes additional information has to be aquired. Trends of the vegetation index can be brought in relation to desertification. Land-use changes can be analysed with the aid of spectral reflectances and, if high resolution images (such as from aircraft) are available, of the planting pattern. If the radiative properties of the components covering the surface are known, it is possible to determine by "spectral decomposition" their fractional coverage. Spectral reflectances combined with a measure of surface albedo will inform us, if the net solar flux at the surface changed. The effective surface temperature adds the information about the net infrared flux and by this allows to estimate the total net radiation at the surface (if the sky emission can assumed to be constant). The temperature data in addition inform us whether vegetation species may be under stress. Microwave sensors are capable to provide information about the soil moisture in the uppermost soil layer and if this information is combined with soil-water models and precipitation data series also soil moisture profiles can be estimated. Combining further these various bits of information one can approach the question of the surface energy fluxes and evaporation. With an elaborate method Bastiaanssen (1995) and Bastiaanssen et al. (1997) were able to determine these fluxes for a test area. Roerink and Menenti (1999) and Menenti et al. (1989) simplified the method for quasi-operational application using the fact that if albedo is low and if at the same time the temperature is low, then evaporation is close to potential evaporation and if on the other hand temperature and visible reflectance are high then there cannot be much vegetation and the sensible heat flux will be maximum which is close to the difference between the downwelling radiation flux minus the soil heat flux. In the temperature versus albedo diagram the line for the maximum sensible heat flux starts high and slopes down because with increasing albedo less energy becomes available and the surface temperature sinks. The line of the maximum latent heat flux starts low and gently slopes up because with increasing albedo and decreasing net radiation the evaporative flux becomes smaller which reduces the cooling effect and the surface temperature slightly rises. The distances of a pixel in this diagram from these two lines allows to compute the ratio between the sensible and latent heat fluxes (the Bowen Ratio) and by solving the energy budget equation the heat fluxes can be computed. The soil heat flux can be estimated with the experience gained from field experiments. A cloudy atmosphere makes budget estimates much more difficult but if there is no precipitation one can assume that the partitioning of the net radiation flux does not change under clouds and if one can infer the net radiation flux at the ground also the absolute magnitude of the heat fluxes (and evaporation) can be derived. By building up this tree of information it finally becomes feasible to draw a fairly good picture of the land-surface processes at a large spatial scale.

The new generation of satellites will in addition provide more information about the composition of the atmosphere then so far is possible. Trace constituent concentrations are becoming available for the troposphere and improved information about the water vapour distribution is expected which will help to explain and to remove discrepancies in the modeling of this quantity.

With a number of ground measurements at "Anchor Stations" this information can be validated and the algorithms be improved which are used for the evaluation of the data. If furthermore the analysis of the measurements made in space can be continued for long time periods trends and extents of changes could be analysed for large areas. The intensive evaluation and use of observations made from space combined with research at "Anchor Stations" will support the evaluation of existing ground based data and may ring in a first phase of a basin wide Mediterranean research area.

6 The Concept of Mediterranean Anchor Stations

The field experiment EFEDA carried out in Spain, 1991, covering an area of about 10^4 km^2 in Castilla - La Mancha lasted not much longer than one month (June). This was just the time when the land-surface processes switched from the wet to the dry regime but it was too short to include processes in the study which have much longer response times such as some hydrological processes and their impacts on vegetation. It was therefor desirable to carry on with at least a reduced number of observations for a much longer time period. At the ground only a few stations could further be operated but additional information could be deduced from satellite data. EFEDA included an intensive comparison program between quantities measured at the ground and information inferred from satellites. Because of the strong seasonal variability of surface parameters and the large variety of ecosystems found in the Mediterranean area it was at that time concluded that a number of stations at which if not permanently then from time to time surface and atmospheric state variables are measured which are also assessed from space would tremendously help to improve the accuracy of the of the data derived from the satellite measurements and would protect from misinterpretations which could occur if the area is not enough known to the analyser of remote sensing data.

It was therefore proposed to arrange for a limited number of sites which should be established in areas representative for the various Mediterranean environments and climatic zones. Very important would be to measure such quantities which are not routinely measured by operational services such as all six important components of the radiation (solar, reflected, longwave up- and downwelling, PAR and UV), sensible and latent heat fluxes, optical depth of the atmosphere, water vapour amount of the atmosphere, soil moisture, surface emissivity, vegetation cover and state and if there is a requirement additional special quantities of importance for the region. The area should be large enough to fill a number of high resolution satellite image pixel to make a thorough comparison. Then these high resolution pixel could be averaged over the pixel size of medium resolution instruments such as on board of meteorological satellites. These medium resolution images have to be used to produce day-by-day data series which are needed to remove the influence of clouds and to obtain continuous data series which high resolution imagery cannot provide.

These "Anchor Stations" should also offer the opportunity to perform limited field experiments in case new sensors respectively the algorithms used to evaluate the measurements with new sensors have to be calibrated and their performane tested. They may also serve for training courses. To date there are so far only few and not completely equipped stations which could serve as Anchor Stations. Their number should greatly be enhanced and each country around the Mediterranean sea should establish at least one of these stations which should build a network.

A Tunesian - Italian initiative recently took up this idea of an anchor station network for trasect from south Tunesia to the Ligurian coast. Such a transect indeed would cut through a number of representative Mediterranean ecosystems all of them with different problems and probably different responses to climate change. It would provide the basis for studies of the north-south gradient across the Mediterranean sea.

Along this transect satellites would monitor the "green wave" and its dependence on the large scale circulation and the NAO. The variability of surface energy and water vapour fluxes would be estimated from changes of spectral reflectances, emissivity, and surface temperatures. It can also be observed - if the time series becomes long enough - how land-use changed over the years and whether desertification is progressing. Ground based studies would then try to find out which had been the causes for these changes if man-made, do they have climatic or economical reasons, and what are the causes for desertification. The entrainment of the operating flux measuring stations in Italy would in addition open the field for studies into the relationship between biomass production as sensed by the vegetation index and carbon dioxide and water vapour fluxes as well as leaf area index measured at the ground. In Tunisia studies of the variability of the borderline between desert and semi-desert belong into the category of gradient studies as the comparison between the situation in Sardegna and Corsica would be - which clearly can be seen in satellite images. Finally the transect reaches the area near the Alps which is strongly influenced by events like the Mistral and the Genova cyclone. The simultaneous assessment of the water vapour concentrations and of the aerosol content as well as its properties is obligatory because these data are not only of interest for climate research but are in addition indispensable for the accurate correction of the satellite data.

The variability of the central part of the Mediterranean seems not to be strongly related to the large scale oscillations like the NAO and the MO. To study the impacts of these phenomena on the ecosystems an east - west transect would be necessary for which a re-activation of the sites used in follow-on activities of EFEDA would provide a first string which spans from Spain (Castilla-La Mancha/Valencia over Tuscany (Radicondoli) and Crete (Messara valley) to Israel (Beersheva). Sites in Turkey, Lebanon, Egypt, the Balkan, Morocco and islands should be added. A West - East research program could bridge over to the exemplary project "Degradation of the Drylands of Asia" of the Center for Environmental Remote Sensing, Chiba University, Japan, which ends in Turkey. The natural cross-over point of both transects would be the island of Lampedusa where presently integrative and multinational atmospheric research is performed at an Italian research station.

The oceanographic community is getting organized in the context of the *Mediterranean Forecasting System*, an EU-MAST Project (Pinardi and Flemming, 1998). The goal is the prediction of the 3-D physical sea state and related marine biochemical components at time scales of weeks to months by implementation of an automatic monitoring and nowcasting/forecasting modelling system. 25 institutions of Mediterranean countries and some extra-Mediterranean Institutions are involved in this activity. A pilot project started in September 1998. Satellite data may contribute to this project by providing sea surface temperatures, sea level height, water pollution, and information about surface currents.

This oceanographic system combined with the network of land stations and basin-wide satellite observations would be an important European contribution ti the Global Climate Observation System (GCOS).

7 Conclusions

There have been and exists a large number of activities in the Mediterranean basin which could, if properly combined, provide an exemplary core of a monitoring and research area for a climatic as well as with respect to anthropogenic disturbances sensitive region and a contribution to GCOS. Additional fields of research would be attached as the requirements for specific expertises increases. Different approaches to reach this goal could be thought of. The most simple but probably not very effective way would be organize periodically meetings at which all involved scientists and organizations report their results and by this communication constitute a network for research and applications (dissemination of research results) in the Mediterranean area. A central unit would have to integrate and to disseminate the combined knowledge by periodic reports as is the usage of IPCC. A second way would be that a few large institutions take the lead. There exist a number of well equipped research institutions throughout the Mediterranean specialized on different topics which could form jointly with some involved extra-Mediterranean research institutes a network of centres of excellence aiming at an exchange and integration of data as well as modelling the Mediterranean sub-system. Alternatively a new multinational Mediterranean Research Institute could be established into which this information is flowing and amalgamated and where information for the user communities is extracted and disseminated. On the other hand nothing has a more integrative power than a joint field experiment. In other regions of the world where comparable problems in understanding the system exist, GEWEX experiments have been carried out to improve not only the knowledge about the system but also to intensify the inter-disciplinary communication. Much relevant experience have been gained in Europe with BALTEX that could be transferred to the Mediterranean. Such an experiment would need the contribution of national operational services and a free exchange of data.

Anchor stations would become the backbone and basis of a large scale experiment which would be guided and accompanied by the evaluation of existing data. It has been demonstrated that observations from space and ground based remote sensing equipment would be the tools to bridge over the areas between experimental sites. Presently the Italian - Tunisian initiative is starting to build a transect between Tunisia and the South of France to study environmental changes with satellite data. Equally interesting would be a West - East transect including islands to study the Mediterranean Oscillation, the synergies between the western and eastern part of the basin respectively changes that occur in the western countries as compared to the situation in the Near East which may be related to the anticyclic variability manifested in the Mediterranean Oscillation. Last not least the scale has even to be enlarged as the dependency of the Mediterranean water budget on processes that occur in the North Atlantic and Indian Oceans gets into view.

It may be advisable to start the development of experiment plans separately within the four areas defined in this paper entraining as many national agencies as possible and useful. But then the synergies between the different approaches should be developed, "Anchor Stations" should be selected that can be used jointly, field campaigns have to be synchronized, and deliberations about the assimilation and use of the data must be started in a comprehensive way. For this task it needs a strong integrating body.

References

Bastiaanssen WGM (1995) Regionalization of surface flux densities ans moisture indicators in composite terrain. PhD Thesis. Wageningen Agricultural University Report 109, DLO-Winand Staring Centre, Wageningen, The Netherlands

Bastiaanssen WGM, Pelgrum H, Droogers P, de bruin HAR, Menenti M (1997) Area-average estimates of evaporation, wetness indicators and top soil moisture during two golden days in EFEDA. Agricultural and Forest Meteorolology 87:119-137

Ellis JS, Vonder Haar TH, Levitus S, Oort AH (1978) The annual variation in the global heat balance of the Earth. J. Geophys. Res. 84:1958-1962

Menenti M, Bastiaanssen WGM, van Eick D, Abi Ele karim MA (1989) Linear relationshipsbetween surface refklectance and temperature and their appliocation to map evaporation of groundwater. Adv. Space Res. 9, No.1: 165-176

Oort AH (1971) The observed annual cycle in the Meridional transport of atmospheric energy. Journal Atm. Sciences 28:325-339

Oort AH, Vonder Haar TH (1976) On the observed annual cycle in the ocean-atmosphere heat balance over the northern hemisphere. Journal Phys. Oceanography 6:781-800

Peixoto JP, Oort AH (1982) The atmospheric branch of the hydrologic cycle and climate. In: Street-Perrott et al. (Eds.) Variations in the global water budget. D. Reidel, Hingham, Mass.

Roerink GJ, Menenti M (1999) Surface energy fluxes: From point to regional scale. In: Synthesis of Change Detection Parameters into a Land-surface Change Indicator for Long term Desertification Studies in the Mediterranean Areas (RESYSMED). Final report, Contract ENV4-CT97-0683, Firenze and München

Vonder Haar TH, Suomi VE (1971) Measurements of the Earth's radiation budget from satellites during a five-year period. Journal Atm. Sciences 28:305-314

WCRP (1990) Global Energy and Water Cycle Experiment (Scientific Plan). WCRP-40, WMO/TD-No. 376

WCRP (1992) Scientific plan for the GEWEX Continental-scale International Project (GCIP). WCRP-67 WMO/TD-No. 461

Appendix
MED-HYCOS

According to an information by Tommaso Abrate, WMO, the MED-HYCOS project is organized and operates as follows:

1 The first phase is being executed by WMO with the financial support of the World Bank. Twenty countries of the Mediterranean basin are participating in this project. The Pilot Regional Centre (PRC) is located in Montpellier in France, hosted by the IRD (Research Institute for Development, formerly ORSTOM).

2 The major output of the first phase has been the establishment of the Mediterranean Hydrological Information System (MHIS), which has been developed by the PRC with the contribution of experts seconded from participating countries. The following types of information are available through the MHIS:
   • Information on the project (project document, progress reports, meeting reports, etc), on its status of implementation and the PRC staff, including seconded experts;
   • Information on the participating NHSs (descriptions of the NHS, national focal points, other staff involved in the project activities);
   • Information on the stations forming the project network (location, channel characteristics, hydrological regime, length of the historical series, quality and quantity of the information available);
   • The regional data bank
   • Information about water resources availability and policy in the countries;

3 The most important feature of the MHIS is represented by the regional data bank and the related tools for data retrieval and display. The regional database contains data from 94 stations, including 31 equipped with DCPs. These DCPs are located as follows: Albania (2), Bosnia and Herzegovina (1), Bulgaria (3) Croatia (2), Cyprus (3), Italy (6), Jordan (1), Malta (2), FYR Macedonia (1), Morocco (2), Slovenia (1), Turkey (3), Tunisia (4). Data from conventional stations are updated weekly or monthly, based on the information supplied by participating countries. Data from DCP are retrieved three-hourly from EUMETSAT Web site.

4 Three software applications have been developed for accessing the database:
   • MED-DAT: allows tabular and graphic representation of data, overlapping of diagrams from different years or different stations, automated routine download process of data from EUMETSAT;
   • MED-MAP: allows access to data through a cartographic interface, providing basic GIS facilities;
   • MED-CLIM: allows areal analysis and cartographic representation of climatological series.